APPLIED CHEMISTRY AND CHEMICAL ENGINEERING

Volume 2

Principles, Methodology, and Evaluation Methods

APPLIED CHEMISTRY AND CHEMICAL ENGINEERING

Volume 2

Principles, Methodology, and Evaluation Methods

Edited by

A. K. Haghi, PhD
Lionello Pogliani, PhD
Devrim Balköse, PhD
Omari V. Mukbaniani, DSc
Andrew G. Mercader, PhD

AAP APPLE ACADEMIC PRESS

Apple Academic Press Inc.
3333 Mistwell Crescent
Oakville, ON L6L 0A2 Canada

Apple Academic Press Inc.
9 Spinnaker Way
Waretown, NJ 08758 USA

Library and Archives Canada Cataloguing in Publication

Applied chemistry and chemical engineering / edited by A.K. Haghi, PhD, Devrim Balköse, PhD, Omari V. Mukbaniani, DSc, Andrew G. Mercader, PhD.
Includes bibliographical references and indexes.
Contents: Volume 1. Mathematical and analytical techniques --Volume 2. Principles, methodology, and evaluation methods --Volume 3. Interdisciplinary approaches to theory and modeling with applications --Volume 4. Experimental techniques and methodical developments --Volume 5. Research methodologies in modern chemistry and applied science.
Issued in print and electronic formats.
ISBN 978-1-77188-515-7 (v. 1 : hardcover).--ISBN 978-1-77188-558-4 (v. 2 : hardcover).--ISBN 978-1-77188-566-9 (v. 3 : hardcover).--ISBN 978-1-77188-587-4 (v. 4 : hardcover).--ISBN 978-1-77188-593-5 (v. 5 : hardcover).--ISBN 978-1-77188-594-2 (set : hardcover).
ISBN 978-1-315-36562-6 (v. 1 : PDF).--ISBN 978-1-315-20736-0 (v. 2 : PDF).-- ISBN 978-1-315-20734-6 (v. 3 : PDF).--ISBN 978-1-315-20763-6 (v. 4 : PDF).-- ISBN 978-1-315-19761-6 (v. 5 : PDF)
1. Chemistry, Technical. 2. Chemical engineering. I. Haghi, A. K., editor

| TP145.A67 2017 | 660 | C2017-906062-7 | C2017-906063-5 |

Library of Congress Cataloging-in-Publication Data

Names: Haghi, A. K., editor.
Title: Applied chemistry and chemical engineering / editors, A.K. Haghi, PhD [and 3 others].
Description: Toronto ; New Jersey : Apple Academic Press, 2018- | Includes bibliographical references and index.
Identifiers: LCCN 2017041946 (print) | LCCN 2017042598 (ebook) | ISBN 9781315365626 (ebook) | ISBN 9781771885157 (hardcover : v. 1 : alk. paper)
Subjects: LCSH: Chemical engineering. | Chemistry, Technical.
Classification: LCC TP155 (ebook) | LCC TP155 .A67 2018 (print) | DDC 660--dc23
LC record available at https://lccn.loc.gov/2017041946

Apple Academic Press also publishes its books in a variety of electronic formats. Some content that appears in print may not be available in electronic format. For information about Apple Academic Press products, visit our website at **www.appleacademicpress.com** and the CRC Press website at **www.crcpress.com**

ABOUT THE EDITORS

A. K. Haghi, PhD

A. K. Haghi, PhD, holds a BSc in Urban and Environmental Engineering from the University of North Carolina (USA), an MSc in Mechanical Engineering from North Carolina A&T State University (USA), a DEA in applied mechanics, acoustics and materials from the Université de Technologie de Compiègne (France), and a PhD in engineering sciences from the Université de Franche-Comté (France). He is the author and editor of 165 books, as well as of 1000 published papers in various journals and conference proceedings. Dr. Haghi has received several grants, consulted for a number of major corporations, and is a frequent speaker to national and international audiences. Since 1983, he served as professor at several universities. He is currently Editor-in-Chief of the *International Journal of Chemoinformatics and Chemical Engineering* and the *Polymers Research Journal* and on the editorial boards of many international journals. He is also a member of the Canadian Research and Development Center of Sciences and Cultures (CRDCSC), Montreal, Quebec, Canada.

Lionello Pogliani, PhD

Lionello Pogliani, PhD, was Professor of Physical Chemistry at the University of Calabria, Italy. He studied Chemistry at Firenze University, Italy, and received his postdoctoral training at the Department of Molecular Biology of the C. E. A. (Centre d'Etudes Atomiques) of Saclay, France, the Physical Chemistry Institute of the Technical and Free University of Berlin, and the Pharmaceutical Department of the University of California, San Francisco, CA. Dr. Pogliani has coauthored an experimental work that was awarded the GM Neural Trauma Research Award. He spent his sabbatical years at the Centro de Química-Física Molecular of the Technical University of Lisbon, Portugal, and at the Department of Physical Chemistry of the Faculty of Pharmacy of the University of Valencia-Burjassot, Spain. He has contributed nearly 200 papers in the experimental, theoretical, and didactical fields of physical chemistry, including chapters in specialized books. He has also presented at more than 40 symposiums. He also published a book on the numbers 0, 1, 2, and 3. He is a member of the International Academy of

Mathematical Chemistry. He retired in 2011 and is part-time teammate at the University of Valencia-Burjassot, Spain.

Devrim Balköse, PhD

Devrim Balköse, PhD, is currently a faculty member in the Chemical Engineering Department at the Izmir Institute of Technology, Izmir, Turkey. She graduated from the Middle East Technical University in Ankara, Turkey, with a degree in Chemical Engineering. She received her MS and PhD degrees from Ege University, Izmir, Turkey, in 1974 and 1977, respectively. She became Associate Professor in Macromolecular Chemistry in 1983 and Professor in process and reactor engineering in 1990. She worked as Research Assistant, Assistant Professor, Associate Professor, and Professor between 1970 and 2000 at Ege University. She was the Head of the Chemical Engineering Department at the Izmir Institute of Technology, Izmir, Turkey, between 2000 and 2009. Her research interests are in polymer reaction engineering, polymer foams and films, adsorbent development, and moisture sorption. Her research projects are on nanosized zinc borate production, ZnO polymer composites, zinc borate lubricants, antistatic additives, and metal soaps.

Omari V. Mukbaniani, DSc

Omari Vasilii Mukbaniani, DSc, is Professor and Head of the Macromolecular Chemistry Department of Iv. Javakhishvili Tbilisi State University, Tbilisi, Georgia. He is also the Director of the Institute of Macromolecular Chemistry and Polymeric Materials. He is a member of the Academy of Natural Sciences of the Georgian Republic. For several years he was a member of the advisory board of the *Journal Proceedings of Iv. Javakhishvili Tbilisi State University* (Chemical Series) and contributing editor of the journal *Polymer News* and the *Polymers Research Journal*. He is a member of editorial board of the *Journal of Chemistry and Chemical Technology*. His research interests include polymer chemistry, polymeric materials, and chemistry of organosilicon compounds. He is an author more than 420 publications, 13 books, four monographs, and 10 inventions. He created in the 2007s the "International Caucasian Symposium on Polymers & Advanced Materials," ICSP, which takes place every other two years in Georgia.

Andrew G. Mercader, PhD

Andrew G. Mercader, PhD, studied Physical Chemistry at the Faculty of Chemistry of La Plata National University (UNLP), Buenos Aires, Argentina, from 1995–2001. Afterwards he joined Shell Argentina to work as Luboil, Asphalts and Distillation Process Technologist, as well as Safeguarding and Project Technologist. His PhD work on the development and applications of QSAR/QSPR theory was performed at the Theoretical and Applied Research Institute located at La Plata National University (INIFTA). He received a post-doctoral scholarship to work on theoretical-experimental studies of biflavonoids at IBIMOL (ex PRALIB), Faculty of Pharmacy and Biochemistry, University of Buenos Aires (UBA). He is currently a member of the Scientific Researcher Career in the Argentina National Research Council, at INIFTA.

Applied Chemistry and Chemical Engineering, 5 Volumes

Applied Chemistry and Chemical Engineering,
Volume 1: Mathematical and Analytical Techniques
Editors: A. K. Haghi, PhD, Devrim Balköse, PhD, Omari V. Mukbaniani, DSc, and Andrew G. Mercader, PhD

Applied Chemistry and Chemical Engineering,
Volume 2: Principles, Methodology, and Evaluation Methods
Editors: A. K. Haghi, PhD, Lionello Pogliani, PhD, Devrim Balköse, PhD, Omari V. Mukbaniani, DSc, and Andrew G. Mercader, PhD

Applied Chemistry and Chemical Engineering,
Volume 3: Interdisciplinary Approaches to Theory and Modeling with Applications
Editors: A. K. Haghi, PhD, Lionello Pogliani, PhD, Francisco Torrens, PhD, Devrim Balköse, PhD, Omari V. Mukbaniani, DSc, and Andrew G. Mercader, PhD

Applied Chemistry and Chemical Engineering,
Volume 4: Experimental Techniques and Methodical Developments
Editors: A. K. Haghi, PhD, Lionello Pogliani, PhD, Eduardo A. Castro, PhD, Devrim Balköse, PhD, Omari V. Mukbaniani, PhD, and Chin Hua Chia, PhD

Applied Chemistry and Chemical Engineering,
Volume 5: Research Methodologies in Modern Chemistry and Applied Science
Editors: A. K. Haghi, PhD, Ana Cristina Faria Ribeiro, PhD, Lionello Pogliani, PhD, Devrim Balköse, PhD, Francisco Torrens, PhD, and Omari V. Mukbaniani, PhD

CONTENTS

LIST OF CONTRIBUTORS

R. A. Ahmedova
Institute of Polymer Materials of Azerbaijan, National Academy of Sciences, S.Vurgun Str, 124, AZ 5004 Sumgait, Azerbaijan

Hikmat A. Ali
Chemistry Department, College of Science, University of Basrah, Basrah, Iraq. Email: hikmatali42@ gmail.com

K. T. Archvadze
Food Industry Department, Georgian Technical University, 77 Kostava, 0175 Tbilisi, Georgia. E-mail: keti987@mail.ru

Afaq M. Aslanbayli
Institute of Petrochemical Processes, Azerbaijan National Academy of Sciences, 30, Khodjaly av., AZ 1025 Baku, Azerbaijan

I. R. Chachava
Food Industry Department, Georgian Technical University, 77 Kostava, 0175 Tbilisi, Georgia

Peter Duchovič
Department of Science, VIPO a.s., Gen. Svobodu 1069/4, 958 01 Partizánske, Slovakia

Miguel A. Esteso
Departamento de Química Física, Facultad de Farmacia, Universidad de Alcalá, 28871 Alcalá de Henares (Madrid), Spain

R. Gakhokidze
Department of Chemistry, Ivane Javakhishvili Tbilisi State University, Ilia Chavchavadze Ave., 0128 Tbilisi, Georgia

Abasgulu Guliyev
Institute of Polymer Materials of Azerbaijan, National Academy of Science, S.Vurgun str., AZ 5004 Sumgait, Azerbaijan. E-mail: abasgulu@yandex.ru

A. M. Guliyev
Institute of Polymer Materials of Azerbaijan, National Academy of Sciences, Baku, Azerbaijan

K. G. Guliyev
Institute of Polymer Materials of Azerbaijan, National Academy of Sciences, Baku, Azerbaijan

Nemat A. Guliyev
Institute of Petrochemical Processes, Azerbaijan National Academy of Sciences, 30, Khodjaly av., AZ 1025 Baku, Azerbaijan

Ts. D. Gulverdashvili
Institute of Polymer Materials of Azerbaijan, National Academy of Sciences, Baku, Azerbaijan

Gulara N. Hasanova
Institute of Petrochemical Processes of Azerbaijan, National Academy of Sciences, Khojaly Ave., 30, AZ 1025 Baku, Azerbaijan

Ch. H. İsmaylova
Institute of Polymer Materials of Azerbaijan, National Academy of Sciences, S.Vurgun Str, 124, AZ 5004 Sumgait, Azerbaijan

Nazil F. Janibayov
Institute of Petrochemical Processes of Azerbaijan, National Academy of Sciences, Khojaly Ave., 30, AZ 1025 Baku, Azerbaijan. E-mail: j.nazil@yahoo.com

Ajith J. Jose
Department of Chemistry, St. Berchmans College (Autonomous), Changanassery 686101, Kerala, India

Peter Jurkovič
Department of Science, VIPO a.s., Gen. Svobodu 1069/4, 958 01 Partizánske, Slovakia

Alaa S. Khalaf
Chemistry Department, College of Science, University of Basrah, Basrah, Iraq

Moayad N. Khalaf
Chemistry Department, College of Science, University of Basrah, Basrah, Iraq

Sh. Maghsoodlou
Department of Textile Engineering, University of Guilan, Rasht, Iran. E-mail: sh.maghsoodlou@gmail.com

S. B. Mamedli
Institute of Polymer Materials of Azerbaijan, National Academy of Sciences, Baku, Azerbaijan

B. A. Mamedov
Institute of Polymer Materials of Azerbaijan, National Academy of Sciences, S.Vurgun Str, 124, AZ 5004 Sumgait, Azerbaijan

S. S. Mashaeva
Institute of Polymer Materials of Azerbaijan, National Academy of Sciences, S.Vurgun Str, 124, AZ 5004 Sumgait, Azerbaijan

Ján Matyašovský
Department of Science, VIPO a.s., Gen. Svobodu 1069/4, 958 01 Partizánske, Slovakia. E-mail: jmatyasovsky@vipo.sk

T. I. Megrelidze
Food Industry Department, Georgian Technical University, 77 Kostava, 0175 Tbilisi, Georgia

Raniero Mendichi
Istituto per lo Studio delle Macromolecole, Consiglio Nazionale delle Ricerche, Milano, Italy

George Meskhi
Samtskhe-Javakheti State University, Faculty of Engineering, Agrarian and Natural Sciences, 106 Rustaveli str., 0800 Akhaltsikhe, Georgia. Email: george.meskhi@yahoo.com

Fizuli A. Nasirov
Institute of Petrochemical Processes, Azerbaijan National Academy of Sciences, 30, Khodjaly av., AZ 1025 Baku, Azerbaijan. E-mail: fizulin52@rambler.ru; fizuli_nasirov@yahoo.com

Ahmed M. Omer
Laboratory of Bioorganic Chemistry of Drugs, Institute of Experimental Pharmacology and Toxicology, SK-814 04 Bratislava, Slovakia

Sukanchan Palit
Department of Chemical Engineering, University of Petroleum and Energy Studies, Bidholi via Premnagar, Dehradun 248007, Uttarakhand, India. E-mail: sukanchan68@gmail.com, sukanchan92@gmail.com

Sevda R. Rafiyeva
Institute of Petrochemical Processes of Azerbaijan, National Academy of Sciences, Khojaly Ave., 30, AZ 1025 Baku, Azerbaijan

Gafar Ramazanov
Department of Science, Sumgait State University, Badalbayli Street, AZ 5008 Sumgait, Azerbaijan

M. Luisa Ramos
Coimbra Chemistry Centre, Department of Chemistry, University of Coimbra, 3004-535 Coimbra, Portugal

Ana C. F. Ribeiro
Coimbra Chemistry Centre, Department of Chemistry, University of Coimbra, 3004-535 Coimbra, Portugal. E-mail: anacfrib@ci.uc.pt

Daniela F. S. L. Rodrigues
Coimbra Chemistry Centre, Department of Chemistry, University of Coimbra, 3004-535 Coimbra, Portugal

Jozef Rychlý
Polymer Institute, Slovak Academy of Sciences, Slovakia

Maysa M. Sabet
Polymer Materials Research Department, Advanced Technologies and New Materials Research Institute (ATNMRI), City of Scientific Research and Technological Applications (SRTA-City), New Borg El-Arab City, P.O. Box: 21934 Alexandria, Egypt.
Laboratory of Bioorganic Chemistry of Drugs, Institute of Experimental Pharmacology and Toxicology, SK-814 04 Bratislava, Slovakia

Ján Sedliačik
Department of Science, Technical University in Zvolen, Masaryka 24, 960 53 Zvolen, Slovakia

Rita Shahnazarli
Institute of Polymer Materials of Azerbaijan, National Academy of Science, S.Vurgun str., AZ 5004 Sumgait, Azerbaijan

N. Sidamonidze
Department of Chemistry, Ivane Javakhishvili Tbilisi State University, Ilia Chavchavadze Ave., 0128 Tbilisi, Georgia. E-mail: sidamonidzeneli@yahoo.com

Anamika Singh
Department of Botany, Maitreyi College, University of Delhi, Delhi, India. E-mail: arjumika@gmail.com

Rajeev Singh
Department of Environmental Studies, Satyawati College, University of Delhi, Delhi, India

Ladislav Šoltés
Institute of Experimental Pharmacology and Toxicology, Slovak Academy of Sciences, 841 04 Bratislava, Slovakia

Dmitro L. Starokadomsky
Chuiko Institute of Surface Chemistry, National Academy of Sciences (NAS), Kiev, Ukraine. E-mail: stard3@i.ua

Heru Susanto
Department of Information Management, College of Management, Tunghai University, Taiwan. E-mail: heru.susanto@lipi.go.id
Department of Computational Science, The Indonesian Institute of Sciences, Jakarta, Indonesia

L. V. Tabatadze
Food Industry Department, Sukhumi State University, Ana Politkobskaia 9, 0186 Tbilisi, Georgia

Tamer M. Tamer
Polymer Materials Research Department, Advanced Technologies and New Materials Research Institute (ATNMRI), City of Scientific Research and Technological Applications (SRTA-City), New Borg El-Arab City, P.O. Box: 21934 Alexandria, Egypt. E-mail: ttamer85@gmail.com
Laboratory of Bioorganic Chemistry of Drugs, Institute of Experimental Pharmacology and Toxicology, SK-814 04 Bratislava, Slovakia

Jince Thomas
School of Chemical Sciences, Mahatma Gandhi University, Kottayam, Kerala, India

Dominika Topoľská
Institute of Experimental Pharmacology and Toxicology of SAS, Slovak Academy of Sciences, Bratislava, Slovakia

R. R. Usmanova
Department of Chemical Science, Ufa State Technical University of Aviation, Ufa 450000, Bashkortostan, Russia

Katarína Valachová
Institute of Experimental Pharmacology and Toxicology of SAS, Slovak Academy of Sciences, Bratislava, Slovakia

R. Vardiashvili
Department of Chemistry, Ivane Javakhishvili Tbilisi State University, Ilia Chavchavadze Ave., 0128 Tbilisi, Georgia

Luis M. P. Verissimo
Coimbra Chemistry Centre, Department of Chemistry, University of Coimbra, 3004-535 Coimbra, Portugal
Departamento de Química Física, Facultad de Farmacia, Universidad de Alcalá, 28871 Alcalá de Henares (Madrid), Spain

Runcy Wilson
School of Chemical Sciences, Mahatma Gandhi University, Kottayam, Kerala, India
Department of Chemistry, St. Cyrils's College, Kilivyal, Adoor 691529, Kerala, India

G. E. Zaikov
Department of Polymer Science, Emanuel Institute of Biochemical Physics, Russian Academy of Sciences, Moscow 119991, Russia

LIST OF ABBREVIATIONS

1,2-SPBD	syndiotactic 1,2-polybutadiene
3D	three-dimensional
AC	active centers
ACO	ant colony optimization
ACR	allylcarbinyl radicals
AIBN	azobisisobutyronitrile
AlkXh-Co	cobalt alkyl xanthogenate
AN	acrylonitrile
ANN	artificial neural networks
ANOVA	analysis of variance
AOPs	advanced oxidation processes
AP	4-aminophenol
APIs	application programming interfaces
BR	butyl rubber
BSA	bovine serum albumin
CCPS	para-(2-carboxy) cyclopropyl styrene
CH	collagen hydrolysate
COD	chemical oxygen demand
CPCR	cyclopropylcarbinyl radical
DEAC	diethyl aluminum chloride
DMB	2,3-dimethyl-1,3-butadiene
DM	decision maker
DMF	N,N-dimethylformamide
DoE	design of experiment
DTP	dithiophosphorylated
EAOPs	electrochemical AOPs
EAs	evolutionary algorithms
EAs	evolutionary strategies
EMO	evolutionary multiobjective optimization
ENR	epoxidized NR
EP	epoxy polymer
EP	evolutionary programming
FD	formaldehyde
FFD	full-factorial design

FGSs	functionalized graphene sheets
FO	forward osmosis
FTIR	Fourier transform infrared spectroscopy
GA	genetic algorithms
GO	graphene oxide
GPC	gel permeation chromatograph
HA	hyaluronan
HA	hyaluronic acid
HFCS	high-fructose corn syrup
HYALs	hyaluronidases
i.Sil	initial silica
IBP	ibuprofen
MA	methyl acrylate
MD	molecular dynamics
MED	multieffect distillation
MIDO	mixed-integer dynamic optimization
MMT	montmorillonite
mod.Sil	modified silica
MOEAs	multiobjective evolutionary algorithms
MOGA	multiobjective generic algorithm
MPE	modified polyethylene
MSF	multistage flash
NF	nanofiltration
NR	natural rubber
OAP	oligoaminophenols
o-DCB	o-dichlorobenzene
ODDT	Open Drug Discovery Toolkit
ODH	oxidative dehydrogenation
PAN	polyacrylonitrile
PDB	Protein Data Bank
PG	polymer grade
PMC	paramagnetic centers
PMMA	polymethyl methacrylate
PVA	polyvinyl alcohol
QSAR	quantitative structure–activity relationship
RO	reverse osmosis
S/N ratio	signals-to-noise ratios
SA 2	Screen Assistant 2
SBR	styrene–butadiene rubber
SCRs	structurally conserved regions

SEC	size-exclusion chromatographic
SEC-MALS	size-exclusion chromatography–multiangle light scattering
SEM	scanning electron microscope
SI	swarm intelligence
SPSA	simultaneous perturbation stochastic approximation
TAA	trialkylaluminum
TBT	tributyltin
TEA	triethylaluminum
UASB	upflow anaerobic sludge blanket
UF	ultrafiltration
UF	urea–formaldehyde
VCP	vinylcyclopropanes
WBOS	Weissberger biogenic oxidative system
XNBR	carboxylated acrylonitrile butadiene rubber

PREFACE

The chapters are divided into four sections: Polymer Chemistry and Technology, Bioorganic and Biological Chemistry, Nanoscale Technology, and Other Selected Topics. The volume will be valuable chemical engineers and researchers in many fields.

Applied Chemistry and Chemical Engineering, 5-Volume Set includes the following volumes:

- Applied Chemistry and Chemical Engineering,
  Volume 1: Mathematical and Analytical Techniques
- Applied Chemistry and Chemical Engineering,
  Volume 2: Principles, Methodology, and Evaluation Methods
- Applied Chemistry and Chemical Engineering,
  Volume 3: Interdisciplinary Approaches to Theory and Modeling
  with Applications
- Applied Chemistry and Chemical Engineering,
  Volume 4: Experimental Techniques and Methodical Developments
- Applied Chemistry and Chemical Engineering,
  Volume 5: Research Methodologies in Modern Chemistry and
  Applied Science.

PART I
Polymer Chemistry and Technology

CHAPTER 1

POLYMER–GRAPHENE NANOCOMPOSITES: A SMART MATERIAL FOR NEXT-GENERATION APPLICATIONS

AJITH J. JOSE[1,*], RUNCY WILSON[2,3], and JINCE THOMAS[3]

[1]*Department of Chemistry, St. Berchmans College (Autonomous), Changanassery 686101, Kerala, India*

[2]*Department of Chemistry, St. Cyrils's College, Kilivyal, Adoor 691529, Kerala, India*

[3]*School of Chemical Sciences, Mahatma Gandhi University, Kottayam, Kerala, India*

Corresponding author. E-mail: ajithjamesjose@gmail.com

CONTENTS

ABSTRACT

Graphene, a one-atom-thick sheet of carbon atoms with a two-dimensional hexagonal lattice structure, has attracted intense attention because of its unique properties such as fast charge carrier mobility, high thermal conductivity, and large surface area with potential applications in energy, catalysis, and sensing. The use of these materials in polymer as a reinforcement material found a dramatic enhance in the overall properties of the host material. The combination of the amazing physical properties of graphene and the ability to disperse in various polymer matrices has led to emerge of a new era of polymer nanocomposites. This chapter discusses about the use of graphene for making polymer nanocomposite and their impact on various polymers.

1.1   INTRODUCTION

The composites are solid multiphase materials formed through the combination of more than one material with different structural, physical, and chemical properties compared to their individual components. This makes composites different from the other multicomponent systems such as blends and alloys. In composites, one phase is continuous and is called matrix, while the other is filler material which make the dispersed phase. Composite materials based on the nature of matrix phase can be divided into polymeric, ceramic, and metallic composites. Usually, the filler phase is embedded to the host matrix phase to make a composite which has properties far from either phase alone. Polymers often have advantages over other materials such as metals and ceramics. They are widely used in various technical applications because of their unique advantages such as ease of production, lightweight, and ductility. However they have lower mechanical, modulus, and strength properties compared to that of metals and ceramics. The commercial importance of polymers and their increasing use, results to the continuous demand for improvement in their properties to meet the necessary conditions. By the composite technology, polymer properties are improved while maintaining their lightweight and ductile nature.[1,2] The recent advances in nanoparticle synthesis have unquestionably accelerated the growth of the composite industry. The capacity to synthesize and characterize atomic-level particles has produced a new generation of high-performance fillers and new class of materials, "polymer nanocomposites." Polymer nanocomposites are commonly defined as the combination of a polymer matrix and additives that have at least one dimension in the nanometer range. The additives can be one-dimensional, such as nanotubes and fibers, two-dimensional, which include layered clay minerals or graphene sheets, or three-dimensional, including spherical particles.

Graphene is a single layer of carbon atoms arranged in a honeycomb lattice which represents the two-dimensional carbon allotrope. Graphene is considered as the mother of all carbon based materials of all other dimensionality, for example, the 0D bukyminister, the 1D carbon nanotube, and the 3D graphite. Geim and Novoselov have provided the first demonstration for successful isolation of graphene.[3] Graphene has the superior mechanical, thermal, and electronic transport properties, and the defect-free graphene especially possesses a Young's modulus as high as about 1 TPa, breaking stress of over 100 GPa, thermal conductivity of about 5000 W/mK, and electrical conductivity of about 7000 S/cm.[4–7] The recent development of graphene-based materials produced by scalable chemical routes got a significant interest to establish effective dispersion techniques for and investigate the use of graphene-based materials for elastomer reinforcement.[8–11] In this chapter, we will discuss the graphene-based polymer nanocomposites and their limited properties.

1.2 MORPHOLOGY

To harness the fantastic properties, graphene has been considered to be incorporated into polymers, to prepare graphene-filled nanocomposites, which may offer a novel and intriguing nanostructured materials for various applications. The dispersion of nanometer-sized particles in the polymer matrix has a significant impact on the properties of nanocomposites. Interfacial bonding between graphene and the host polymer dictates the final properties of the graphene reinforced polymer nanocomposite. Pristine graphene is not compatible with organic polymers and does not form homogeneous composites. The surface modification of graphene is an essential step for obtaining a molecular level dispersion of individual graphene in a polymer matrix.[12]

In general, surface modification of nanoparticles is carried out by either chemical or physical methods. Chemical methods involve modification either with modifier agents or by grafting polymers. Silane coupling agents are the most used type of modification agents. Surface modification based on physical interaction is usually implemented by using surfactants or macromolecules adsorbed onto the surface of nanoparticles. The principle of surfactant treatment is the preferential adsorption of a polar group of a surfactant to the surface of nanoparticle by electrostatic interaction. A surfactant reduces the interaction between the nanoparticles within agglomerates by reducing the physical attraction and easily incorporated into polymer matrix. Thus, surface modification promotes the surface hydrophobicity of nanoparticles.[13]

There were lots of methods in the literature available for preparing polymer/graphene nanocomposites. In the case of natural rubber (NR) latex–graphene nanocomposites, the nanocomposites preparation can be mainly divided into two classes: (1) solution treatment (latex mixing): graphene dispersed in an appropriate solvent and is intermixed with NR latex (NRL) to form a homogeneous mixture, and then followed by the coagulation of the latex or solution casting method is employed; (2) Mechanical mixing (conventional) methods: graphene powder is directly blended with a solid NR matrix by an internal mixer or a two-roll mill process.[14–20]

1.3   MECHANICAL PROPERTIES

The major requirement of polymer nanocomposites is to optimize the balance between the strength/stiffness and the toughness as much as possible. Therefore, it is usually necessary to characterize the mechanical properties of the nanocomposites from different viewpoints. Tensile test is the most widely used method to evaluate the mechanical properties of the resultant nanocomposites, and accordingly Young's modulus, tensile strength, and the elongation at break are three main parameters obtained.[21] There are several possible reasons for this, which include enhanced specific area of graphene platelets, improved mechanical interlocking/adhesion at the nanofiller matrix interface, and the two-dimensional geometry of graphene platelets

The mechanical properties of NR reinforced with exfoliated graphene oxide (GO) nanoplatelets shows an excellent increase in the Young's modulus of the composites even at low filler loading, and it is mentioned that there exists a strong interfacial interaction between GO and NR. It is also shown that the Young's modulus increases with the strain rate and the dissipation increases for increased GO content because of the increased friction caused by the presence of high-aspect-ratio GO platelets.

She et al. reported that the self-assembly depresses restacking and agglomeration of grapheme oxide (GO) sheets and leads to homogenous dispersion of GO within epoxidized NR (ENR) matrix. The formation of hydrogen bonding interface between ENR and GO demonstrates a significant reinforcement for matrix polymer, when compared with pure ENR. The composite with 0.7 wt% GO loading receives 87% increase in tensile strength and 8.7-fold increases in modulus at 200% elongation after static in situ vulcanization.[22] Xing et al. reported that the well dispersed graphene nanosheets have strong interfacial interaction with styrene–butadiene rubber (SBR), resulting in improvement of the mechanical property of SBR.[23] It was found that nearly 11-fold increase in the tensile strength is attained upon addition of 7 phr of graphene. Ramanathan et al. reported the substantial improvement of the mechanical and thermal properties of polymethyl methacrylate (PMMA) with the addition of 1 wt% of thermally reduced graphene. The level of improvement in the mechanical properties (tensile strength and Young's modulus) is greater than the improvement by incorporation of 1 wt% of single-walled carbonnanotube (SWCNT) and expanded graphite.[24]

1.4 ELECTRICAL PROPERTIES

The most promising aspects of graphene-based materials are their potential for use in device and other electronics applications, owing to their high electrical conductivity. "Paper" materials made of stacked graphene platelets have been reported to exhibit conductivities as high as 35,100 S/m, and such highly conductive materials, when used as fillers, may increase the bulk conductivity of an otherwise insulating polymer by several orders of magnitude. In order for a nanocomposite with an insulating matrix to be electrically conductive, the concentration of the conducting filler must be above the electrical percolation threshold, where a conductive network of filler particles is formed.[25, 26]

The vulcanized graphene/NR composites with a conductive segregated network exhibit good electrical conductivity, water vapor permeability, and high mechanical strength. The composite exhibits a percolation threshold of ~0.62 vol% and a conductivity of 0.03 S/m at a content of 1.78 vol%, which is ~5 orders of magnitude higher than that of the composites made by conventional methods at the same loading fraction. Al-Hartomy et al. studied the influence on the concentration of graphene nanoparticles in NR on the dielectric and microwave properties in a wide frequency range (1–12 GHz). Results recommended graphene as a filler for NR-based composites to afford specific dielectric and microwave properties, especially when their loading with much more expensive carbon nanotubes is not possible.[27]

1.5 RHEOLOGY

Rheological properties have important implications in many and diverse applications. The applications of rheology are very important in the areas of industries involving metal, plastic, and many other materials. The rheological investigations will provide a mathematical description of the viscoelastic behavior of material under study. Also, study of nanocomposite rheology is important for the understanding of processing operations but it may also be used to examine nanocomposite microstructure. In linear viscoelastic rheology measurements, the low-frequency moduli may provide information on the platelet dispersion; for instance, the presence of a low-frequency storage modulus plateau is indicative of rheological percolation due to formation of a "solidlike" elastic network of filler. Ozbas et al. investigated the effects of functionalized graphene sheets (FGSs) on the mechanical properties and strain-induced crystallization.[28] The results revealed that the neat-NR exhibits strain-induced crystallization around a strain of 2.25, while incorporation of 1 and 4 wt% FGS shifts the crystallization to strains of 1.25 and 0.75, respectively. Xing et al. found that the well dispersed graphene nanosheets (GE) have a strong interfacial interaction with NR. The results revealed that the addition of low content of graphene can remarkably increase the tensile strength and the initial tensile modulus of NR. With incorporation of as low as 0.5 phr of GE, a 48% increase in the tensile strength and an 80% increase in the initial tensile modulus are achieved without sacrificing the ultimate strain. Dynamic mechanical measurement indicates that the storage modulus of the nanocomposites is greatly enhanced with addition of GE, while the loss tangent peak is depressed due to the reduced mobility of the rubber molecules.[29]

1.6 BARRIER EFFECT

The incorporation of inert and nonporous fillers into a polymer nanocomposites results in an increase in barrier property because filler particles lowers both solvent solubility and diffusivity within the polymer. Solvent transport in nanocomposites proceeds by a solution-diffusion mechanism in which the permeability (p) is given by $S \times D$, where S and D denote the solubility and diffusivity of the permeating species, respectively. The solubility provides a measure of interaction between the polymer matrix and penetrant molecules, whereas the diffusivity describes molecule mobility, which is normally governed by the size of the penetrant molecule as it winds its way through the permanent and transient voids afforded by the free volume of the nano-composites. Therefore, barrier property is to be strongly dependent on the amount of free volume in the polymer matrix.[30,31] Nanofillers are believed to increase the barrier by creating a "tortuous path" that retards the progress of penetrant molecules through the matrix.

The barrier property of graphene/polymer composites is affected strongly by the aspect ratio, dispersion and orientation of the graphene nanosheets, the graphene nanosheets/polymer interface, and the crystallinity of the polymer matrix. Due to its high aspect ratio and high electronic density of the carbon rings, graphene is able to repel atoms and gas molecules, and therefore has a very low solubility of gases. Scherillo et al. noted the self-assembling of reduced GO platelets in NR matrix as an interconnected network results in improved gas barrier properties. It was also noted that the nanocomposite structure results to be much more effective than homogeneous dispersion of graphene platelike particles at low graphene loadings.[32] Kanget et al. fabricated GO/carboxylated acrylonitrile butadiene rubber (GO/XNBR) nanocomposites with high mechanical and gas barrier properties environ-ment-friendly latex cocoagulation method. The results indicated that the gas permeability of the nanocomposite is significantly lower than that of the matrix. The high mechanical properties and low gas permeability of the nanocomposite are correlated to the homogeneous dispersion of the GO sheets and strong interfacial interactions, which facilitate the load transfer from XNBR to the GO sheets.[33]

1.7 THERMAL PROPERTIES

Thermal stability of polymer nanocomposites is estimated from the weight loss upon heating which results in the formation of volatile products. Ther-mogravimetric analysis demonstrates the thermal stability of the nanocom-posites, while differential scanning calorimeter determines the thermal transition behavior. The degradation behavior of polymers is commonly evaluated in terms of three parameters: (1) the onset temperature, considered as the temperature at which the system starts to degrade; (2) the degradation temperature, considered as the temperature at which maximum degradation rate occurs; and (3) the degradation rate, seen in the derivative weight loss as a function of temperature curve. Generally, the incorporation of nanometer-sized inorganic particles into the polymer matrix enhances thermal stability by acting as a superior insulator and mass transport barrier to the volatile products generated during decomposition.[34]

Yang et al. prepared a layer-aligned poly(vinyl alcohol) (PVA)/graphene nanocomposites in the form of films by reducing graphite oxide in the polymer matrix. The results indicate that graphene is dispersed on a molecular scale and aligned in the PVA matrix and there exists strong interfacial interactions between both components mainly by hydrogen bonding, which are responsible for the change of the structures and properties of the PVA/graphene nanocomposites such as the increase in T_g and the decrease in the level of crystallization.[35,36]

1.8   APPLICATIONS

Graphene is two-dimensional carbon nanofiller with a one-atom-thick planar sheet of sp^2 bonded carbon atoms that are densely packed in a honeycomb crystal lattice.[38] It is regarded as the "thinnest material in the universe" with tremendous application potential. Graphene is predicted to have remarkable properties, such as high thermal conductivity, superior mechanical properties, and excellent electronic transport properties. The superiority of graphene nanoplatelets over carbon nanotubes in terms of properties is related to their high specific surface area, wrinkled (rough) surface, as well as the two-dimensional (planar) geometry of graphene platelets.[39] These intrinsic properties of graphene have generated enormous interest for its possible implementation in a myriad of devices. These include future generations of high-speed and radio frequency logic devices, thermally and electrically conducting reinforced nanocomposites, ultrathin carbon films, electronic circuits, sensors, and transparent and flexible electrodes for displays and solar cells.[37]

High surface area carbons are considered as excellent electrode materials for supercapacitors because they have an attractive combination of properties, including no toxicity, wide availability, excellent electrical conductivity, high chemical and thermal stability, and relatively low cost. However, activated carbon suffers from low electrical conductivity and high resistance to ion transport resulting from its complex pore structure, consisting of variable pore sizes, pore structures, and pore lengths. The net result is that supercapacitors of activated carbons produce small specific capacitance restricting their successful application as high power density supercapacitor electrode materials. An alternative form of carbon called graphene offers the potential to utilize carbon but in a structure that overcomes the limitations of activated carbon. Graphene is one-atom-thick carbon sheet. It can be produced as single sheets of graphene or stacks of graphene arranged into platelets. Single graphene sheets and few-layered graphene platelets have remarkable properties such as high surface area, superior stiffness, strength, thermal and electrical conductivity, electronic transport properties, chemical and thermal inertness, etc. Thus, they fulfill requirements for not only supercapacitor electrode materials but also other energy storage/generation devices.

The active cathode in metal/air batteries consists of carbons, catalysts, and binders. Carbon aerogel, activated carbon, carbon black, carbon nanotube, and carbon foam have been investigated for use in lithium/air batteries and it has been known that high surface area carbon with high meso/macropore volume and large pore size increases the storage capacity of Li/Air battery. Nanosized thin grapheme (NTG) with high open surface area may also a good candidate to achieve high-energy storage capacity.[40]

Conducting polymer/graphene composites can also be used as electrode materials in a range of electrochromic devices. The polymer/graphene flexible electrode has some commercial applications in LEDs, transparent conducting coatings for solar cells and displays. The other commercial applications of graphene polymer composites are: lightweight gasoline tanks, plastic containers, more fuel efficient aircraft and car parts, stronger wind turbines, medical implants, biomedical applications, such as ultraminiaturized low-cost sensors for the analysis of blood and urine, and sports equipment. The discovery of graphene as a nanofiller has opened a new dimension for the production of lightweight, low-cost, and high-performance composite materials for a range of applications.[41]

1.9 CONCLUSIONS

Recently, graphene has attracted both academic and industrial interest because it can produce a dramatic improvement in properties at very low filler content. Graphene- and polymers-based nanocomposites show promising growth in technology and applications. However, a few key challenges must be addressed and resolved to realize the potential of graphene-based nanocomposites regarding synthesis methods, costs, and applications. For instance, if we consider the physical synthesis methods such as sonication, exfoliation, and cleaving of GO, the resultant product (graphene) can have a reduced aspect ratio, which can drastically degenerate the reinforcement, bonding interactions, and thermal and electrical properties of both the graphene and the nanocomposites. From the present review, it is clearly evident that both graphene and its derivatives have demonstrated their potential as promising candidates as reinforcements for high-performance nanocomposites. A large amount of research publications in the past 5 years signifies the importance of graphene that might surpass silicon research in the development of microelectronics. The possible applications of graphene-based material include transparent flexible electrodes, graphene/polymer composites for mechanical parts, energy storage, sensors and organic electronics. Graphene/polymer composites have showed the lowest percolation threshold for electrical conductivity and improved mechanical, thermal, and gas barrier properties. However, the core issues such as the homogeneous distribution of individual graphene platelets, their orientation, connectivity, and interface bonding with matrix still require more investigation. The main hurdles with any device fabrication using reduced grapheme oxide (RGO) including defects at atomic level, the folding/wrinkling of RGO, and the over lapping of RGO at macroscale require continuous research endeavors. Due to the high diversity, properties, and advantages of graphene, a multitude of nanocomposite-based applications have been envisioned to be practical. These multifunctional graphene composites coupled with affordable cost will soon be seen in the global market.

KEYWORDS

- graphene
- polymer
- nanocomposite
- carrier mobility
- thermal conductivity

REFERENCES

1. Jordan, J.; Jacob, K. I.; Tannenbaum, R.; Sharaf, M. A.; Jasiuk, I. *Mater. Sci. Eng. A* **2005**, *393*(1), 1–11.
2. Thomas, S.; Stephen, R. *Rubber Nanocomposites Preparation, Properties and Applications*; Wiley: Singapore, 2010.
3. Geim, A. K.; Novoselov, K. S. The Rise of Graphene. *Nat. Mater.* **2007**, *6*, 183–191.
4. Novoselov, K. S.; Geim, A. K.; Morozov, S. V.; Jiang, D.; Zhang, D. Y.; Dubonos, S. V.; Grigorieva, I. V.; Firsov, A. A. Electric Field Effect in Atomically Thin Carbon Films. *Science* **2004**, *306*, 666–669.
5. Zhao, Q.; Nardelli, M. B.; Bernholc, J. Ultimate Strength of Carbon Nanotubes: A Theoretical Study. *Phys. Rev. B* **2002**, *65*(144105), 1–5.
6. Zandiatashbar, A.; Lee, G. H.; An, S. J.; Lee, S.; Mathew, N.; Terrones, M. *Nat. Commun.* **2014**, *5*(3186), 1–9.
7. Lee, C.; Wei, X.; Kysar, J. W.; Hone, J. Measurement of the Elastic Properties and Intrinsic Strength of Monolayer Graphene. *Science* **2008**, *321*(5887), 385–388.
8. Balandin, A.; Ghosh, S.; Bao, W.; Calizo, I.; Teweldebrhan, D.; Miao, F. Superior Thermal Conductivity of Single-layer Graphene. *Nano Lett.* **2008**, *8*(3), 902–907.
9. Bolotin, K. I.; Sikes, K. J.; Jiang, Z.; Klima, M.; Fudenberg, G.; Hone, J. Ultra High Electron Mobility in Suspended Graphene. *Solid State Commun.* **2008**, *146*(9–10), 351–355.
10. Novoselov, K. S.; Geim, A. K.;Morozov, S. V.; Jiang, D.; Zhang, Y.; Dubonos, S. V. Electric Field Effect in Atomically Thin Carbon Films. *Science* **2004**, *306*(5696), 666–669.
11. Geim, A. K.; Novoselov, K. S. The Rise of Graphene. *Nat. Mater.* **2007**, *6*, 183–191.
12. Giannini, L.; Citterio, A.; Galimberti, M.; Cozzi, D. In *Rubber–Clay Nanocomposites*; Galimberti, M. Ed.; Wiley and Sons: New York, 2011; pp 127–144.
13. Zou, H.; Wu, S.; Shen, J. *Chem. Rev.* **2008**, *108*(9), 3893–3957.
14. Lin, Y.; Jin, J.; Song, M. *J. Mater. Chem.* **2011**, *21*, 3455–3461.
15. Vuluga, D.; Thomassin, J. M.; Molenberg, I.; Huynen, I.; Gilbert, B.; Jerome, C.; Detrembleur, C. *Chem. Commun.* **2011**, *47*, 2544–2546.
16. Ozbas, B.; O'Neill, C. D.; Register, R. A.; Aksay, I. A.; Prud'homme, R. K.; Adamson, D. H. *J. J. Polym. Sci. Part B Polym. Phys.* **2012**, *50*, 910–916.
17. Zhan, Y. H.; Wu, J. K.; Xia, H. S.; Yan, N.; Fei, G. X.; Yuan, G. *Macromol. Mater. Eng.* **2011**, *296*, 590–602.
18. Potts, J. R.; Shankar, O.; Du, L.; Ruoff, R. S. *Macromolecules* **2012**, *45*, 6045–6055.
19. Li, C.; Feng, C.; Peng, Z.; Buszman, P.; Trznadel, S.; Linke, A.; Jüni, P. *Polym. Compos.* **2013**, *34*, 88–95.
20. Hernández, M.; Bernal, M. M.; Verdejo, R.; Ezquerra, T. A.; López-Manchado, M. A. *Compos. Sci. Technol.* **2012**, *73*, 40–46.
21. Stanier, D. C.; Patil, A. J.; Sriwong, C.; Rahatekar, S. S.; Ciambella, J. The Reinforcement Effect of Exfoliated Graphene Oxide Nanoplatelets on the Mechanical and Viscoelastic Properties of Natural Rubber. *Compos. Sci. Technol.* **2014**, *95*, 59–66.
22. She, X.; He, C.; Peng, Z.; Kong, L. Molecular-level Dispersion of Graphene Into Epoxidized Natural Rubber: Morphology, Interfacial Interaction and Mechanical Reinforcement. *Polymer* **2014**, *55*(26), 6803–6810.
23. Xing, W.; Tang, M.; Wu, J.; Huang, G.; Li, H.; Lei, Z.; Li, H. Multifunctional Properties of Graphene/Rubber Nanocomposites Fabricated by a Modified Latex Compounding Method. *Compos. Sci. Technol.* **2014**, *99*, 67–74.

24. Ramanathan, T.; Abdala, A. A.; Stankovich, S.; Dikin, D. A.; Herrera-Alonso, M.; Piner, R. D.; Adamson, D. H.; Schniepp, H. C.; Chen, X.; Ruoff, R. S.; Nguyen, S. T.; Aksay, I. A.; Prud'homme, R. K.; Brinson, L. C. 2008.

25. Zhan, Y.; Lavorgna, M.; Buonocore, G.; Xia, H. Enhancing Electrical Conductivity of Rubber Composites by Constructing Interconnected Network of Self-assembled Graphene with Latex Mixing. *J. Mater. Chem.* **2012**, *22*(21), 10464–10468.

26. Tapas, K.; Sambhu, B.; Dahu, Y.; Nam, K. H.; Saswata, B.; Joong, L. H. *Prog. Polym. Sci.* **2010**, *35*, 1350–1375.

27. Al-Hartomy, O. A.; Al-Ghamdi, A.; Dishovsky, N.; Shtarkova, R.; Iliev, V.; Mutlay, I.; El-Tantawy, F. Dielectric and Microwave Properties of Natural Rubber Based Nanocomposites Containing Graphene. *Mater. Sci. Appl.* **2012**, *3*(7), 453–459.

28. Bulent, O.; Shigeyuk, T.; Benjamin, S.; Hsiao, C.; Benjamin, R.; Richard, A.; Ilhan, A. A.; Prud'homme, R. K.; Douglas, A. H. Strain-induced Crystallization and Mechanical Properties of Functionalized Graphene Sheet-filled Natural Rubber. *J. Polym. Sci. B Polym. Phys.* **2012**, *50*(10), 718–723.

29. Xing, W.; Wu, J.; Huang, G.; Li, H.; Tang, M.; Fu, X. Enhanced Mechanical Properties of Graphene/Natural Rubber Nanocomposites at Low Content. *Polym. Int.* **2014**, *63*(9), 1674–1681.

30. Kalaitzidou, K.; Fukushima, H.; Drzal, L. T. Multifunctional Polypropylene Composites Produced by Incorporation of Exfoliated Graphite Nanoplatelets. *Carbon* **2007**, *45*(7), 1446–1452.

31. Patel, N. P.; Zielinski, J. M.; Samseth, J.; Spontak, R. J. Effects of Pressure and Nanoparticle Functionality on CO_2-selective Nanocomposites Derived from Crosslinked Poly(ethylene Glycol). *Macromol. Chem. Phys.* **2004**, *205*, 2409–2419.

32. Scherillo, G.; Lavorgna, M.; Buonocore, G. G.; Zhan, Y. H.; Xia, H. S.; Mensitieri, G.; Ambrosio, L. Tailoring Assembly of Reduced Graphene Oxide Nanosheets to Control Gas Barrier Properties of Natural Rubber Nanocomposites. *Appl. Mater. Interfaces* **2014**, *6*(4), 2230–2234.

33. Kang, H.; Zuo, K.; Wang, Z.; Zhang, L.; Liu, L.; Guo, B. Using a Green Method to Develop Graphene Oxide/Elastomers Nanocomposites with Combination of High Barrier and Mechanical Performance. *Compos. Sci. Technol.* **2014**, *92*, 1–8.

34. Ganguli, S.; Roy, A. K.; Anderson, D. P. *Carbon* **2008**, *46*, 806–817.

35. Yang, X.; Li, L.; Shang, S.; Tao, X. M. Synthesis and Characterization of Layer-aligned Poly(vinyl Alcohol)/Graphene Nanocomposites. *Polymer* **2010**, *51*(15), 3431–3435.

36. Yu, A.; Ramesh, P.; Itkis, M. E.; Bekyarova, E.; Haddon, R. C. *J. Phys. Chem. C* **2007**, *111*, 7565–7569.

37. Song, Z.; Xu, T.; Gordin, M. L.; Jiang, Y. B.; Bae, I. T.; Xiao, Q.; Wang, D. Polymer–Graphene Nanocomposites as Ultrafast-charge and Discharge Cathodes for Rechargeable Lithium Batteries. *Nano Lett.* **2012**, *12*(5), 2205–2211.

38. Wang, G.; Shen, X.; Wang, B.; Yao, J.; Park, J. Synthesis and Characterisation of Hydrophilic and Organophilic Graphene Nanosheets. *Carbon* **2009**, *47*(5), 1359–1364.

39. Hyunwoo, K.; Ahmed, A. A.; Christopher, M. W. *Macromolecules* **2010**, *43*, 6515–6530.

40. Hyunwoo, K.; Yutaka, M.; Christopher, M. W. *Chem. Mater.* **2010**, *22*, 3441–3450.

41. Yongzheng, P.; Tongfei, W.; Hongqian, B.; Lin, L. *Carbohydr. Polym.* **2011**, *83*, 1908–1915.

CHAPTER 2

MODIFICATION OF WASTE POLYSTYRENE FOOD CONTAINERS AND STUDY OF ITS EFFICIENCY AS AN OIL SPILL CLEANUP

HIKMAT A. ALI*, MOAYAD N. KHALAF, and ALAA S. KHALAF

Chemistry Department, College of Science, University of Basrah, Basrah, Iraq

Corresponding author. E-mail: hikmatali42@gmail.com

CONTENTS

ABSTRACT

In the present study, the waste PS food containers were recycled and modified with acyl chloride and maleic anhydride. The final products were characterized by FTIR and ^{1}HNMR, and then their efficiency was studied as oil spill cleanup at different weight and at different time. Then the ability of crude oil recovery from sorbent materials and use of sorbent materials once more were studied.

2.1   INTRODUCTION

Oil is one of the important sources of energy in the modern industrial world, which has to be transported from the source of production to many places across the globe through oceans and in land transport. During transportation, the chance of oil spillage over the water body occurs due to accidents or by deliberate action during wartime that causes severe environmental pollution.[1,2] Oil spill is the release of a liquid petroleum hydrocarbon into the environment, especially in marine areas due to the human activity into the ocean or coastal waters, but spills could also occur on land. Oil spills may also be due to releases of crude oil offshore platforms, drilling rigs, and wells, as well as spills of refined petroleum products.[3,4] An important part of the field of oil spill control is the analysis of oil in various media. Oil analytical techniques are a necessary part of the scientific, environmental, and engineering aspects of oil spills. Analytical techniques are used extensively in environmental assessments of fate and effects. Laboratory analysis can provide information to help identify an oil if its source is unknown or what its sources might be. With a sample of the source oil, the degree of weathering and the amount of evaporation or biodegradation can be determined for the spilled oil. Properties of an ideal sorbent material for oil spill cleanup include oleophilicity and hydrophobicity, high oil sorption capacity, low water pickup (high oil/water selectivity), and high buoyancy. The sorbent materials can be categorized into three major classes: inorganic mineral products, organic natural products, and synthetic organic products.[5-7]

The present study aims to recycle waste PS polymer by modifying it with different organic active groups and using the new modified polymer as oil spill cleanup.

2.2 EXPERIMENTAL TECHNIQUES

The recycling of waste polystyrene (PS food container) carried out by using chemical reactions such as malieted and acylation reactions. The waste PS food containers were cleaned up by washing it with distilled water and with acetone to remove any contamination and then dried. Subsequently, they were cut into chips of ~1 cm² sizes.

2.2.1 MALTED AND ACYLATION REACTIONS

The waste PS reacted with maleic anhydride and acyl chloride using 1:5 weight ratios, respectively. The reaction was carried out in the presence aluminum chloride as catalyst in dichloromethane (DCM) as a solvent catalyst under nitrogen atmosphere and stirrer at room temp. for 24°C. At the end of reaction, filtration is carried out to separate the precipitate from the mixture medium. The precipitate obtained was purified and washed with distilled water and then with 0.5 M HCl. The final product was dried in a vacuum oven at 60°C and weighted.[8-10] The COD of product compounds is A and B, respectively.

SCHEME 2.1   Chemical equation of preparation compound A.

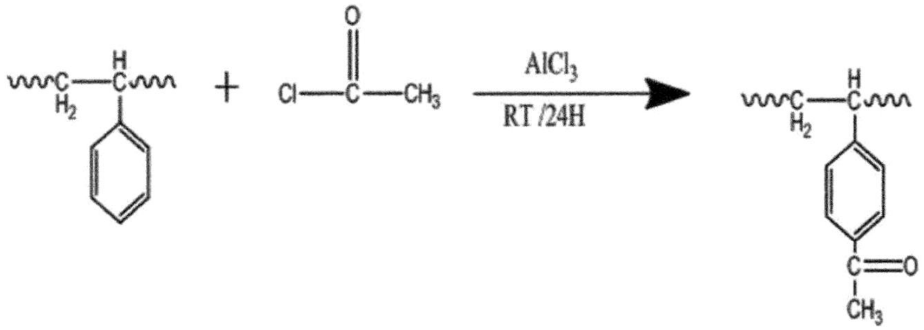

SCHEME 2.2   Chemical equation of preparation compound B.

The chemical structure of the waste PS and the new compounds (A) and (B) were characterized by FTIR spectroscopy as a sold disk by using KBr powder for sold sample.[11,12] From Figure 2.1 and Table 2.1, the FTIR spectra of waste PS food container gave a peak at 3050 cm^{-1} for stretching C–H of aromatic ring and two peaks at 2975 and 2850 cm^{-1} for asymmetric and symmetric CH$_2$, respectively. The peak was at 1600 cm^{-1} for stretching C=C of aromatic ring and two peaks at 1000 and 800 cm^{-1} for in plane bending C–H and out of plane bending C–H of aromatic ring, respectively. The two peaks at 1450 and 750 cm^{-1} for scissoring CH$_2$ and rocking CH$_2$ respectively.

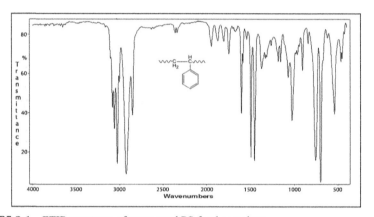

FIGURE 2.1   FTIR spectrum of compound PS food container.

TABLE 2.1   The Important Characteristics FTIR Bands and Their Location for Waste PS food Container.

Ar–H Stre (cm^{-1})	CH2 Asy (cm^{-1})	CH2 Sy (cm^{-1})	C=C Stre (cm^{-1})	C–H In plane (cm^{-1})	C–H Out of plane (cm^{-1})	C–H Sciss (cm^{-1})	CH$_2$ Rock (cm^{-1})
3050	2925	2850	1600	1000	800	1450	725

From Figure 2.2 and Table 2.2, the FTIR spectra of compound (A) gave a peak at 3415 cm^{-1} for stretching OH of carboxylic group and a peak at 1690 cm^{-1} for stretching carbonyl group. The peak at 1640 cm^{-1} for aliphatic carbon double bond and two peaks at 1400 and 690 cm^{-1} for =CH in plane bending and out of plane bending, respectively. From Figure 2.3 and Table 2.3, the FTIR spectra of compound (B) gave two peaks at 2830 and 2887 cm^{-1} for asymmetric and symmetric CH$_3$ and a peak at 1700 for stretching carbonyl group. The two peaks at 1400 and 1415 cm^{-1} for asymmetric bending and symmetric bending, respectively.

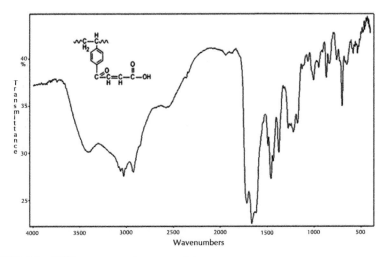

FIGURE 2.2   FTIR spectrum of compound A.

TABLE 2.2   The Important Characteristics FTIR Bands and Their Location for Prepared Compound A.

Comp	O–H Stre (cm⁻¹)	Ar–H Stre (cm⁻¹)	CH2 Asy (cm⁻¹)	CH2 Ay (cm⁻¹)	C=O Stre (cm⁻¹)	C=C Stre (cm⁻¹)	C=C* Stre (cm⁻¹)	=C–H In plane (cm⁻¹)	=C–H Out of plane (cm⁻¹)
J	3415	3055	2887	2800	1690	1580	1640	1400	690

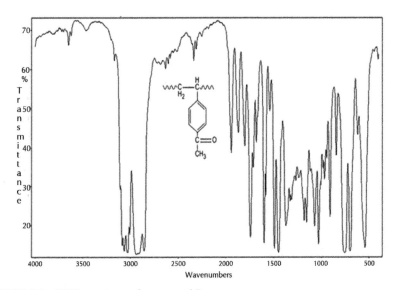

FIGURE 2.3   FTIR spectrum of compound B.

TABLE 2.3 The Important Characteristics FTIR Bands and Their Location for Prepared Compound B.

Comp	Ar–H Stre (cm⁻¹)	CH₃ Asy cm⁻¹	CH₃ Sy (cm⁻¹)	C=O Stre (cm⁻¹)	C=C Stre (cm⁻¹)	C–H In plane (cm⁻¹)	C–H, Out of plane (cm⁻¹)	CH₃ Asy-ben (cm⁻¹)	CH₃ Sy-ben (cm⁻¹)
K	3027	2830	2887	1700	1550	860	750	1490	1415

A further confirmation for formation of amino derivative from waste *poly(ethylene terephthalate)* (PET) was determined from ¹H NMR analysis.[12,13] The ¹HNMR spectroscopy was used for characterization of waste PS food containers by using chloroform (Sigma-Aldrich, 99.9% atom D) as a solvent. From Figure 2.4, the waste PS gave the peaks at 1.4 and 1.9 ppm for methylene (d) and (c), respectively. The peaks at 6.6 and 7.2 ppm for proton (a) and (b) of aromatic ring, respectively. The peak at 7.4 ppm for solvent. After modification of waste PS food containers, the compound (A) was characterized by using dimethylformaide (DMF) (Sigma-Aldrich, 99.9% atom D) as a solvent. From Figures 2.5 and 2.6, the compound A gave new doublet peaks at 7.5 and 8.4 ppm for methylene (b) and (a), respectively. The peak at 9.5 ppm is singlet for OH (c), the peaks at 2.75, 2.9, and 8 ppm for solvent. The compound B was characterized by using chloroform (Sigma-Aldrich, 99.9% atom D) as a solvent. From Figure 2.7, the compound B gave a new singlet peak at 0.9 ppm for methyl group (a).

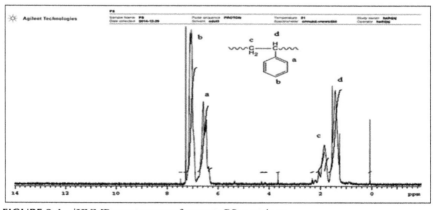

FIGURE 2.4   ¹HNMR spectroscopy for waste PS container.

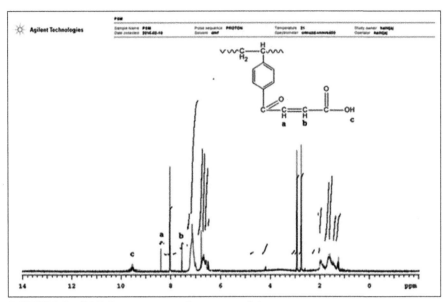

FIGURE 2.5 ¹HNMR spectroscopy for compound A.

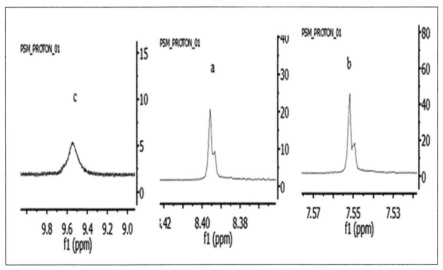

FIGURE 2.6 Expansion of ¹HNMR spectroscopy for compound A.

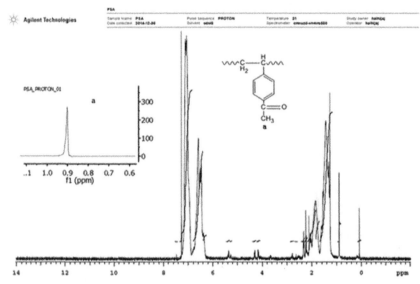

FIGURE 2.7   ¹HNMR spectroscopy for compound B.

2.3   EXPERIMENTAL PROCEDURE

The spill oil cleanup tests were carried out under the standard method (American Society for Testing and Materials F726-06). About 0.5 g of compound (L) was added to the oil spill and was stirred for 5 min to become a homogenous surface. After 1 h, the mixture was picked up and put on a filter paper to remove the excess crude oil before weighing it on a balance.[14,15] Oil absorbency (g/g) was calculated by the weight ratio between the absorbed oil and the original dried materials. The same procedure to oil spill cleans up by using compounds A and B, as shown in Figure 2.8.

FIGURE 2.8   Oil spill cleanup.

2.4   STUDY EFFECT INCREASING WEIGHT OF SORBENT MATERIALS ON OIL SPILL UPTAKE WITH CONSTANT TIME

The effect increasing weight studied by using different weight of sorbent materials (A and B) at constant time (1 h) and room temperature (27°C). This study carried out at the same procedure above (Section 3.3) as shown in Figure 2.9.

FIGURE 2.9   Effect of increasing weight of compounds on oil uptake.

In the present study, the general sorption mechanisms include absorption. The oil uptake (g) for sorbent materials obtained from the following equation:

$$\text{Oil uptake} = m_0 - m_s. \tag{2.1}$$

The oil sorption capacities for these sorbent materials obtained from the following equation:[16]

$$Q = \frac{m_0 - m_s}{m_s} \tag{2.2}$$

where Q is the oil sorption capacity (g/g), m_0 is the total mass of wet sorbent after the oil drained for 10 s, and m_s is the mass of the sorbent before sorption (g). The wet sorbent was taken out using a nipper, and then weighed. Each sample with a same sorption time was measured three times independently, and the average value was calculated. All of the oil sorption measurements were conducted at 26°C.

The oil uptake value (q) was calculated using eq 2.1. The q values of sorbent material increases with the increase in the weight of sorbent material as shown in Figures 2.10 and 2.11. When it reaches to optimum q values as shown in Table 2.4, the q values have liner relation with increase in the weight of sorbent material.[17]

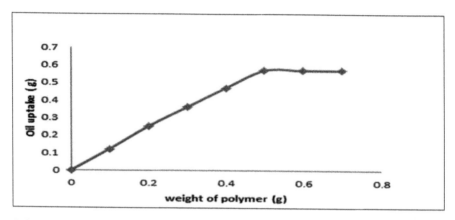

FIGURE 2.10   Effect of increasing weight on oil spill cleanup for compound A.

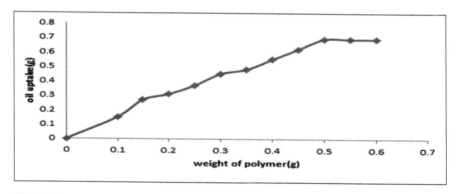

FIGURE 2.11   Effect of increasing the time on oil spill uptake for compound A.

TABLE 2.4   Optimum Weight and Optimum Oil Uptake of Sorbent Material.

Comp	Optimum weight (g)	Optimum oil uptake (g)	Oil absorption capacity (g/g)	Molecular weight (Da)
A	0.5	0.57	1.140	47480
B	0.5	0.67	1.340	46895

2.5   STUDY EFFECT OF INCREASING THE TIME ON OIL SPILL UPTAKE WITH CONSTANT WEIGHT OF POLYMERS

The effect of increasing time of absorption studied by using different times with constant weight (optimum weight) of sorbent material at room temperature 27°C for compounds A and B. This study is carried out at the same procedure in Section 2.3. From Figure 2.12 to Figure 2.13, we observe the oil uptake increasing with increasing the time of absorption. When it reaches to optimum values of (q), as shown in Table 2.5, the q values have liner relation with increase in the time of absorption. Absorbent materials are attractive for some applications because of the possibility of collection and

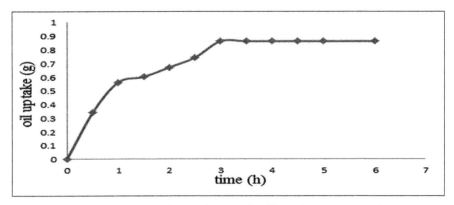

FIGURE 2.12   Effect of increasing the time on oil spill uptake for compound B.

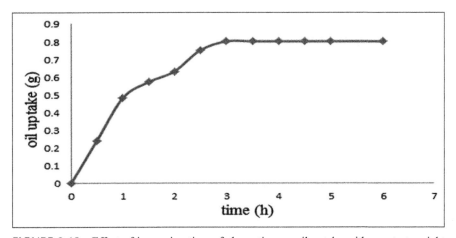

FIGURE 2.13   Effect of increasing time of absorption on oil uptake with constant weight (0.5 g) of compound B.

complete removal of the oil from the oil spill site. The good absorbent materials should include hydrophobicity and oleophilicity,[18] high uptake capacities, high rate of uptake, retention over time, oil recovery from absorbents, and the reusability. The oil absorption capacity depend on several factors such as the hole in structure of polymer, polar groups exhibit hydrophilic and oleophobic properties and cross-linking density (swelling ability), lower cross-linking density higher swelling, and absorption capacity and molecular weight, high molecular weight exhibits. The modified compounds (A and B) differ in molecular weight, polar group and holes, or cross-linking density in the backbone of polymers, therefor these compounds different oil absorption capability. The compound (A) is of high molecular weight (Table 2.5), high polar group and contains lower cross-linking or holes in their structure, therefore it exhibited high oil absorption capability Figure 2.12. The absorption capacity of sorbent materials followed arrangement (A and B) from higher (A) to lower (B) oil absorption capability as shown in Table 2.5. This difference in oil absorption capability of sorbent materials is due to difference in their molecular weight, polar group, and structure.[19] The optimum time and optimum oil uptake were shown in Table 2.5.

TABLE 2.5 Optimum Time and Optimum Oil Uptake of Sorbent Material.

Comp	Optimum time (h)	Optimum oil uptake (g)	Oil absorption capacity (g/g)	Molecular weight (Da)
A	3	0.86	1.720	47480
B	3	0.80	1.600	46895

2.6   CRUDE OIL RECOVERY FROM SORBENT MATERIALS

After oil spill treatment, the crude oil was recovered by washing the sample (sorbent materials loaded with crude oil) with toluene to dissolve crude oil and then filtered; the precipitate(sorbent material) was washed several times with toluene and dried to be used once more. The filtrate (crude oil) was transferred into rotary where it was evaporated to remove the solvent.

2.7   STUDY OIL SPILL UPTAKE EFFICIENCY BY THE PURIFIED COMPOUNDS

The oil uptake efficiency with the time was studied at constant weight for compounds A and B after purification from one absorption process.

From these results, the purified compounds showed good absorption capability to be used once more as sorbent materials for oil spill cleanup as shown in Figures 2.14 and 2.15. The results showed that the compound A has more absorption capability to use it once more than compound B (Figs. 2.16–2.18).

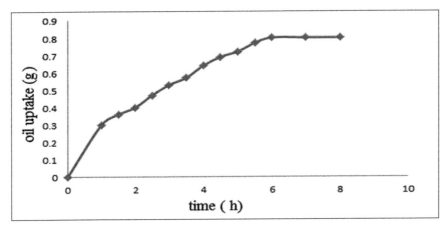

FIGURE 2.14 Oil spill uptake efficiency for purified compound A with constant weight (0.5 g).

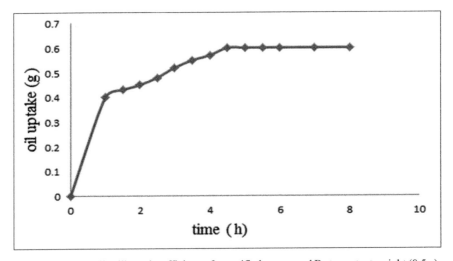

FIGURE 2.15 Oil spill uptake efficiency for purified compound B at constant weight (0.5 g).

FIGURE 2.16   Oil spill uptake.

FIGURE 2.17   Absorption process of oil spill.

FIGURE 2.18   After absorption process.

2.8 CONCLUSIONS

The main conclusions of the present study could be summarized in the following points:

1. The waste PS as a good material can be used in several applications.
2. New modification polymer was synthesized and characterized by FTIR and NMR spectroscopy.
3. The new modification polymer A and B showed high ability for oil spill cleanup.
4. The crude oil can be recovered from sorbent materials after absorption processes.
5. The sorbent materials can be used once more after crude oil recovery.

KEYWORDS

- **waste polymer**
- **oil spill cleanup**
- **FTIR**
- **HNMR**
- **sorbent materials**

REFERENCES

1. Hussein, M.; Amer, A. A.; Sawsan, I. I. Oil Spill Sorption Using Carbonized Pith Bagasse: Trial for Practical Application. *Int. J. Environ. Sci. Tech.* **2008,** *5,* 233–242.
2. Adebajo, M. O.; Frost, R. L.; Kloprogge, J. T.; Carmody, O.; Kokot, S. Porous Materials for Oil Spill Cleanup: A Review of Synthesis and Absorbing Properties. *J. Porous Mater.* **2003,** *10,* 159–170.
3. Zhu, K.; Shang, Y.; Sun, P.; Li, Z.; Li, X.; Weli, J.; Wang, K.; Wu, D.; Cao, A.; Zhu, H. Oil Spill Cleanup from Sea Water by Carbon Nanotube Sponges. *Front. Mater. Sci.* **2013,** *7,* 170–176.
4. Ramirez, C. E.; Batchu, S. R.; Gardinali, P. R. High Sensitivity Liquid Chromatography Tandem Mass Spectrometric Methods for the Analysis of Dioctyl Sulfosuccinate in Different Stages of an Oil Spill Response Monitoring Effort. *Anal. Bioanal. Chem.* **2013,** *405,* 4167–4175.
5. Iqbal, M. Z.; Abdala, A. A. Oil Spill Cleanup Using Grapheme. *Environ. Sci. Pollut. Res.* **2013,** *20,* 3271–3279.

6. Gong, Y.; Zhao, X.; Cai, Z.; O'Reilly, S. E.; Hao, X.; Zhao, D. A Review of Oil, Dispersed Oil and Sediment Interactions in the Aquatic Environment: Influence on the Fate, Transport and Remediation of Oil Spills. *Mar. Pollut. Bull.* **2014,** *79,* 16–33.

7. Burns, K. A. Analytical Methods Used in Oil Spill Studies. *Mar. Pollut. Bull.* **1993,** *26,* 68–72.

8. Li, Q.; Zhang, R.; Li, J.; Yan, X.; Wang, L.; Gong, F.; Su, Z.; Ma, G. In Situ Inhibitor (HCl) Removal Promoted Heterogeneous Friedel–Crafts Reaction of Polystyrene Microsphere with Lewis Acids Catalysts. *J. Mol. Catal. A Chem.* **2013,** *370,* 56–63.

9. Donchak, V.; Harhay, K. Synthesis of Fluorinated Polystyrene. *Chem. Chem. Technol.* **2008,** *2*(1), 11–14.

10. Furniss, B. S.; Hannaford, A. J.; Smith, P. W. G.; Tatchell, A. R. *Practical Organic Chemistry Text Book, Chapter 6*; John Wiley & Sons Inc.: New York, 1986; pp 1005–1016.

11. Stuart, B. *Infrared Spectroscopy: Fundamentals and Applications*; Wiley & Sons, Ltd, 2004. ISBNs: 0-470-85427-8 (HB); 0-470-85428-6 (PB).

12. Mistry, B. D. *Handbook of Spectroscopic Data Chemistry*; Oxford Book Company, 2009. ISBN: 978-81-89473-86-0.

13. Gauglitz, G.; Vo-Dinh, T. *Handbook of Spectroscopy*; Wiley-VCH Verlag GmbH & Co. KGaA, 2003. ISBN 3-527-29782-0.

14. Zhu, H.; Qiu, S.; Jiang, W.; Wu, D.; Zhang, C. Evaluation of Electrospun Polyvinyl Chloride/Polystyrene Fibers as Sorbent Materials for Oil Spill Cleanup. *Environ. Sci. Technol.* **2011,** *45,* 4527–4531.

15. J. Lin, J.; Shang, Y.; Ding, B.; Yang, J.; Yu, B. B.; Al-Deyab, S. S. Nanoporous Polystyrene Fibers for Oil Spill Cleanup. *Mar. Pollut. Bull.* **2012,** *64,* 347–352.

16. Zheng, M.; Ahuja, M.; Bhattacharya, D.; Clement, T. P.; Hayworth J. S.; Dhanasekaran, M. Evaluation of Differential Cytotoxic Effects of the Oil Spill Dispersant Corexit 9500. *Life Sci.* **2014,** *95,* 108–117.

17. Patoway, M.; Ananthakrishnan, R.; Pathak, K. Superhydrophobic and Oleophilic Barium Sulfate Material for Oil Spill Clean-ups: Fabrication of Surface Modified Sorbent by a One-step Interaction Approach. *Environ. Chem. Eng.* **2014,** *2,* 2078–2084.

18. Adebajo, M. O.; Frost, R. L.; Kloprogge, J. T.; Carmody, O.; Kokot, S. Porous Materials for Oil Spill Cleanup: A Review of Synthesis and Absorbing Properties. *J. Porous Mater.* **2003,** *10,* 159–170.

19. Reynolds, J. G.; Coronado, P. R.; Hrubesh, L. W. Hydrophobic Aerogels for Oil-spill Clean up Synthesis and Characterization. *J. Non-Cryst. Solids* **2001,** *292,* 127–137.

CHAPTER 3

TECHNICAL NOTE ON SURFACE MODIFICATION OF SILICA BY EPOXY RESIN

DMITRO L. STAROKADOMSKY[*]

Chuiko Institute of Surface Chemistry, National Academy of Sciences (NAS), Kiev, Ukraine

[*]*Corresponding author. E-mail: stard3@i.ua*

CONTENTS

ABSTRACT

Epoxy–silica polymer composites were prepared using an epoxy–dyan resin ED20 (Epoxy520) with 0, 2, and 5 wt% of silica PK300 (S = 300 m^2/g) with the original and modified (by initial epoxy resin) surfaces. The composites with epoxy-modified silica exhibited more homogeneous morphology compared with the composites with the neat silica. After curing, filled composites have higher resistance (compared with the unfilled) in aggressive ether acetate ink solvent, and similar resistance to concentrated HNO$_3$. Epoxy modification of the fillers may enhance stability in solvents (e.g., 5 wt% of silica), but reduces resistance to nitric acid (at 4 wt%). Filling by neat silica gives it better thermostability, whereas epoxy-modified silica does not improve it. Filling leads to an increase in adhesion strength characteristics of composite. Modification of the filler can significantly increase the adhesion to steel. Thus, filling by neat silica can be a good method to improve a number of composite characteristics, and epoxy-modification of silica shows an additive effect.

3.1   INTRODUCTION

It is known about the significant effect of silica filling on the structure and performance of epoxy polymer (EP). This effect depends on the nature of silica surface and its interaction with the epoxy resin. Modification of silica filler is a good method to improve the properties of silica-filled composite.

As silica modificators for EPs, more than six leading groups can be selected. There are: silanes[1] and siloxanes;[2] anhydrides and acides;[3] diols,[4] epoxides, and amines (components for EP);[4–6] hydrophobizators;[7] and other complex compounds. It should be expected that the range of promising modifiers for silica is largest. For example, there are silanes and olygoacrylates (their applications have been studied mainly for filled polyacrylates) that shows a positive effect on the properties of EP (our work[8]) It is known that the introduction of different types of silica is able to improve a chemical resistance in corrosive environments[9] and a number of the strength properties of the polymer.[8,11,12]

The influencing theme of epoxy-modified silica in the polyepoxide properties seems slow-studied at this moment. It is known that Porsch et al.[4] successfully modified and researched epoxy and dyol-modified silica. Interaction of polyepoxide with a silica surface during the polymerization was studied by Batler et al.[10]

Therefore, investigation of changes in the properties of the epoxy composite after surface modification of nanosilica by neat epoxy resin was of great interest.

We can assume that the surface modification of SiO_2 by epoxy from alcohol solution forms, attached (through hydrogen bonds and other interactions) to the surface, oligomeric chains, which subsequently could participate in polymerization (Fig. 3.1).

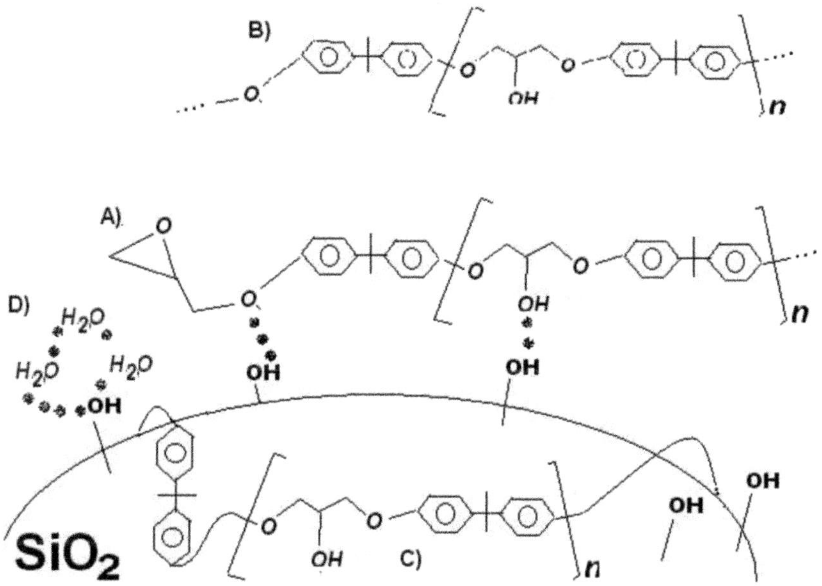

FIGURE 3.1 Schematic representation of the surface layer in system "silica/epoxy." (A) Epoxy oligomer molecules interacting with the surface through hydrogen bonds, (B) oligomer/polymer molecules "in bulk," (C) the oligomer molecule "lying" on the surface of SiO_2, and (D) clusters of surface-adsorbed water molecules.

Effect of filler modification is expected to enhance the reactivity of the surface-structured macromolecules. Also, effect of modifications is expected due to hydrophobic surface of silica and the weakening of its acidity. Immobilization of epoxy molecules on the surface of silica will block or oust clusters of surface-adsorbed water, which acts as inhibitor of polymerization (Fig. 3.1c). This should improve the distribution of silica in the resin. Consequence should be to improve the strength, transparency, homogeneity of the composition. As a result, after curing, it may increase visual aesthetics, strength, and durability of the composite.

3.2   EXPERIMENTAL METHODS AND REACTIVES

Such complex characteristics of composite have been investigated: visual (transparency), adhesion strength, resistance to aggressive environments (swelling, decomposition), and thermal stability.

The silica PK300 (S = 300 m^2/g, treated at 400°C) was taken as filler. Silica was modified (by adsorption from ethanol) by initial epoxy resin Epoxy520 (Czech production). This epoxy resin was filled by 2 and 5 wt% of initial and epoxy-modified silica. After 1 month, compositions were hardened by 1.5 wt% of polyethylene polyamine.

For swelling, the tablets of template (d = 1 cm, h = 0.1–0.2 cm) were immersed in ink solvent Inkwin (ether acetate, see Fig. 3.1) and in concentrated nitric acid. The changes in weight of templates calculated in percent versus its initial mass.

For strength methods were used next tests (according Soviet/Russian standard GOST 14760-59):

(1) *Adhesion strength at moving* of glued fiberglass plates (area of glued surface 3 cm^2)
(2) *Adhesion strength at tearing* of glued steel cylinders (area of glued surface 5 cm^2)

For statistical verification of data, the 5–10 identical templates (according student criteria) of each composite were tested. The result average F was calculated after excluding of minimal result [$F = (F_1 + F_2 + ... + F_{i-1})/(i - 1)$, where i—number of templates].

3.3   RESULTS AND DISCUSSION

3.3.1   VISUAL EFFECTS

Composites with initial silica (i.Sil) are more turbid and viscous if compared with unfilled polymer (H) and with filled by modified silica (mod.Sil) composites. The viscosity and turbidity of i.Sil composite grows in row 0–2–5 wt% of silica (Fig. 3.2). At 5 wt% of i.Sil, composite becomes ticsotropic. In contrary, for mod.Sil all composites (2 and 5 wt% of mod.Sil) are transparent similar to H-composite (Fig. 3.2). This fact tells about better compatibility of phases in composite (and its better homogenization) after epoxy-modifying of silica.

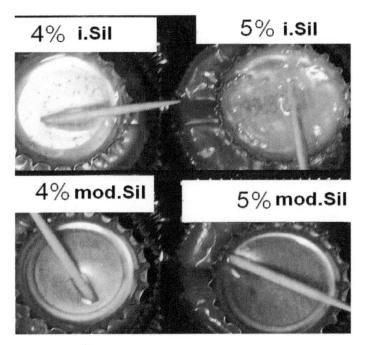

FIGURE 3.2   Templates with 4–5 wt% of silica before hardening (templates with transparent mod.Sil).

Microscopy: The real structure of epoxy–polymer includes many pores and defects (Fig. 3.3). After filling, the number of pores decreases and its size increases (Fig. 3.4B). In addition, microisles of silica agglomerates appear (Fig. 3.4C).

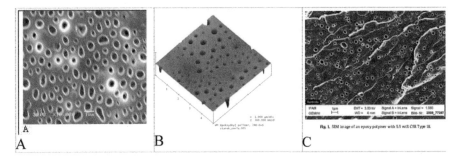

FIGURE 3.3   SEM (A) and AFM (B) images (scale 5 mcm) of unfilled epoxy polymer with 0 wt% of filler, obtained by Sprenger.[7]

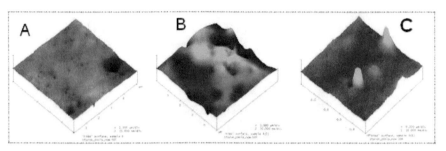

FIGURE 3.4   AFM images of surfaces for composites A—unfilled (scale 5 mcm), B—3.5 wt% of silica (scale 5 mcm); 3.5 wt% of silica (scale 1 mcm).

3.3.2   STABILITY OF COMPOSITES IN AGGRESSIVE MILIEUS

Swelling in aggressive ink solvent: Ink solvent (Fig. 3.5) Inkwin is a phys-ical-aggressive environment for epoxy-glued details of polygraphic systems (plotters). Molecule of solvent can suffer sterical difficulties when incor-porated in nanopores of polymer structure. But for solvent obviously is not difficult to penetrate in micro/mesopores and cavities of structure.

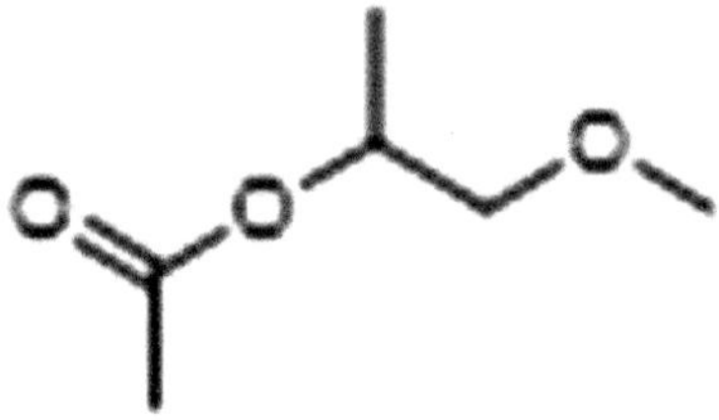

FIGURE 3.5   Formula of ink solvent Inkwin, according our NMR analysis.

After endurance in solvent, the pores, microchannels, and not-hardened boards in templates appears (due to leaching of substances by solvent), the enduring templates become yellow. As a rule, endurance in solvent creates a globular structure of EP, with globules up to 1 mm (Fig. 3.6).

Thus, filling lets decrease the size and number of defects or heterogeneity in EP. They are the least visible in the sample with the unmodified silica (i.Sil), which may indicate the formation of a more solid and dense structure of the composite. The sample with mod.Sil has globules of smaller size than the unfilled sample (Fig. 3.6).

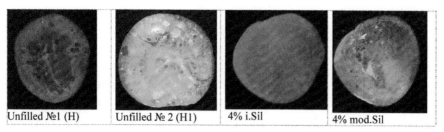

| Unfilled №1 (H) | Unfilled № 2 (H1) | 4% i.Sil | 4% mod.Sil |

FIGURE 3.6 The samples (diameter $d = 0.5 \pm 0.1$ cm) after 24 days of exposure in a solvent. Visible globular formation (samples H, H1, and 4% mod.Sil) and silica agglomerates (in sample 4% i.Sil) in the middle of the samples.

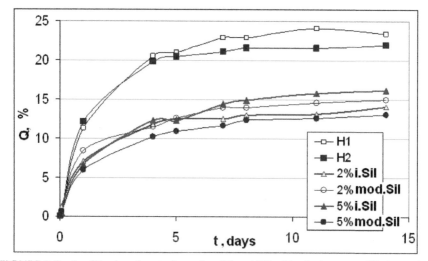

FIGURE 3.7 Swelling in solvent of samples. H1 and H2—identical mass unfilled samples.

Experiment on the swelling of the samples in a solvent showed that the swelling is almost finished before the 15th day of exposure (Fig. 3.7). From Figure 3.7, it follows that 2–5 wt% of silica leads to obvious slowing of swelling in a solvent. The unfilled samples (as we can see, the spread of data on them is negligible) after the first day swells by $Q > 10$ wt%, and after 5 days by $Q > 20\%$. Filling 2 reduces Q by almost half and composite with 5 wt% mod.Sil is maximal resistant sample (Fig. 3.7). Thus, filling leads to plumping of polymer-composite structure of the composite network, that hinders the penetration of the solvent molecules. Modification at certain concentrations (5 wt%) can contribute even more noticeable seal structure of the composite.

Swelling in concentrated nitric acid: In concentrated HNO_3, in all cases, there is a rapid decomposition of the composite (visual degradation of composites in acid is shown in Fig. 3.9). The swelling curves have a classic look and includes two sections—fast (up to 3 days before line 1 in Fig. 3.8) and slow (3–16 days, the line 2 Fig. 3.8). At these stages, acid molecules are introduced into the polymer network with oxidation of the polymer. This is manifested visually by the color change of the samples (with white transparent to brown, Figure 3.9), apparently due to the accumulation of nitrogen dioxide and the decomposition products of the polymer. After 16 days the swelling curve reflects the chaotic spread of results (Fig. 3.8)—apparently due to the competition between the gas release, swelling, and leaching. The addition of silica does not change the parameters of the swelling (Fig. 3.8).

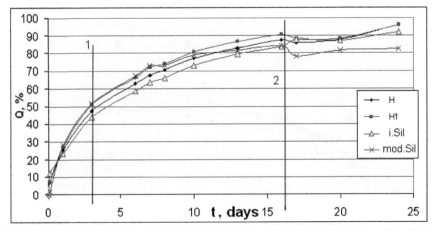

FIGURE 3.8   Swelling of freshly prepared sample—unfilled (H and H1), 4% and 4 wt% of silica in concentrated nitric acid.

Rather, filling by i.Sil makes polymer more resistant to oxidation with nitric acid, probably due to the acidic properties of the filler surface. It follows that surface and aggregation of i.Sil can act as stabilizer of polymer oxidation with nitric acid. Perhaps an important role here belongs to the acidic nature of surface of neat silica and its aggregation.

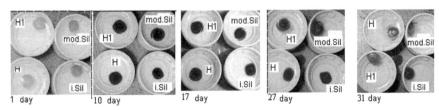

FIGURE 3.9   Shape of the samples in concentrated nitric acid.

Epoxy modification of silica negates the stabilizing properties of silica in epoxy polymer. The appearance of the epoxy modifier on silica surface makes the composite even less stable than the unfilled composite. It is seen by increase in the rate of swelling (at first stage, see "mod.Sil" Fig. 3.8) and a reduction in the lifetime of the acid (Table 3.1 and Fig. 3.9). As can be seen from Table 3.1, the time to complete decomposition of the samples (lifetime) is limited to weeks. This fact suggests the need to find other ways of modifying silica to improve acid resistance.

TABLE 3.1   Lifetime (Number of Days from the Start of Immersion in Acid Until Decomposition) of Samples in Concentrated HNO_3.

	H (unfilled)	i.Sil	Mod.Sil
At $15 \pm 1°C$	29	32	25
At $20 \pm 1°C$	10	14	7

3.3.3   THE ADHESION STRENGTH OF COMPOSITES

The tests of the adhesive shear for gluing fiberglass and tearing for gluing steel molds were conducted. Breaking load (in kg) for unfilled composite was taken as 100%, and effect of filling was calculated as $(100 \pm x)$% versus the load for unfilled. The results are shown in Table 3.2.

TABLE 3.2   Comparative Indicators of the Breaking Load of Composites (in kgf) as % to the Breaking Load of the Unfilled Polymer (100%).

	Unfilled	2% i.Sil	5% i.Sil	2% mod.Sil	5% mod.Sil
Adhesive shear (on fiberglass)	100 (200 kgf)	112	115	110	100
Adhesive tearing (on steel)	100 (450 kgf)	107	109	117	117

As can be seen from Table 3.2, the introduction of neat silica (i.Sil) steadily increases the studied mechanical properties. Adhesive tearing responds a significant increase in the content and modification of the filler (Table 3.2)—is clearly visible sequence of the reinforcement of composites in the series:

H < 2% i.Sil < 5% i.Sil < 2, and 5% of mod.Sil.

But surface modification does not always lead to an increase of the strength (adhesive shear strength, Table 3.2). Thus, we can say that the surface modification of silica improves compatibility with epoxy resin visually (Fig. 3.2), as some real strength indicators (Table 3.2).

3.3.4 HEAT RESISTANCE: A QUALITATIVE VISUAL ANALYSIS

Thermal stability was evaluated qualitatively, by one-side heating on the steel surface with visual fixation of temperature of destructive changes—intensive gassing (smoke emission, boiling) or discoloration (yellowing, browning, carbonization)—during decomposition of samples. Thermal stability of samples for smoke and gas emission reveals the extreme temperatures of their stability as a material. Temperature at the start of degradation (color change from transparent to brown), and the final temperature of decomposition (melting and strong smoke formation) is important to determine the potential risk of fire for polymer products.

TABLE 3.3 Extreme Temperatures by Heating the Samples on the Substrate.

Sample	T (°C)	Effect
Rapid heating (≈50°C/min)		
0 мас% (sample **H**)	270	Boils
4% i.Sil	385	Twitch
4% mod.Sil	310	Boils
Slow heating (20 ± 10°C/min)		
0	290–300	Boils, then at 360°C crack
4% i.Sil	420–440	Without boiling, burn after 440°C
4% mod.Sil	335	Strong twitching; at 235°C in a blackened whiten polymer agglomerates.

For the unfilled sample, in all cases, an active destruction begins at about 280 ± 10°C. Filling by i.Sil gives a better thermal stability of composite. However, here an increased scatter in the data are seen as a clear consequence of the heterogeneity of the structure due agglomeration of silica. Filling by mod.Sil reduces heat resistance versus i.Sil filled as unfilled composite. Thus, filling by i.Sil can improve thermal stability of composite, but modification of silica surface by epoxy resin is not optimal way to its further strengthening.

3.4 CONCLUSIONS

1. Surface modification of silica by epoxy resin for subsequent filling and curing of epoxy resin provides a more homogeneous distribution of filler in the composition. This effect proves a better susceptibility of epoxy-modified silica with epoxy phase.
2. The study of swelling in ether acetate ink solvent showed a significant slowdown of swelling after filling. The most effective was the introduction of 5 wt% of epoxy modified and 2 wt% of neat silica. This fact suggests the more seal polymer structure after filling. Exposure of solvent also reveals pores, globules, and not-harden zones in the structure of polymer. Filling leads to a visual decrease in the number and size of these regions.
3. Epoxy-modification of silica makes the composite less resistant to degradation in the concentrated nitric acid. This may indicate an important role of the unmodified silica surface to increase the acid resistance of the filled polymer.
4. It is often multidirectional impact of surface-modification on the adhesion properties of composite. Test on the adhesion strength to fiberglass (shear) and steel (tearing) have shown the possibility of composites amplification by 2–5 wt% filing of neat silica ($S = 300$ m²/g). Epoxy modification of silica surface improves adhesion strength to steel (growth on 15–20% for modified silica versus 7–9% for neat silica), but does not improve the adhesion strength to fiberglass (growth 12–15% for neat silica and 0–10% for modified silica). This may show a better susceptibility of composite "epoxy resin–neat silica" with fiberglass due to self-organized aggregate chains of silica in composite structure.
5. Analysis of the thermal stability shows that filling by 4 wt% of neat silica can significantly improve the heat resistance of the sample (growth up to 20–30%). But epoxy-modification of silica reduces this effect (no more than 10%). This may indicate an important role of the unmodified silica surface to increase the thermoresistance of the filled polymer.

KEYWORDS

- filling
- silica
- modification
- epoxy
- composite
- adhesion
- swelling
- thermostability

REFERENCES

1. Chonkaew, W. Modifications of Epoxy Resins for Improved Mechanical and Tribological Performances and Their Effects on Curing Kinetics. Dissertation Prepared for the Degree of Ph.D., University of North Texas, May 2008, 187, p 3.
2. Taylor, A. C.; Bray, D. J.; Guild, F. J.; Hsieh, T. H.; Kinloch, A. J.; Masania, K. The Toughness of Epoxides Modified with Silica Nanoparticles. http://www.adhesionsociety.org/wp-content/uploads/2013-Annual-Meeting-Abstracts/Taylor_The_2013.pdf (accessed 2016).
3. Barabanova, A.; Shevnin, P.; Prahina, T., et al. Nanocomposites from Epoxy Resin and SiO$_2$ Particles. *Visokomolekularnie Soedinenia* [in Russian], **2008**, *50A*(7), 1242–1254.
4. Porsch, B. Epoxy- and Diol-modified Silica: Optimization of Surface Bonding Reaction. http://dx.doi.org/10.1016/0021-9673(93)80385-L (accessed 2016).
5. Meador, M. A. B.; Fabrizio, E. F.; Ilhan, F.; Dass, A.; Zhang, G.; Vassilaras, P.; Chris Johnston, J.; Leventis, N. Cross-linking Amine-modified Silica Aerogels with Epoxies: Mechanically Strong Lightweight. Porous Materials. http://pubs.acs.org/doi/pdf/10.1021/cm048063u (accessed 2016).
6. Chen, M.-H.; Chen, C.-R.; Sun, S.-P.; Su, W.-F. Low Shrinkage Light Curable Nanocomposite for Dental Restorative Material. *Dent. Mater.* **2006**, *22*, 138–145.
7. Sprenger, S. Epoxy Resins Modified with Elastomers and Surface-modified Silica Nanoparticles. *Polymer* **2013**, *54*, 4790–4797.
8. Starokadomsky, D. L.; Telegeev, I. G. *Modification of Nanosilica by Acrylates for Improving of Strength and Stability of Epoxy-Silica Compositions*, 3th Intern. Conf. Nanostructured Materials NANO-2012: Belarus-Ukraine-Russia, November 2012, Minsk, Belarus, p 447.
9. Starokadomsky, D. L.; Telegeev, I. G. Resistance of Epoxy Polymer with 2–5 wt% Nanosilica in Aggressive Acid Medium. *Open J. Polym. Chem.* **2012**, *2*(3), 117–123. DOI: 10.4236/ojpchem.2012.23016.

10. Batler, J.; Becker, N.; Ziehmer, M.; Thomassey, M.; Zielinski, B.; Muller, U.; Sanctuerry, R. Interactions Between Silica Nanoparticles and an Epoxy Resin Before and During Network Formation. *Polymer* **2009,** *50*(14), 3211–3219.

11. Hsieh, T. H.; Kinloch, A. J.; Masania, K.; Sohn Lee, J.; Taylor, A. C.; Sprenger, S. The Toughness of Epoxy Polymers and Fibre Composites Modified with Rubber Microparticles and Silica Nanoparticles. *J. Mater. Sci.* **2010,** *45*(5), 1193–1194.

12. Starokadomsky, D. L. About Influence of Non-modified Nanosilica on Physicomechanical Properties of Epoxy Polymer Composites. *Russ. J. Appl. Chem.* **2008,** *12*, 2045–2051.

CHAPTER 4

PHOTOSENSITIVE COPOLYMERS OF PARA-(2-CARBOXY) CYCLOPROPYL STYRENE WITH 2,3-DIMETHYL-1,3-BUTADIENE

K. G. GULIYEV*, S. B. MAMEDLI, TS. D. GULVERDASHVILI, and A. M. GULIYEV

Institute of Polymer Materials of Azerbaijan, National Academy of Sciences, Baku, Azerbaijan

*Corresponding author. E-mail: ipoma@science.az

CONTENTS

ABSTRACT

The radical copolymerization of para-(2-carboxy) cyclopropyl styrene (CCPS) with 2,3-dimethyl-1,3-butadiene (DMB) has been carried out. It has been shown that the copolymerization reactions proceed with opening of double bond of vinyl group of CCPS and 1,4-addition of DMB. The copolymerization constants have been determined, and the factors of activity and the polarity of the used monomers have been calculated. It has been shown that the synthesized copolymers are capable for structurization under action of UV-irradiation and show the high photosensitive properties.

4.1 INTRODUCTION

The prospects of use of polyfunctional reactive polymers, in particular, for creation of the negative photoresists stipulate an interest to their synthesis.[1–4] One of the basic solutions of this problem is the synthesis, homo- and copolymerization of functionalized styrene monomers.[5–7]

In this connection, the purpose of this study was the preparation of polymer light-sensitive compounds for creation of new negative photoresists. The copolymerization reaction of para-(2-carboxy) cyclopropyl styrene (CCPS) with 2,3-dimethyl-1,3-butadiene (DMB) has been selected as the object of investigation.

4.2 EXPERIMENTAL DETAILS

The polyfunctional monomer of CCPS has been synthesized according to methodology presented in Ref. 8.

CCPS has been identified by the methods of elemental, chemical, IR-, and PMR-spectroscopic analyses. Its purity, determined by a method of gas–liquid chromatography, was 99.9%.

The copolymerization reaction of CCPS with DMB was carried out in ampoules with benzene solution in the presence of 0.2 mol % of azobisisobutyronitrile (AIBN) from total amount of comonomers at 70°C. The forming copolymers were purified by twofold precipitation from benzene solutions by methanol and dried at 30°C in vacuum (15–20 mm merc.c.).

The IR-spectra of monomers and for thin copolymer films were obtained using spectrometer UR-20 and PMR-spectra using spectrometer BS-487B (Tesla, 80 MHz) in the deuterated chloroform solution. The contents of carboxyl groups and internal double bonds in composition of copolymers have been determined by methods presented in Ref 9 and method of ozonization on apparatus ADS-4, respectively.

For investigation of photochemical structurization of copolymers, their 2–10% solutions were prepared and applied on glass substrate (K-8) by a size of 60 × 90 mm. Putting was carried out by a method of centrifugation. The layer was dried for 10 min at room temperature, for 20 min at 40–45°C and pressure 10 mm merc.c. As a source of UV-irradiation, a mercuric lamp DPT-220 (current strength 2.2 A) was used, at a distance from source of radiation—15 cm, supposed exponometer shutter rate—720 mm/h, and exposition time—5–20 s. The fraction of insoluble polymer was determined in relation of its mass to the initial film mass.

4.3   RESULTS AND DISCUSSION

The results of elemental and chemical (determination of content of carboxyl groups and double bond) analyses of various fractions of joint polymerization products of CCPS with DMB are practically identical, that is, during their copolymerization the homopolymers of the used monomers are not formed. At the same time, the values of these indices are essentially differed for copolymers prepared at various molar ratios of comonomers.

In the IR-spectra of copolymers of CCPS with DMB (CCD), the absorption band at 1035–1045 cm^{-1} characteristic for three-membered carbon cycle is kept. In addition, there are fixed the absorption bands corresponding to valence vibrations of benzene ring (1450, 1500, and $1570–1600^{-1}$), nonplanar deformation vibrations of two neighboring aromatic C–H bond (830–835 cm^{-1}), valence vibrations of C–H bond in $-CH_2-$, $>CH-$, and $-CH_3$ groups (2930–2940, 2860–2970, and 2965–2970 cm^{-1}) and valence vibrations of internal double bond (1660–1670 cm^{-1}). It also should be noted that on the total spectral contour in the field of 2800–3000 cm^{-1} three maxima at 2975, 2935, and 2867 cm^{-1}, belonging to $\delta_{as}(CH_3)$, $\delta_{as}(CH_2)$, and $\delta_s(CH_3)$ are, respectively, isolated. The absorption bands in the field of 3500–3600 cm^{-1} can be referred to the valence vibrations of associated hydroxyl groups.

The relative intensities of absorption band of determined groups and bonds in the IR-spectra of copolymers prepared at various molar ratios of CCPS with DMB are differed. In particular, with growth of content of CCPS in the initial monomer mixture the relative intensity of absorption band of cyclopropane cycle (1035–1045 cm^{-1}) is increased, and in a case of growth of the molar fraction of DMB in the initial mixture the absorption bands at 1660–1670 cm^{-1} become comparatively more intensive.

In PMR-spectra of CCD, the resonance signals of benzene (6.65–7.94 ppm) and cyclopropane (0.68–1.75 ppm) rings, and methyl and methylene groups (1.6 and 2.0 ppm, respectively) are fixed.

The above-mentioned spectral data show that at radical copolymerization of CCPS with DMB an opening of double bond of vinyl group of CCPS takes place, and cyclopropane group remains unaffected as a side group.

One can conclude that although CCPS is the polyfunctional compound, its copolymerization with DMB is realized on the following scheme with formation of the reactive polyfunctional copolymers:

For elucidation of dependence of composition of the copolymers on composition of the initial mixture of monomers, the copolymerization reaction was carried out at various molar ratios of CCPS (M_1) and DMB (M_2). The mole fractions of the corresponding links m_1 and m_2 have been found by determination of content of carboxyl groups and internal double bond (Fig. 4.1, Tables 4.1 and 4.2). On the basis of these results, the copolymerization constant values of CCPS with DMB have been determined, and on scheme Q–e the factors of their activity and polarity have been calculated. It is seen from Table 4.1 that $r_2 > r_1$, that is, DMB is the more active monomer than CCPS at joint copolymerization. In all depths of conversion of comonomers, the soluble copolymers are prepared, that is, during copolymerization of CCPS with DMB; the processes of structurization of the prepared copolymers do not take place.

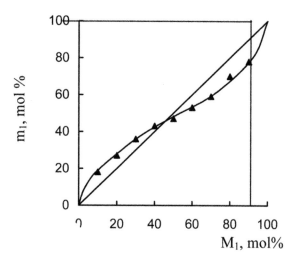

FIGURE 4.1 Dependence of molar composition of links CCD on molar content of CCPS.

TABLE 4.1 Conditions of Carrying Out of the Copolymerization Reaction of CCPS with DMB and Compositions of Copolymers (t—70°C, initiator AIBN).

Content of monomers in the initial mixture (mol%)		Composition of copolymer (mol%)		Content of double bond (%)	Content of COOH groups (%)
M_1	M_2	m_1	m_2		
90	10	78.59	21.41	3.11	21.4
80	20	68.45	31.55	4.90	19.9
70	30	59.27	40.73	6.71	18.4
60	40	52.64	47.46	8.26	17.2
50	50	47.80	52.20	9.44	16.2
40	60	42.33	57.67	10.9	15.0
30	70	36.10	63.90	12.7	13.5
20	80	27.15	72.85	15.8	11.0
10	90	17.52	82.48	19.7	7.84

$r_1 = 0.31, r_2 = 0.43, Q_1 = 1.04, e_1 = -0.39; Q_2 = 5.86, e_2 = -1.81$

The calculated values of parameters Q and e for CCPS, their comparison with analogous styrene parameters ($Q = 1, e = -0.8$) evidence about distance influence of substituents in cyclopropane ring on general electron state of all molecules of CCPS. The spectral data and results of our previous investigations[10] indicate to availability of uniform system of conjugated bonds in molecules of paracyclopropyl substituted styrenes. As a result of this, an electronic nature of substituent of cyclopropane ring shows an influence on electron density of vinyl group and changes activity of monomer during radical homo- and copolymerization.

The synthesized copolymers are well soluble in benzene, acetone, dioxane, and other organic solvents.

The found copolymerization constant values allowed to obtain information about microstructure of copolymers (Table 4.2).[11] It is observed that with increase in share of links m_1 in composition of copolymers, a block length of LM_2 is decreased and at approximately equal contents of both links the maxima of blocking and alternation of links are observed. The synthesized copolymers, basically, consist of short blocks m_1 and m_2 in content of links m_1, approximately from 40 to 70 mol%. The relatively long blocks of links m_2 are formed in considerable excess of m_2 in the initial mixture of monomers.

An availability of strongly absorbing groups in macromolecule of CCD allowed to use them as the light-sensitive base of photoresists. It has been established that under action of UV-irradiation the synthesized copolymers CCPS with DMB are enough easily converted into insoluble form.

TABLE 4.2   Average Blocks Length of Links of CCPS and DMB in Compositions of Copolymers.

Composition of copolymers (mol%)		Microstructure		
m_1	m_2	LM_1	LM_2	R
78.59	21.41	3.79	1.05	41.34
59.27	40.73	1.72	1.18	68.78
47.80	52.20	1.31	1.43	72.99
36.10	63.90	1.13	2.0	63.77
17.52	82.48	1.03	4.87	33.87

LM_1 and LM_2—average blocks length of links m_1 and m_2, respectively, and R—Harwood coefficient.

The spectral analysis of copolymers showed that after UV-irradiation depending on process duration, the absorption bands characteristic for cyclopropane ring (1035–1045 cm^{-1}) and internal double bond (1660–1670 cm^{-1}) are decreased or completely disappear. A decrease of intensity of the absorption bands of above-mentioned groups in a case of irradiation of polymer films in the initial stage of UV-irradiation is not strongly differed and in irradiation duration >30 min their absorption bands are practically disappear. Indeed, in the spectrum of CCD after irradiation it was not succeeded to detect the absorption bands characteristic for cyclopropane ring and double bond. It is apparent that these groups take part in the photochemical reactions leading to structurization of copolymer (Table 4.3). Consequently, an availability of possibility of proceeding of these reactions stipulates the higher inclination to CCD to the photochemical structurization. As follows from data of Table 4.4, the synthesized copolymers possess high lithographic indices and can be used as the photosensitive base of photoresists of negative type.

TABLE 4.3   Data of Structurization of Copolymers of CCD (IV) After UV-irradiation.

Copolymer	Content of links (mol%)		Quantity of insoluble fraction (%) at irradiation time (min)				
	m_1	m_2	1	2	6	10	12
I	17.52	82.48	20	52	68	87	97
II	27.15	72.85	18	45	62	83	93
III	36.10	63.90	16	40	60	80	92
IV	47.80	52.20	15	33.5	55.6	88.6	91

TABLE 4.4   Properties of Copolymers CCD (IV).

Copolymer*	Concentration of CCD in benzene (%)	Microdeficiency (def./cm²)	Film thickness (mkm)	Exposition time (s)	Photosensitivity (cm²/J)
	2.5	0.15	15	20	56.4
I	5.0	0.18	15	20	58.2
	7.5	0.20	20	20	55.1
	10	0.25	20	20	52.3
	2.5	0.15	15	20	54.7
II	5.0	0.15	15	20	57.1
	7.5	0.18	15	20	56.3
	10	0.25	20	20	52.1
	2.5	0.15	15	20	59.6
III	5.0	0.17	15	20	57.4
	7.5	0.18	15	20	55.8
	10	0.20	15	20	53.5
	2	0.15	15	15	53.5
IV	5	0.17	15	15	54.8
	8	0.20	20	20	52.5

*Compositions of copolymers I–IV are presented in Table 4.2.

The method of preparation of the soluble polyfunctional copolymers developed by us allows to synthesize the samples of CCD with various content of substituted cyclopropane rings and internal double bond, that is, varying synthesis conditions one can regulate purposefully the composition and consequently, physical-mechanical, adhesion, and other properties of copolymers. It has been revealed that with growth of share of substituted cyclopropane fragments the strength and adhesion properties of CCD are increased: adhesion strength is changed in the range of 5.6–16.4 MPa, depending on composition of copolymer. At the same time, an enrichment of CCD with links m_2 is accompanied by improvement of elastic properties. Such characteristics of copolymers allow to prepare the glossy films with low microdeficiency from them.

Thus, the copolymers possessing good film-forming and adhesion properties, high photosensitive and strength indices in combination with elasticity and stability for cyclic loadings, necessary in formation of photoresist films, have been synthesized.

KEYWORDS

- radial copolymerization
- para-(2-carboxy) cyclopropyl styrene
- 2,3-dimethyl-1,3-butadiene
- monomers
- photosensitive properties

REFERENCES

1. De Boer, C. D. *J. Polym. Sci. Polym. Lett. Ed.* **1973,** *11*(1), 25–27.
2. Vayner, A. Y.; Dyumaev, K. M. *Chem. Ind.* **1989,** *7*, 483–487.
3. Decker, C. *Prog. Polym. Sci.* **1996,** *21*, 593–650.
4. Vayner, A.Y.; Dyumaev, K. M.; Likhacheva, I. A., et al. *Dokl. RAS* **2004,** *396*(3), 362–365.
5. Guliyev, K. G.; Ponomaryeva, G. Z.; Guliyev, A. M. *J. Appl. Chem. Russ.* **2006,** *79*(3), 497–500.
6. Guliyev, K. G.; Ponomaryeva, G. Z.; Guliyev, A. M. *High Mol. Comp. Russ. B.* **2007,** *49*(8), 1577–1581.
7. Guliyev, K. G. *J. Appl. Chem. Russ.* **2008,** *81*(4), 636–639.

8. Guliyev, K. G.; Nazaraliyev, K. G.; Guliyev, A. M. *Azerb. Chem. J.* **1999,** *1*, 87–90.
9. Toroptseva, A. M.; Belogorodskaya, K. V.; Bondarenko, V. M. *Laboratory Session on Chemistry and Technology of High-molecular Compounds*. L.: Khimiya. 1972; p 125.
10. Guliyev, K. G.; Ponomaryeva, G. Z.; Mamedli, S. B.; Guliyev, A. M. *J. Struct. Chem. Russ.* **2009,** *50*(4), 720–722.
11. Zilberman, E. N. *High-molecular Compd. Russ. B* **1979,** *21*(1), 33–36.

CHAPTER 5

POLYMERIC MATERIALS AND THEIR USE IN AGRICULTURE

K. T. ARCHVADZE[1,*], T. I. MEGRELIDZE[1], L. V. TABATADZE[2], and I. R. CHACHAVA[1]

[1]*Food Industry Department, Georgian Technical University, 77 Kostava, 0175 Tbilisi, Georgia*

[2]*Food Industry Department, Sukhumi State University, Ana Politkobskaia 9, 0186 Tbilisi, Georgia*

Corresponding author. E-mail: keti987@mail.ru

CONTENTS

ABSTRACT

It is planned to conduct researches on improvement of heliotechnological designs with use of the latest technologies—infrared radiators and polycarbonate materials for improvement of quality of the dried-up production. Offered heliodrying from the device are intended for small-scale country enterprises. Use of heliodrying which allows to begin drying of agricultural production in any weather conditions, and also, having begun drying, not to interrupt because of adverse conditions.

5.1 INTRODUCTION

Polymer materials are widely used in the food industry and agriculture. Georgia is rich in agricultural products. Mineral resources of fruits, berries, and vegetables in some areas of our country are not used to the desired efficiency. Therefore, the use helioinstallations in drying agricultural products can significantly increase the volume of domestic production of dried fruits. The use of polymeric materials, those of the polycarbonates in particular, for heliodrying apparatus gives fine results.

Due to the high strength and toughness (250–500 kJ/m^2), polycarbonates are used as construction materials in various industries. Polycarbonates are produced from various polymers.[1–3]

Polycarbonate—Synthetic thermoplastic polymer, one type of polyester of carbonic acid and dihydric alcohols having the general formula [–ORO–C(O)–]n, where R—aliphatic or aromatic residue. Of the greatest industrial importance are aromatic polycarbonates (Figs. 5.1 and 5.2), especially polycarbonate based on bisphenol A, because of the availability of bisphenol A is synthesized by condensation of phenol and acetone.

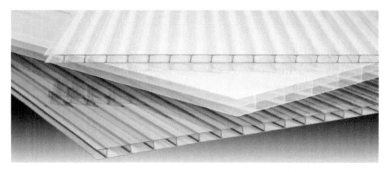

FIGURE 5.1 Polycarbonate.

FIGURE 5.2   Greenhouse polycarbonate.

5.2   EXPERIMENTAL

5.2.1   MATERIALS

Polycarbonate, a large heliodrying apparatus, the convective heliodrying apparatus, fruits, and vegetables.

5.2.2   OBJECTS OF STUDY

Heliodrying apparatus, polycarbonate, food products, fruits, and vegetables.

Products dried in heliodrying installations with a polycarbonate coating have better consumer properties than products obtained by natural air-drying. As the results of experiments for vitamin composition of the products before and after drying, the loss of vitamin 'C' in the heliodrying apparatus(Figs. 5.3 and 5.4) is 2–3 times lower than in case of the natural drying in the open air. The drying rate compared with conventional drying in the open air increases by 3–5 times, and in case of drying grass it increases 24 times. These dried fruits are eco-friendly and are of high quality without additions for improving color, flavors, and preservation capacity. Retentivity of agricultural products prepared in heliodrying establishments is better than that of the products dried by natural drying.

FIGURE 5.3   A large heliodrying apparatus.

FIGURE 5.4   The convective heliodrying apparatus.

5.3   RESULTS AND DISCUSSION

We present the results of drying herbs and dogwood dried in the heliodryer.

5.3.1   ANALYSIS OF EXPERIMENTAL DATA

As can be seen from the results of experiments (Table 5.1), the drying of medicinal herbs in the heliodryer took 4 h; natural drying in the shade lasted for 4 h. The drying process of medicinal plants in the heliodryer was 24 times faster than in case of the natural drying of herbs in the shade (Fig. 5.5).

TABLE 5.1 Experimental Results.

	Product Name Medicinal Plants	
	T/C	U/C
Initial mass (g)	720	710
Final mass (g)	212	213
Weight reduction (%)	70.5	70
Drying time (h)	4	96

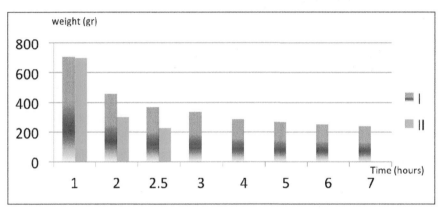

FIGURE 5.5 The change in the mass of raw material (plum) for days: I. Change in the mass of raw materials if dried by natural drying in the open air; II. Change in the mass of raw material when dried in the heliodryer.

5.3.2 DRYING PLUMS

For efficiency of drying agricultural products, it is recommended to dry them in solar dryer with polycarbonate surface. Use of solar dryer, as studies have shown, reduces the drying time, improves the preservation of flavor, nutrients and taste, provides sterility of products, and facilitates storage of the dried product (there is no damage to products and it lasts longer than usual). The experiments have shown that the use of simple solar dryers in small farms and urban residents provides savings of costs, physical labor, and it spares the environment (there is no heat generation and carbon dioxide emission into the environment). The consumer receives an environmentally clean and high-quality food. The use of high temperatures (electric driers and ovens) for the preparation of dried fruits often leads to the destruction of vitamins. Drying of raw by natural–solar–air methods in outdoors usually takes a long time, which is also not the best way; it also affects the quality of the product and its vitamin content. However, drying in H/D devices, as shown by analysis of dried products, is the best, because it preserves vitamin composition, and consumer taste and quality products as high as possible.[4–7]

Proposed solar dryers with polycarbonate cover are combined dryers. In windy weather, a dried product cannot be left in the open air because of the strong wind and raw materials are covered, but in this devices drying in windy conditions is not less intense than in hot, if we rotate H/D to the direction of the wind. High speed drying is explained due to the strengthening of the convective motion. Wind creates an effect similar to a fan at the entrance of solar dryer, and on the exit end of the tube.

In the proposed plants, different agricultural products may be dried under direct sunlight, and without it. For this purpose, several options are available in the design. The solar dryer installation temperature is higher than in the environment (in the sun) from about 10°C to 35°C, depending on the coating of drying chamber. Convective solar dryer is proposed for urban residents. This device is small in size. Bulky solar dryer device mainly is better for small-scale farming.

The drying process in the solar dryer machines ensures the sterility of products. While drying of agricultural products in solar dryer, materials can be left in the machine for a time from several hours to several days, depending on the product to be dried without tracking of:

- The process of drying (no need to turn over the product)
- The purity of raw materials to be dried. Solar dryer is closed in the process of drying, the product is not exposed to contamination by dust and rainfall, by dew and it is not damaged by birds or insects
- Weather conditions. In the rain and windy weather, the open air-dried product is removed to indoors, while solar dryer proceeds of drying products

Because of reduction in drying time, the product is less susceptible to color changes in comparison with the solar air-drying method that makes a product of higher quality. This material is better preserved, determining their nutritional and biological value (sugar, vitamins, and others). Drying in solar dryer also has other economic advantages, since it uses free energy source—the sun—drying time is significantly reduced, increases productivity, and reduces production costs.

Some dried fruits contain preservatives which may act as allergens in sensitive individuals. A dried fruit prepared in solar dryer proposed, no preservatives, colorings, and flavorings. Studies have shown that in the early stages of natural drying in rainy weather, the product can become moldy and be unusable. Drying process should start in sunny weather. Solar dryer (especially with ventilation) makes it possible to start drying in all weather conditions.

Note that during storage, the products that were dried in solar dryer had more resistant to molds than products dried naturally in the open air. Solar dryer can be located on any platform, oriented to the south to make the most of the radiant flux of solar energy. The higher the air temperature in the environment, the greater the temperature difference between the air of solar dryer and the environment. The unit is lightweight and easy to transport, during the day, and it can move and rotate relative to the sun. Offered dryers are simple to manufacture and can be widely used in large and small farmers and private farms.

- Thus, considering everything mentioned above, the proposed solar dryer has a number of advantages over traditional methods, namely:
- Not complicated to produce (they are easy to make and repair using local materials)
- Availability during the usage
- Relatively inexpensive (compared with other types of solar dryers)
- Effective
- Save the qualitative characteristics of fruit and vegetables
- Inexpensive to use
- Reduced drying steps
- Reduced drying time

KEYWORDS

- **polycarbonate**
- **heliodrying apparatus**
- **drying**
- **product**
- **infrared radiators**

REFERENCES

1. Zuev, V. V.; Uspenskaia, M. V.; Olehnovich, A. O. *Physics and Chemistry of Polymers Proceeding.* (In Russian), SPb. ITMO, St. Petersburg, Russia, 2010.
2. Semchikov, Y. D. *Macromolecular Compounds* (In Russian); Academy: Moskva, 2003.
3. Plaksin, Y. M.; Malakhov, H. N.; Vladimir, L. *Processes and Devices for Foodstuffs Processing,* 2nd Ed., (Rev. and Add); Colossus: Moskva, 2007.
4. Chagin, O. V.; Kokin, N. R.; Pastine, V. V. *Equipment for Drying Foodstuffs*; Ivan. Chemical. Primary Process. Univ.: Ivanovo, 2007, p 138.
5. Kazantseva, N. S. *Commodity Food Products. Textbook for High Schools*; M.: Dashkov & Co.
6. *Journal Fast Canning*; OOO News Express. Special Issue No. 06, 2011.
7. Kondrashova, E. A.; Konik, N. V.; Peshkov, T. A. *Commodity Food Products*; M.: Alpha-book: USA, 2009.

CHAPTER 6

ELECTROCONDUCTIVE POLYMER MATERIALS ON THE BASIS OF OLIGO-4-AMINOPHENOL

R. A. AHMEDOVA, S. S. MASHAEVA, CH. H. İSMAYLOVA, and B. A. MAMEDOV*

Institute of Polymer Materials of Azerbaijan, National Academy of Sciences, S. Vurgun Str., 124, AZ 5004 Sumgait, Azerbaijan

Corresponding author. E-mail: ipoma@science.az

CONTENTS

ABSTRACT

By oxidative polycondensation of 4-aminophenol (AP), the polyconjugated oligomers, including aromatic links with hydroxyl and amine functional groups have been synthesized. These oligomers show solubility and melt-ability, thermal stability, semiconductivity, and paramagnetism, and notice-able activity during reactions characteristic for aromatic hydroxyl and amine groups. They have been used as the active additions to industrial polymers and for preparation of composition materials with high physical–mechan-ical, heat–physical, and electric properties.

6.1   INTRODUCTION

The compositions of polyfunctional polyconjugated macromolecular compounds with thermoplasts, resins, and fibers are used in creation of sensors, converters, and smart materials in textile industry.[1–4] The purposeful change and management of electric properties of such material in maintaining of solubility and meltability is the actual and important task. In connection with this, we have synthesized the oligoaminophenols (OAP), including reactive phenol hydroxyl and aromatic amine groups in each elementary link, which was used further as the active addition for preparation of ther-mostable electroconductive polymer compositions of special purpose.

6.2   EXPERIMENTAL

The synthesis of OAP was carried out in three-necked flask by volume 250 mL, equipped with thermometer, mechanical mixer, and reflux condenser. In total, 10.9 g (0.1 mol) 4-aminophenol (AP), 85 g 30% solution of hydrogen peroxide (0.3 mol H_2O_2), and 70 mL distilled water was placed into flask. The oxidative polycondensation reaction was carried out at 363K for 4 h, at intensive stirring of reaction mixture. The synthesized OAP was purified from residue of monomer with washing of hot distilled water and dried at 373K in vacuum (13.3 Pa) to constant mass.

The OAP oxidation was carried out in glass reactor that was equipped with magnetic stirrer, thermometer, and bubbler. Furthermore, 10.7 g (0.1 mol) OAP and 72 mL C_2H_5OH was loaded in a reactor and after achieving of given temperature through system, it was transmitted with dry and puri-fied oxygen at the rate of 5.6 L/h. After reaction completion, the reaction mixture was subjected to the filtration; the prepared precipitate was washed by ethanol and dried in vacuum (13.3 Pa) at 313–323K to constant mass. The molecular weight of OAP samples was determined by a method of gel-permeating chromatography.[5]

The IR-spectra of OAP were taken for thin oligomer films applied on monocrystals NaCl; the EPR-spectra of OAP samples were taken on spectrometer PE-1306. DFPH and 2, 2, 6, 6-tetramethyl-4-oxypyridine-1-oxyl were used as standards.[6,7]

The electrical measurements have been carried out by the common technique on constant current by means of amplifier B 3-16, and on alternating current by means of bridge R-571—at low frequencies and cube meter—in the field of high frequencies (5×10^4–3×10^7 Hz).

The rubber mixtures on the basis of nitrile rubber (SNR-40) or butyl rubber (BR) and the synthesized oligomer compounds were made by means of laboratory rollers. On the rollers, firstly rubber was plasticized and then the necessary components, including oligomer compound were added in definite sequence and stirred. Depending on peculiarity of components of mixture, the rollers' temperature was regulated: in a front roll in the range of 303–313K and in an end roll in the range of 343–348K.

Observing the sequence of introduction of the components, the making of rubber mixture was realized as follows: (1) BR—0; stearin tech.—3 min; captax—5 min, thiuram—7 min, zinc oxide + oligomer—10 min; sulphur—15 min, general time of mixing—20 min, rolling temperature—313–323K, (2) SNR-40—0; stearin tech.—15 min; captax—17 min; zinc oxide—19 min; carbon black DG + oligomer—21 min; sulphur—34 min; general time of mixing—41 min; vulcanization temperature—423K.

Then by means of hydraulic press PG-63, the prepared mixtures were pressed in special press-forms with the plates with thickness of 1.5–2.0 mm. With use of standard knives, form prepared plates were used to cut the samples of desired shape and sizes for determination of physical–mechanical (tensile strength, specific elongation, residual deformation), electrical, and other properties on common methods.

The strength and deformation properties of the samples were determined by breaking machine PM-250.

6.3   RESULTS AND DISCUSSION

The preparation of OAP was carried out by the oxidative polycondensation reaction of AP in the presence of hydrogen peroxide and macroaminophenoxyl radicals with various concentrations of the paramagnetic centers (PMC) that have been synthesized by oxidation of OAP by oxygen in alcohol-alkaline medium.

The synthesized samples of OAP are powders of dark brown or black color, well soluble in polar organic solvents, and melting under loading at 383–408K depending on conditions of synthesis. Their composition and structure has been established by the methods of elemental, chemical, and IR-spectral analyses, and the following molecular weight indices were established by a method of gel-permeating chromatography (Table 6.1):

$$\overline{M_w} = 540–880, \overline{M_n} = 410– 550, \overline{Mw} / \overline{Mn} = 1.32–1.60 \qquad (6.1)$$

As observed from Table 6.1, the temperature growth, concentration of monomer and oxidizer, ratio of monomer:oxidizer, and also process duration in definite intervals intensifies the oxidative polycondensation of AP. The oxidative polycondensation reaction of AP has a first order on monomer and H_2O_2.

The results of elemental analysis and determination of content of hydroxyl groups testify that they are practically identical for synthesized oligomers and AP. This indicates to the fact that during AP oilgomerization the dehydration reaction does not occur and the ether bonds are not formed. Indeed, in the IR-spectra of the oligomers, the absorption band of C–O–C group at 1230 cm 1600–1800 is not detected. The elemental composition of oligomers on the basis of p-aminophenol is not practically differed and is close to the calculation values of elemental composition of aminophenol links (C—67.29%, H—4.67%, and N—13.08%). A content of hydroxyl groups in their composition is changed in the range of 14.0–16.3%.

In the IR-spectra of the oligomers, a wide absorption band in the field of 3200–3550 cm^{-1} is detected (valence vibrations of associated hydroxyl groups), and the absorption bands in the field of 1230 cm^{-1} correspond to their deformation vibrations. An availability of benzene ring in the composition of the oligomers is confirmed by three maximums in the fields of

TABLE 6.1 Synthesis Conditions and Some Indices of Oligoaminophenol.

1-NP (mol/L)	H_2O_2 (mol/L)	T (K)	Time (h)	Yield of oligomer (%)	Nitrogen content (%)	Content of OH groups (%)	Molecular weight indices		
							$\overline{M}_w$	$\overline{M}_n$	$\overline{M}_w / \overline{M}_n$
1.0	1.0	343	4	38.3	13.02	14.0	540	410	1.32
1.0	1.0	353	4	62.4	12.56	14.2	630	440	1.43
1.0	1.0	363	4	78.6	12.78	14.5	670	445	1.51
1.0	1.0	368	4	83.5	12.87	14.7	710	480	1.48
1.5	1.5	368	4	85.1	12.82	14.9	730	500	1.46
1.0	2.0	368	4	91.2	12.35	15.6	820	540	1.52
1.0	3.0	368	4	96.7	12.08	16.3	880	550	1.60
0.5	0.5	368	4	70.8	–	–	–	–	–
2.0	2.0	368	4	78.3	–	–	–	–	–
1.3	1.3	368	3	66.8	–	–	–	–	–
1.3	1.3	368	5	87.0	–	–	–	–	–
1.3	3.9	368	4	97.2	–	–	–	–	–

1590–1600; 1500–1510, and 1470–1480 cm^{-1}; the absorption bands in the field of 750–770 and 835–840 cm^{-1} (relatively the intensive absorption bands in the field of 835–840 cm^{-1} indicate the availability of the isolated C–H groups, while that of 750–770 cm^{-1} characterizes two neighboring aromatic CH groups) correspond to nonplanar deformation vibrations of C–H groups. Thus, the synthesized oligomers consist of aminophenol links combining through C–C bond of aromatic nuclei of AP, advantageously in *p*-positions relatively hydroxyl group.

p-OAP

It has been shown that the OAP samples show the electron exchange activity. Their alkaline and aqueous solutions intensively absorb the molecular oxygen. The oxidation rate constant values of OAP are sufficiently high (0.92×10^{-2}–$3.8 \cdot 10^{-2}$ min^{-1} at 303–323K, E = 62.4 kJ/mol in methanol). Consequently, these oligomers can perform a function of antioxidant in compositions increasing heat and thermal stability.

OAP and its oxidized samples (macroaminophenoxyl radicals) show the paramagnetic (concentration of the PMC ~ 7.0×10^{17}–9.2×10^{18} spin/g) and semiconductor (σ_0 ~ 10^{-8}–10^{-4} Ω^{-1} cm^{-1} at 298K, E = 1.34–1.67 eV) properties. Moreover, the growth of concentration of PMC in the composition of OAP by 1–2 orders leads to a noticeable increase of their electrical conductivity.

A possibility of increase of the electrical conductivity of OAP with growing concentration of PMC in its composition has been used for creation of the antisatic polymer compositions on the basis of thermoplasts by introduction of 5–15% OAP in their composition with the following treatment of material surface by aqueous or alcohol solution of alkali. OAP powders of thermoplasts was thoroughly mixed in ball mill and from this mixture, 2.6 kg/cm^2 was loaded at 423K; the discs with diameter of 0.05 m and thickness 0.001–0.002 m were obtained; the obtained disc was dipped for 15–20 s in 1% alkali solution in water or ethanol. It was sustained on air at room temperature for 30–35 min and dried at 323–333K. Then, the measurement of ρ_v material was carried out. As observed from Table 6.2, ρ_v after inclusion of OAP into composition of thermoplasts is essentially decreased.

TABLE 6.2 Values of Specific Surface Resistance of the Composition Materials on the Basis of Thermoplasts and OAP.

OAP (%)	Thermoplast (%)	On the basis of HPPE	ρ_v (Ohm·cm)	
			On the basis of PP	On the basis of PS
5	95	0.81×10^8	2.5×10^8	4.8×10^8
		(7.3×10^8)	(9.1×10^8)	(9.8×10^8)
10	90	3.5×10^8	8.1×10^8	9.5×10^8
		(6.4×10^8)	(1.2×10^7)	(2.7×10^7)
15	85	8.6×10^8	9.8×10^8	2.8×10^8
		(3.7×10^7)	(6.0×10^7)	(8.5×10^7)

In brackets: the values of ρ_v after treatment of composition surface by water solution of alkali.

The synthesized OAP samples have been used as the active addition for preparation of rubber mixtures of butyl (BR) and butadiene nitrile (BNR-40) rubbers. In this case, the rubber mixtures have been produced by standard recipe of ingredients with only difference that instead of carbon black (partially or completely) OAP (from 25 to 50 mass p. per 100 mass p. of rubber) is used.

It has been shown that an introduction of OAP in composition of rubber mixtures, instead of carbon black, leads to the increase of tensile strength and decrease of specific elongation of modulus of elasticity of the prepared rubbers. For example, for rubbers prepared by mixture vulcanization on the basis of BR including 15 mass p. OAP instead of carbon black, the tensile strength is increased to 22.5–23.5 MPa, specific elongation reaches 610–672%, and modulus of elasticity in elongation to 200% is decreased from 9.0–9.8 to 6.8–8 MPa. The values of specific volume electrical conductivity reach $(2.8–5.8) \times 10^{-7} \ \Omega^{-1} \ cm^{-1}$. For rubbers prepared by mixture vulcanization on the basis of BNR-40, including 11.3 mass p. OAP instead of carbon black, a tensile strength is increased to 23.8–26.7 MPa, specific elongation is 410–530%, modulus of elasticity in elongation to 200%—8.8–14.6 MPa, and the values of specific volume electrical conductivity reach $(4.5–6.2) \times 10^{-7} \ \Omega^{-1} \ cm^{-1}$. Along with this, there is an increase in thermal stability and durability of the prepared rubbers, which has been apparently connected with structural peculiarity of OAP (availability of benzene rings in chain of the developed system of polyconjugation stipulates a high thermal stability, and an availability of hydroxyl and amine groups in aromatic rings—antioxidant activity).

Since the OAP samples with various PMC content show a high electrical conductivity, their joint use with electroconductive one allows to prepare the rubbers with specific volume conductivity 10^{-8}–10^{-6} Ω^{-1} cm^{-1}. Moreover, a growth of the OAP content from 22.5–45.0 mass p (rubber mass) instead of carbon black and also an increase of PMC concentration in the composition of OAP lead to the increase of the specific electrical conductivity of the prepared rubbers. An effect of percolation is reached in OAP content ~26.5 mass p and ~23.5 mass p for rubbers prepared from BNR-40 and from BR, respectively.

6.4 CONCLUSION

By oxidative polycondensation of AP in the presence of aqueous solution of hydrogen peroxide, the polyfunctional, polyconjugated oligomers have been synthesized, the macromolecules of which consist of aminophenol links. They possess solubility, meltability, thermal stability, paramagnetism, and semiconductivity, and also high reactivity in the electron-exchange reactions and in interaction with epoxide compounds and anhydrides of carboxylic acids. It has been shown that an introduction of oligo-4-aminophenol into composition of thermoplasts (HPPE, PP, and PS) in a quantity of 5–15% leads to a noticeable decrease of specific volume resistance of the prepared compositions (to 3.7×10^7 Ω cm—for HPPE, 5.0×10^7 Ω^{-1} cm—for PP, and 8.5×10^7 Ω cm—for PS). The oligo-4-aminophenol has been used as the active addition to industrial butyl and nitrile rubbers. It has been established that a partial or complete substitution of carbon black by oligo-4-aminophenol in the standard rubber mixtures, on the basis of BR and BNR-40, leads to the formation of electroconductive rubbers.

ACKNOWLEDGMENT

The work has been executed with support of scientific-research project No. 24 "SOCAR SDF Fund."

KEYWORDS

- **electroconductive polymer compositions**
- **aminephenols**
- **oligoaminophenols**
- **thermal stability**
- **paramagnetism**

REFERENCES

1. Pud, A.; Ogurtsov, N.; Korzhenko, A.; Shapoval, Q. *Prog. Polym. Sci.* **2003,** *28,* 1701–1758.
2. Kaverinsky, V. S.; Smekhove, F. M. *Electric Properties of Lacquer Materials and Coatings*; M.: Khimiya: Moscow, Russia, 1990; p 158.
3. Valipour, A. Y.; Modhaddam, P. N.; Mamedov, B. A. *Life Sci. J.* **2012,** *9*(4), 409–421.
4. Huang, J. Synthesis and Applications of Conducting Polymer Polyanline Nanofibres. *Pure Appl. Chem.* **2006,** *78*(1), 15–27.
5. Ragimov, A. V.; Mamedov, B. A.; Gasanova, S. G. *Polym. Int.* **1997,** *43*(4), 343–348.
6. Mamedov, B. A.; Shahnazarli, R. Z.; Rzayev, R. S., et al. *J. Chem. Chem. Eng.* **2011,** *5*(9), 832–840.
7. Mashaeva, S. S.; Mamedov, B. A.; Agaev, N. M., et al. *Process. Petrochem. Oil Refin.* **2008,** *34*(2), 60–69.

CHAPTER 7

MODIFICATION AND APPLICATION POSSIBILITIES OF BIOCOMPOSITES BASED ON THERMOPLASTIC COLLAGEN

JÁN MATYAŠOVSKÝ[1,*], JÁN SEDLIAČIK[2], PETER JURKOVIČ[1], PETER DUCHOVIČ[1], and LADISLAV ŠOLTÉS[3]

[1]*Department of Science, VIPO a.s., Gen. Svobodu 1069/4, 95801 Partizánske, Slovakia*

[2]*Department of Science, Technical University in Zvolen, Masaryka 24, 96053 Zvolen, Slovakia*

[3]*Institute of Experimental Pharmacology and Toxicology, Slovak Academy of Sciences, 84104 Bratislava, Slovakia*

Corresponding author. E-mail: jmatyasovsky@vipo.sk

CONTENTS

ABSTRACT

This chapter studies the possibility of modified collagen in order to reduce formaldehyde (FD) emissions in wood-based products bonded with urea-formaldehyde (UF) resin in detail.

7.1 INTRODUCTION

Many types of synthetic plastics do not decompose naturally, which has caused the serious environmental pollution from the waste polymers. Biodegradable polymers are polymers which can completely degrade under the action of biological organisms and environmental weather conditions. This type of biopolymer is completely biodegradable; consequently, it is effective for the volume reduction of synthetic polymers waste by partial degradation.

Collagen is a renewable material, and therefore, there is an effort to constantly improve existing products and find new options for their application and processing. Many possibilities of effective and ecologic processing and application of leather wastes tanned and untanned to different products have already been analyzed.[1,5,7] Collagen is the main component of animal tissue mostly of leather, bones, and tendons and belongs among technically the most important fibrous proteins.[6]

Haroun (2010)[2] modified polyethylene (MPE) with protein, collagen hydrolyzate (CH). There was noted that CH easily blends with MPE, but as other biopolymers, it also has effects on the original mechanical properties of the MPE. The CH addition in the blend significantly increases the biodegradation rate. The effect of CH on MPE biodegradability has been investigated. About 53% biodegradation is observed, after 24 days, when the polymer is blended with 5% CH and about 63% biodegradation is found in the case of polymer blended with 20% CH. Although MPE/CH thermoplastic film with 40% content of CH have shown better performance in biodegradation, the mechanical strength properties were rather poor in this case.

At the development of new collagen based materials, one possible solution is the production of plastic matrix with the component, which is capable of gradual degradation under external conditions. The combination of bio- and synthetic polymers gives space to achieve a favorable economic, commercial, and environmental performance–biodegradable component is the part of raw materials from leather and food industry. One of the ways to modify properties and process ability of collagen is plasticization and compounding by suitable types of polymers. Films were made by blow extrusion technology from prepared granules and their parameters and optimal concentrations of modifying agents were tested.[4] Thermoplastic adhesive was prepared by collagen modification and tested with high strength and flexibility of the adhesive bonds. Within the experimental measurements, basic research on the antibacterial stabilizing by the colloidal silver was also carried on. Thermooxidative stability of different materials in their consolidation and stabilization was tested by differential scanning calorimetry.[8–10]

This chapter investigates the possibility of modified collagen to reduce formaldehyde (FD) emissions in wood-based products bonded with urea-formaldehyde (UF) resin. Previous results of laboratory tests have confirmed that collagen is a suitable modifier for the treatment of condensation time, viscosity, reducing FD emissions, and increasing the strength of the bond.[3] Life time and condensation time of UF resin were tested and studied by determining the kinetics of condensation reactions of dimethylolurea with urea and aldehyde.[11]

This research describes the possibility to prepare biodegradable polymer composites and films based on collagen, development of hot-melt adhesive. The aim of experiments was verifying the possibility of preparing organic polymer biocomposite on the compounding extrusion line. As the mixture is homogenized at the melting point of the polymer matrix during compounding, limiting factor in the choice of polymers is the melting temperature, to avoid thermal degradation of collagen.

7.2 EXPERIMENTAL PART

7.2.1 BIODEGRADABLE POLYMER FOILS

Collagen is necessary to be chemically modified for its further application. Copolymers of ethylene vinyl acetate and polyamide 6/12 were used as the polymer matrix for collagen modification. Individual compounds were homogenized during the compounding process. Prepared granules were tested at the foil production by blowing technology. Physical and mechanical parameters of foils were evaluated as strength, elongation, permeability to water vapor, and biodegradability. Obtained results are perspective mainly from the view of controlled biodegradation. Fourier transform infrared spectroscopy (FTIR) spectra were processed at FTIR–ATR Nicolet Avatar 330 and scanning electron microscope (SEM) images were taken at VEGA II LMU by Tescan.

Used materials:

- EVAc—ethylene vinyl acetate copolymer Evatane 1080 and Evatane 24-03
- Polyamide 12
- coPA 6/12—copolymer polyamide, Microfine
- Collagen—laboratory prepared CH, Vipocol KHN
- Modifier I—based on polyvalent amines in granular form
- Modifier II—based on polyvalent alcohols
- Modifier III—silver sulfate

7.2.2 COLLAGEN BASED HOT-MELT ADHESIVE

This investigation was focused on the research and preparation of non-FD thermoplastic hot-melt adhesive based on collagen biopolymer. Parameters of adhesive can be varied by selecting appropriate modifying additive, for example, elongation of the film, hydrophilicity, hydrophobicity, viscosity, melting, and solidification points, and so on.

7.2.3 COLLAGEN AS MODIFIER OF UF ADHESIVE MIXTURES

Natural modifiers of UF adhesives were raw collagen materials produced from waste of the leather industry. To modify the CH, there were applied: urea, di-aldehydes, glycerol, and so on.

Marking of collagen samples:

- Collagen No. 1—CH modified with urea and glutaraldehyde
- Collagen No. 2—CH modified with urea
- Collagen No. 3—CH, prepared from waste from leather industry

Rotary cut veneer sheets of beech wood (*Fagus sylvatica* L.) free from defects with dimensions of 300 × 300 mm, thickness of 1.7 mm, and moisture content of approximately 6% were used for the experiments. Three-layer experimental plywood panels were laboratory prepared using commercial UF resin KRONORES CB 1639F (Diakol Strážske, s.r.o., Slovakia). The UF resin (solid content of 66% and viscosity Ford 4/20 of 120 s) was uniformly spread on one side of the veneers in amount of 160 g/m². Veneers were then laid up and hot-pressed in a laboratory press using the pressing temperature of 125°C, pressure of 1.8 MPa, and time of 5 min. Shear strength of plywood samples was tested according to EN 314-1[13] and 314-2.[14]

7.3   RESULTS AND DISCUSSION

7.3.1   BIODEGRADABLE POLYMER FOILS

Experiments were focused on the preparation of polymer composites based on ethylene vinyl acetate and polyamide copolymers with the collagen biopolymer. The impact of these components on the physical and mechanical properties and biodegradation of polymer foils was investigated. Series of samples were prepared with the content of collagen and modifier from 20% to 60% and vinyl acetate content of 18%, 24%, and 28%.

Based on obtained results, it was found that composites based on EVAc polymer matrix containing 30–50% of a mixture of collagen and modifier 80:20 are able to process by blow extrusion technology. Figure 7.1 presents the sample of 50% EVAc copolymer + 50% modified collagen with the thickness of 0.09 mm.

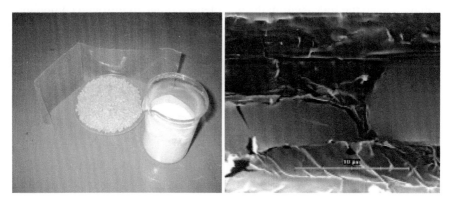

FIGURE 7.1   Collagen powder and granules (EVAc + collagen) and EVAc foil modified by collagen and scanned by SEM.

7.3.2   BIODEGRADATION OF POLYMER BIOCOMPOSITES

Prepared foil sheets based on EVAc and collagen were placed into the soil and the effect of soil culture and water on the degradation of the material was investigated. Figure 7.2 presents the biodegradability of foils and the process of their chemical decomposition.

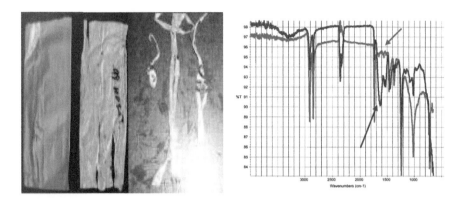

FIGURE 7.2   EVAc foil with collagen and testing their degradation after 24 h and 45 days in the soil. FTIR spectra of EVAc foil degraded after 24 h—red curve and foil tested after 45 days of degradation in the soil—green curve.

As indicated by FTIR spectra, the action of soil and/or water degraded the collagen from the foil—within the wavelength range of nitrogen compounds from 1550 to 1650 cm^{-1}, no peaks have appeared (green line), which would correspond to the presence of collagen in the material. For the implementation, there have been proposed polymer composites and films with the content of vinyl acetate and modified collagen:

- Foil EVAc + 20% of modified collagen
- Foil EVAc + 40% of modified collagen

Technical parameters of the foils prepared by blowing extrusion technology are shown in Table 7.1.

TABLE 7.1   Technical Parameters of Prepared Foils.

Sampling	Tensile strength at break (MPa)	Relative elongation at break (%)	Permeability to water vapor (g.m^{-2}.24h^{-1})
Foil EVAc + col 20%	10.5	462.1	10.2
Foil EVAc + col 40%	7.6	74.7	16.1

Further experiments were aimed on polyamide PA 12 and copolymer PA 6/12, which has lower temperature of melting, approx. 140–180°C. Considering the sensitivity of polyamides to the moisture, collagen has been dried prior compounding at 150°C for 4 h. Foils based on PA 12 and copolymer PA 6/12 were prepared by extrusion on a laboratory Brabender followed by cooling on the rollers, SEM photos are in Figure 7.3. Strength of the foils achieved values from 9.9 to 24.4 MPa and the elongation from 4.9% to 193.7%.

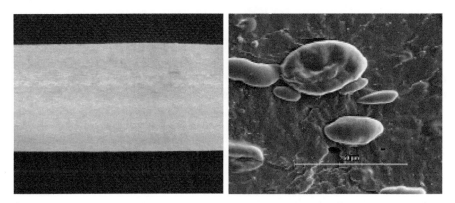

FIGURE 7.3   Foil of copolymer PA 6/12 modified by collagen and scanned by SEM.

7.3.3   COLLAGEN MODIFICATION WITH COLLOID SILVER

The morphology of collagen samples with colloid silver was examined by dual SEM microscope FEI Quanta 3D. The metallization of base sheet was realized by sputtering of Au in a thin layer with the thickness of 3 nm.

SEM images of the particles of colloidal silver with a collagen adhesive illustrate the long-term 1 month stability of glue at pH = 5.7, see Figure 7.4. Technical parameters of adhesive at the temperature of 60°C are:

- pH = 6.5 ± 0.5
- Viscosity: 2100–2200 MPa.s
- Dry content matter: 55–60%
- Surface tack: immediate
- Conservation by Ag nanoparticles

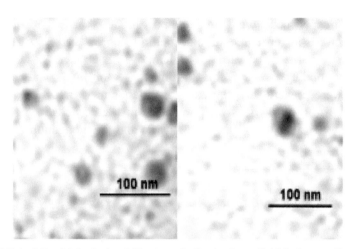

FIGURE 7.4   Microbiological stabilization of adhesive by colloidal silver: collagen + 100 ppm Ag and collagen + 100 ppm Ag (tested after one month).

7.3.4 COLLAGEN-BASED HOT-MELT ADHESIVE

The investigation was oriented on the research and preparation of non-FD thermoplastic hot-melt adhesive based on collagen biopolymer. Biopolymers of animal origin (e.g., technical leather glue, gelatin etc.) were specified by laboratory tests as suitable raw materials for preparation of thermoplastic collagen adhesives, with possible use for book sheets bonding as can be seen in Figure 7.5.

FIGURE 7.5   Thermoplastic collagen adhesive used for gluing of book hard covers.

7.3.5 THE INFLUENCE OF COLLAGEN ON FD EMISSION FROM UF ADHESIVE MIXTURES

The effect of the addition of collagen modifications no. 1, 2, and 3 on the FD content was tested in UF adhesive mixtures hardened at 100°C during the time of 20 min. From the obtained measured values follow, that collagen is the suitable modifier for the reduction of FD in UF-cured adhesive compositions. Collagen no. 1 most significantly reduces emissions of FD, and this effect is increasing with the increasing concentration. The content of FD in reference sample was 0.35 mg/g and in the modified resin only 0.25 mg FD/g as can be seen from Figure 7.6. These results were confirmed on plywood testing, where the content of FD stated according to EN 120[12] was lower down to 50% in comparison with the reference sample.

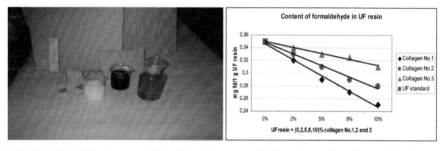

FIGURE 7.6 UF resins modified with collagen applied for plywood bonding and influence of different modifications on the lowering of formaldehyde emission.

7.4 CONCLUSION

Copolymer of ethylene–vinyl acetate, polyamide 12, polyamide copolymer 6/12 and CH were specified for preparation of polymeric biocomposites. Polymer biocomposites and foils with controlled biodegradation were prepared by combining the components of mixtures, with defined physical and mechanical properties.

Obtained results confirmed the possibility to use the modified biopolymer collagen for preparation of ecologic and thermoplastic adhesives. Glued joint performed high strength and flexibility after application of modification plasticization agents based on collagen. Application of colloidal silver has been testing to increase the stability of the adhesive. Industrial tests of gluing of book sheets confirmed the possibility to process adhesive in all types of machines at standard required quality.

Presented research proved the possibility of lowering the FD emission from wood products glued with UF adhesive. Results of laboratory tests confirmed, that collagen prepared from leather waste is a suitable additive for lowering of FD emission from wood products glued with UF adhesive. Tests confirmed the decrease of FD content in plywood in comparison with the reference sample down to 50%.

ACKNOWLEDGMENT

This work was supported by the Slovak Research and Development Agency under the contracts No. APVV-14-0506 "ENPROMO" and APVV-15-0124 "ISOCON", and VEGA agency No. 1/0626/16.

KEYWORDS

- thermoplastic collagen
- biodegradability
- hot-melt adhesive
- formaldehyde
- plywood

REFERENCES

1. Buljan, J.; Reich, G.; Ludvik, J. In *Mass Balance in Leather Processing*, Proceedings of the Centenary Congress of the IULCS, London, 1997; pp 138–156.
2. Haroun, A. A. Preparation and Characterization of Biodegradable Thermoplastic Films Based on Collagen Hydrolyzate. *J. Appl. Polym. Sci.* **2010,** *115*(6), 3230–3237.
3. Matyašovský, J.; Jurkovič, P.; Sedliačik, J.; Novák, I. Possibilities of Application of Collagen Colloid from Secondary Raw Materials as a Modifier of Polycondensation Adhesives. In *News in Chemistry, Biochemistry and Biotechnology: State of the Art and Prospects of Development*; Nova Science Publisher: USA, 2010; pp 85–94.
4. Matyašovský, J.; Sedliačik, J.; Matyašovský, J., Jr.; Jurkovič, P.; Duchovič, P. Collagen and Keratin Colloid Systems with a Multifunctional Effect for Cosmetic and Technical Applications. *J. Am. Leather Chem. Assoc.* **2014,** *109*(9), 284–295.

5. Matyašovský, J.; Sedliačik, J.; Jurkovič, P.; Kopný, J.; Duchovič, P. De-chroming of Chromium Shavings Without Oxidation to Hazardous Cr^{6+}. *J. Am. Leather Chem. Assoc.* **2011,** *106*(1), 8–17.

6. Peterková, P.; Lapčík, L., Jr. Collagen—Properties, Modifications and Applications. *Chem. Listy* **2000,** *94*(6), 371–379.

7. Pünterer, A. Ecological Challenge of Producing Leather. *J. Am. Leather Chem. Assoc.* **1995,** *90*(7), 206–215.

8. Sedliačik, J.; Matyašovský, J.; Šmidriaková, M.; Sedliačiková, M.; Jurkovič, P. Application of Collagen Colloid from Chrome Shavings for Innovative Polycondensation Adhesives. *J. Am. Leather Chem. Assoc.* **2011,** *106*(10), 332–340.

9. Šimon, P.; Kolman, L. DSC Study of Oxidation Induction Periods. *J. Therm. Anal. Calorim.* **2001,** *64*(2), 813–820.

10. Šimon, P. Induction Periods—Theory and Applications. *J. Therm. Anal. Calorim.* **2006,** *84*(1), 263–270.

11. Fukumoto, T.; Thomas, P. S.; Stuart, B. H.; Šimon, P.; Adam, G.; Shimmon, R.; Guerbois, J. P. Estimation of the Storage Life of Dimethylol Urea Using Nonisothermal Accelerated Testing. *J. Therm. Anal. Calorim.* **2012,** *108*(3), 439–443.

12. EN 120: 1995 Wood Based Panels. Determination of Formaldehyde Content. Extraction Method Called the Perforator Method. http://file.yizimg.com/346790/2010121717041621.PDF.

13. EN 314-1: 2005 Plywood. Bonding Quality. Part 1: Test Methods. http://standards.globalspec.com/std/969036/din-en-314-1.

14. EN 314-2: 1998 Plywood. Bonding Quality. Part 2: Requirements. http://apawood-europe.org/official-guidelines/european-standards/individual-standards/en-314-2/.

CHAPTER 8

SYNTHESIS OF SYNDIOTACTIC 1,2-POLYBUTADIENE WITH COBALT ALKYLXHANTHOGENATE + TRIALKYLALUMINUM CATALYTIC DITHIOSYSTEMS

FIZULI A. NASIROV*, NEMAT A. GULIYEV, NAZIL F. JANIBAYOV, AFAQ M. ASLANBAYLI, and SEVDA R. RAFIYEVA

Institute of Petrochemical Processes, Azerbaijan National Academy of Sciences, 30, Khodjaly Av., AZ 1025 Baku, Azerbaijan

Corresponding author. E-mail: fizulin52@rambler.ru; fizuli_nasirov@ yahoo.com

CONTENTS

ABSTRACT

A new highly active and stereo regular Cobalt alkyl xhanthogenate + trialkylaluminum catalytic dithiosystems for obtaining highly crystalline syndiotactic 1,2-polybutadiene have been developed. Activity and stereo regularity of these catalysts in comparison with known cobalt-containing catalytic dithiosystems of butadiene polymerization, dependence of process outcomes on the nature of ligands of cobalt-compounds, catalyst components concentration and ratio, as well as the influence of temperature on catalyst activity and selectivity were studied. These catalysts in toluene solution showed very high activity and stereo regularity in butadiene polymerization: butadiene conversion—50.0–99.0%, 1,2-content—94.0–99.0%, intrinsic viscosity—1.3–3.5 dL/g; crystallinity—63.0–98.0%, syndiotacticity—93.0–98.5%, melting point—175–212°C. The optimal parameters of synthesis of high molecular mass and highly crystalline syndiotactic 1,2-polybutadiene: $[Co] = (1.0–2.0) \cdot 10^{-4}$ mol/L; $[M] = 3.0$ mol/L; Al:Co = (50–100):1; $T = 40–80°C$ was established.

Highly crystalline syndiotactic 1,2-polybutadiene (1,2-SPBD) has higher izod strength, lower initial modulus, and lower thermal distortion temperature than polypropylene in spite of the higher melting point. Owing to the reactive vinyl group, the reaction of 1,2-SPBD (cross linking, cyclization and introduction of new functional groups) can easily proceed by the action of peroxide, radiation, or cationic reagents. The swelling resistances are good in contact with aliphatic, aromatic, or halogenated solvents. The 1,2-SPBD is chemically stable against water, base, and acids such as acetic acid and hydrogen chloride.

8.1   INTRODUCTION

The basic application areas of syndiotactic 1,2-polybutadiene (1,2-SPBD), the thermoplastic resin, allowing to obtain the whole spectrum of the products possessing elastomeric and thermoplastic properties, are: preparation of rubber mixes, in particular, for manufacture of tires, packing polymer-film materials; the various compositions, used in such scopes as a microcapsule for the medical purposes, ceramics, semipermeability membranes, adhesives, synthetic leather, oil resistant tubes, coatings for semiconductor devices, no woven materials, and so on; carbon fibers, possessing increased durability at a stretching.[1–10]

Syndiotactic 1,2-polybutadienes has been synthesized using various catalysts based on compounds of Ti, Cr, Pd, Co, V, Fe, and Mo. Research results, relating to the catalysts and processes of syndiotactic 1,2-polybutadiene synthesis, have been described in many literature materials and patents.[8–15]

We have developed the new nickel- or cobalt-containing catalytic dithiosystems for the polymerization of butadiene[16,17]; and in this chapter are presented results of polymerization of butadiene to the photodegradable syndiotactic 1,2-polybutadiene in the presence of a new Cobalt alkyl xanthogenate (AlkXh-Co) + Trialkylaluminum (TAA) catalytic dithiosystems. The characteristics of synthesized dithio derivaties are shown in Table 8.1.

8.2 EXPERIMENTAL

8.2.1 MATERIALS

Polymerization was conducted with toluene as a solvent. After pre drying over metallic Na for 24 h, toluene was distilled over metallic Na and preserved under Na (stored in the calibrated reservoir under metallic Na as well).

The butadiene, supplied by Efremov Synthetic Rubber Plant (Russian Federation), was analyzed by gas chromatography and found to have the purity 99.5% (wt).

Triethylaluminum (TEA) and Diethylaluminumchloride (DEAC) were obtained from Red Kino plant (Moscow, Russian Federation) with a minimum purity of 85.0% (wt) in benzene solution and were used as received. The TEA and DEAC diluted to 10.0% (vol.) in dry toluene and were drawn from a crown-sealed beverage bottle kept in a glove box.

8.2.2 POLYMERIZATION PROCEDURE

Where necessary, manipulations were carried out under dry, oxygen-free inert argon or nitrogen gas in a glass reactor of 50–200 mL capacity or in Schlenk-type apparatus with appropriate techniques and gas tight syringes. The desired volume of toluene, monomer, AOC, and cobalt components solutions were added to the reactor continuously from the calibrated glass reservoir with stirring and temperature control was initiated. In a typical experiment, the reaction vessel, such as glass ampoule or dilatometer, was evacuated whilst hot, allowed to cool and then filled with dry, oxygen-free argon or nitrogen.

TABLE 8.1 Physico-Chemical and Analytical Data of Used Cobalt Dithiocompounds.

Symbol	Chemical formula	Nomenclature	Yield, % (Mass)	Melting point (°C)	Molecular mass, Found/ Calculated	Element composition (%), found/calculated			
						C	H	P	S
1	2	3	4	5	6	7	8	9	10
EXh-Co	$\left[\begin{array}{c} C_2H_5O \\ C_2H_5O \end{array}\rangle C \stackrel{\displaystyle\backslash\!\!\!\!\backslash S}{\diagdown S}\right]_2 Co$	Cobalt di-ethylxhan-thogenate	98	96	390.0/391.46	15.3/15.34	2.5/2.58	—	16.3/16.38
i-PrXh-Co	$\left[\begin{array}{c} C_3H_7O \\ C_3H_7O \end{array}\rangle C \stackrel{\displaystyle\backslash\!\!\!\!\backslash S}{\diagdown S}\right]_2 Co$	Cobalt di-iso-pro-pylxhanthogenate	96	85	446.5/447.57	18.75/18.79	3.1/3.15	—	14.3/14.33
BuXh-Co	$\left[\begin{array}{c} C_4H_9O \\ C_4H_9O \end{array}\rangle C \stackrel{\displaystyle\backslash\!\!\!\!\backslash S}{\diagdown S}\right]_2 Co$	Cobalt di-butylxhan-thogenate	99	72	505.0/503.68	21.4/21.46	3.55/3.60	—	12.7/12.73
HeXh-Co	$\left[\begin{array}{c} C_6H_{13}O \\ C_6H_{13}O \end{array}\rangle C \stackrel{\displaystyle\backslash\!\!\!\!\backslash S}{\diagdown S}\right]_2 Co$	Cobalt di-hexylx-hanthogenate	95	50	615.0/615.89	25.31/25.35	4.25/4.26	—	10.4/10.41
OXh-Co	$\left[\begin{array}{c} C_8H_{17}O \\ C_8H_{17}O \end{array}\rangle C \stackrel{\displaystyle\backslash\!\!\!\!\backslash S}{\diagdown S}\right]_2 Co$	Cobalt octylxhanthogenate	98	35	730.0/728.11	28.0/28.04	4.7/4.71	—	8.8/8.81

The usual order of reaction components addition was: solvent, cobalt component, AOC (at −78°C) and finally monomer—butadiene. All polymerizations were conducted at 0–100°C. After polymerization the polymerizate was poured to ethanol or methanol and the polymerization reactions were terminated. The precipitated polymer was washed several times with ethanol (methanol). The polybutadiene was dried at 40°C in a vacuum to constant weight.

8.2.3 MOLECULAR MASS AND STRUCTURE DETERMINATIONS

The viscosity of dilute solutions of syndiotactic 1,2-polybutadiene was measured with a Ubbelohde type viscometer in tetraline at 135°C or in o-dichlorobenzene (o-DCB) at 140°C at the concentration of 0.2 g/dL. The intrinsic viscosity [η] was estimated by double extrapolation η_{sp}/C and $\ln\eta_{rel}/C$ to $C \to 0$, where η_{sp} is the specific viscosity, "C" is the solution concentration (g/dL), and η_{rel} is the relative viscosity.[6,20]

The molecular mass of high molecular mass and highly crystalline syndiotactic polybutadiene was determined by viscosimetry method with relationships:

$$\left[\varsigma\right]^{135}_{(tetralin)} = \left[\varsigma\right]^{140}_{(o\text{-}DCB)} = 9.41\cdot10^{-5}\cdot M^{0.854}.$$

The molecular masses (Mw and Mn) and molecular mass distribution (Mw/Mn) were measured by a gel permeation chromatograph (GPC), constructed in Czech Republic with a 6000A pump, original injector, R-400 differential refractive index detector, styragel columns with nominal exclusion of 500, 10^3, 10^4, 10^5, 10^6. The GPC was operated at a flow rate of 0.8 mL/min with tetraline (or o-dichlorobenzene) as solvent. The sample concentration was kept at about 0.3–0.6% with a sample volume of 100–200 mL. GPC instrument was calibrated according to the universal calibration method by using narrow molecular weight polystyrene standards.[21]

The microstructure of the polybutadiene was determined by IR-spectrometry method (Beckman IR-spectrophotometer).

The ^{1}H- and ^{13}C-NMR spectras of 1.2-polybutadiene were recorded on a Varian 80A spectrometer (USA) operating at a frequency of 300 (^{1}C) and 75.47 MHz (^{13}C). ^{13}C NMR spectra were measured with broad band proton suppression under the JMOD regime using 2.0% polymer solutions in a mixture of o-dichlorobenzene and perdeuterio-o-dichlorobenzene (3:1). Tetramethylsilane was used as an internal standard.[22,7]

Tactisity and crystallinity of polymer were determined accordingly.[23,24]

8.3 INFLUENCE OF EXPERIMENTAL CONDITIONS ON THE MAIN PARAMETERS OF POLYMERIZATION OF BUTADIENE

8.3.1 INFLUENCE OF LIGANDS IN CO-XHANTOGENATES AND ALUMINUM ORGANIC COMPOUNDS

Apart from the metal chosen for polymerization, the ligand nature is also of high importance. Ligands that have been used for polymerization of butadiene were mainly xhanthogenates and also dietyldithiocarbamate Co for to compare. The results are given in Table 8.2.

It seems from Table 8.2 Co-alkyl xhanthogenates in combination with DEAC allow to obtain high molecular mass polybutadiene with 1,4-cis content—90.0–96.0%. Cobalt- alkyldithio carbamate catalytic system (Co-DEDTC + DEAC) gives high molecular mass 1,4-cis + 1,2-polybutadiene with 1,4-cis content—58.0% and 1,2-content—34.0%.

Only the Cobalt-Alkyl xhanthogenate catalytic dithiosystems (Cobalt-Alkyl xhanthogenates + TEA) give high molecular and highly crystalline syndiotactic 1,2-PBD (1,2-SPBD). In their presence, polymer yield was—93.0–99.0% and intrinsic viscosity (135°C, tetralin)—2.2–3.5, 1,2-content—94–99%, crystallinity—86.0–95.0%, and mp = 175–208°C.

8.3.2 INFLUENCE OF ORGANIC SOLVENTS

The influence of the organic solvents (toluene, benzene, chlorobenzene, methylene chloride, and hexane) on the activity and selectivity of the Co-iPrXh + TEA catalytic dithiosystem was studied at: [Co] = 1.0 · 10^{-4} mol/L, [M] = 3.0 mol/L, Al:Co = 100:1. For that the necessary amount of a particular solvent was mixed with Co-iPrXh followed by the addition of TEA and butadiene into reactor at −78°C. The butadiene polymerization reactions were conducted at 40°C during 120 min.

From the results shown in Table 8.3 it seems that when toluene and benzene were used the polymer yield was 96.0–99.0%. The obtained polybutadiene had 1,2-content—96.0–98.0%, mp = 203–208°C, crystallinity—93.0–95.0%, and syndiotacticity—95.0–97.5%. The known catalytic systems showed very low catalytic activity and stereo regularity in toluene and benzene solutions.

For future investigations, toluene—the solvent largely used in industrial polymerization processes, has chosen by us as optimal solvent in polymerization of butadiene to 1,2-SPBD.

TABLE 8.2 Comparison of Efficiency of Different Cobalt-containing Catalytic Dithiosystems CoX_2 +AOC in the Butadiene Polymerization Process.

Reaction Conditions: [Co] = 2.0 · 10⁻⁴ mol/L, [M] = 3.0 mol/L; Al:Me = 100; T = 25°C; τ = 60 min, Solvent—Toluene.

Item No	Cobalt-containing catalytic system (CoX_2+AOC)	Polymer Yield, % (Mass)	Crystallinity (%)	Syndiotacticity (%)	Melting point (°C)	Intrinsic viscosity, [η] (dL/g)	Molecular mass, M ·(10⁻³) M_w	M_w/M_n	Microstructure (%) 1,4-cis	1,4-trans	1,2-	Yield of polymer (kg PBD/g Co·h)
1	Co-EtXh + DEAC	95	—	—	—	2.8	540	1.85	96	2	2	127
2	Co-BuXh + DIBAC	90	—	—	—	2.5	410	1.65	95	2	3	105
3	Co-EtXh + TEA	95	92	98.5	208	2.2	270	2.1	2	1	99	96
4	Co-iPrXh + TEA	99	95	97.0	205	2.3	255	1.82	1	1	98	115
5	Co-BuXh + TEA	98	93	95.5	200	2.5	240	1.6	3	1	96	125
6	Co-HeXh + TEA	95	88	94.0	185	2.9	330	1.75	5	1	94	108
7	Co-OcXh + TEA	93	86	94.5	175	3.5	485	2.2	6	2	95	112
8	Co-DEDTC + DEAC	58	—	—	—	2.55	600	2.3	58	8	34	44

Note: In experiments 3–7 an intrinsic viscosity was measured in tetralin at 135°C.

Sample 8—Diethyldithiocarbamate Co.

TABLE 8.3 The Influence of Organic Solvents Type on the Conversion of Butadiene, Selectivity and Productivity of Co-iPrXh + TEA Catalytic System.

$[Co] = 1.0 \cdot 10^{-4}$ mol/L, $[M] = 3.0$ mol/L, Al:Co = 100:1, $T = 40°C$, $\tau = 120$ min.

Solvent	Yield of PBD, % (Mass)	$[\eta]_{135}$, dL/g	Crystallinity (%)	Syndiotacticity (%)	Melting Point (°C)	Microstructure (%)		
						1,4-cis	1,4-trans	1,2-
Toluene	99.0	2.5	95	97.5	208	1	1	98
Benzene	96.0	2.3	93	95.0	203	3	1	96
Chlorobenzene	92.0	1.75	91	94.0	194	3	2	95
Methylene chloride	90.0	1.65	90	93.8	185	3	2	95
Hexane	70.0	1.5	76	91.0	190	6	2	92
–	56.0	3.5	93	95.0	195	4	1	95

8.3.3 INFLUENCE OF COBALT CONCENTRATION

In these experiments, polymerization of butadiene was carried out at 40°C and the cobalt concentration was varied within limits $[Co] = (0.2–10.0)$ 10^{-4} mol/L at Al:Co = 100:1 using the Co-iPrXh + TEA catalyst in toluene solvent. The results are shown in Table 8.4.

An increase in cobalt concentration from $0.2 \cdot 10^{-4}$ mol/L up to approximately $10.0 \cdot 10^{-4}$ mol/L resulted in an increasing of the activity of used catalyst. As the concentration of cobalt compound increased also increased the yield of polybutadiene from 50.0% to 100.0% and crystallinity from 68.0% to 95.0%. By this, followed decreasing of 1,2-content from 99.0% to 95.0%, mp from 212°C to 187°C, and intrinsic viscosity from 3.0 dL/g to 1.7 dL/g.

8.3.4 INFLUENCE OF BUTADIENE CONCENTRATION

In the presence of Co-iPrXh + TEA catalytic system in toluene solution, at cobalt concentration—$[Co] = 2.0 \cdot 10^{-4}$ mol/L, and temperature—t = 40°C butadiene concentration was varied out in the range of 1.5–6.0 mol/L. Increase in monomer concentration results increase in polybutadiene yield (97.0–99.0%), intrinsic viscosity (2.0–3.2 dL/g), crystallinity (84.0–87.0%). In this case, the "mp" was decreased from 207°C to 203°C (Table 8.4).

TABLE 8.4 The Influence of Butadiene Polymerization Parameters on the Conversion of Butadiene, Selectivity and Productivity of Co-iPrXh + TEA Catalytic System (Solvent—Toluene).

$[Co] \cdot 10^4$ (mole/L)	[M] (mol/L)	Al: Co	Temperature (°C)	Reaction Duration (min)	Yield of 1,2-SPBD % (Mass)	$[\eta]_{135}$ (dL/g)	Crystallinity (%)	Syndiotacticity (%)	Melting Point (°C)	Microstructure (%) 1,4-cis	1,4-trans	1,2-
0.2	3.0	100	40	180	50	3.0	68	98	212	1	–	99
0.5	3.0	100	40	120	65	2.8	72	98	212	1	–	99
1.0	3.0	100	40	120	95	2.5	76	97	210	1	1	98
2.0	3.0	100	40	60	99	2.3	85	98	205	1	1	98
5.0	3.0	100	40	15	99	2.1	90	96	197	2	1	97
10.0	3.0	100	40	15	99	1.7	95	95	187	3	2	95
2.0	1.5	100	40	60	97	2.0	84	96	207	2	1	97
2.0	6.0	100	40	45	99	3.2	87	98	203	1	–	99
2.0	3.0	10	40	180	65	2.9	96	98	212	1	–	99
2.0	3.0	25	40	90	83	2.7	93	98	208	1	–	99
2.0	3.0	50	40	60	91	2.4	78	97	206	1	1	98
2.0	3.0	150	40	45	99	2.0	73	96	186	3	1	96
2.0	3.0	200	40	30	99	1.5	76	94	175	4	2	94
2.0	3.0	100	0	180	26	3.5	98	97	215	1	–	99
2.0	3.0	100	10	120	45	2.8	93	97	212	1	1	98
2.0	3.0	100	25	60	95	2.4	90	96	209	2	1	97
2.0	3.0	100	55	45	99	2.0	81	95	194	4	1	95
2.0	3.0	100	80	45	99	1.5	68	93	180	4	2	94
2.0	3.0	100	100	45	99	1.3	63	93	177	5	1	94

8.3.5 INFLUENCE OF AL:CO RATIO

The influence of Al:Co ratio in the interval of (10–200):1 on the activity, selectivity, and productivity of catalyst was investigated at [Co] = 2.0 · 10⁻⁴ mol/L, t = 40°C by using the Co-iPrXh + TEA catalyst in toluene solvent. From the results of Table 8.4 it seems, that the increase of Al:Co ratio from 10:1 to 200:1 resulted in an increase of polybutadiene yield (65.0–99.0%) and in decrease of intrinsic viscosity (2.9–1.5 dL/g), 1,2-content (99.0–94.0%), crystallinity (96.0–76.0%), and mp (212–175°C).

8.3.6 INFLUENCE OF TEMPERATURE

The influence of temperature in the interval of 0–100°C on the outcome of the butadiene polymerization reaction was investigated in the presence of Co-iPrXh + TEA catalyst in toluene solvent. As it seems from Table 8.4, an increase in temperature resulted in an increase in polybutadiene yield (26.0–100.0%), as would be expected. This results in decreasing of intrinsic viscosity (3.5–1.3 dL/g), 1,2-content (99.0–94.0%), crystallinity (98.0–63.0%), and melting point (215–177°C).

Experimental results has showed that high activity and stereo regularity of the new Cobalt alkyl xhanthogenate + TAA catalytic dithiosystems allow to receive in toluene solution (the most used solvent in polymer synthesis) the syndiotactic 1,2-polybutadienes in various crystallinity, syndiotacticity, and molecular mass by variation of experimental conditions: nature, types, molar ratio of catalytic system components, and temperature.

The obtained experimental results allow us to establish the optimal parameters for synthesis of high molecular mass and highly crystalline syndiotactic 1,2-polybutadiene.[25–27]

[Co] = (1.0–2.0) · 10^{-4} mol/L; [M] = 3.0 mol/L; Al:Co = (50–100): 1; T = 40–80°C.

Nowadays, the mechanism of formation of 1,4-cis- and 1,2-structures in butadiene polymerization is universally recognized: at least, under influence of most frequently used catalytic systems based on Co-, Ni-, and Ti-compounds, the structures of both types are formed from syn-butenyl groups at η^4-cis-coordination of monomer. 1,4-trans-structures are formed from anti-butenyl groups which can result by syn–anti-isomerization and η^2-trans-coordination of a monomer.[16,17]

In the presence of Cobalt-alkyl xhanthogenates + Alkylaluminumhalogenids catalytic systems, we obtained polybutadiene with 96.0% of 1,4-cis units. Only Cobalt-alkyl xhanthogenate + TAA catalytic systems provide to receive high molecular mass and highly crystalline syndiotactic 1,2-polybutadiene. The complex of the obtained results allowed proposing the mechanism of stereo regulating in dienes polymerization and syndiotactic 1,2-polybutadiene synthesis (Table 8.2).

It is known[16,17,28–34] that catalytic properties of used Ziegler–Natta systems depend both on natures of the catalyst components and on a method of catalyst preparation. Hence, the understanding of the processes proceeding at mixture of complexes of transition metals and metal organic cocatalysts in the presence of a small amount of monomer and resulting in origin of active centers (AC) of coordination–ionic polymerization is extremely important for a choice of adequate model of AC and understanding of the mechanism of their action. In dienes, polymerization between transition metal compound of the catalyst and growing polymer chain form the π-allylic type bond and coordination of 1,3-diene with transition metal can be carried out on cis-η^3-, trans-η^1- and trans-η^2-type.

8.4   CONCLUSION

A new highly active and stereo regular Cobalt-alkyl xhanthogenate + TAA catalytic dithiosystems have been developed. Activity and stereo regularity of these catalysts were studied in comparison with known cobalt-containing catalytic dithiosystems of butadiene polymerization, dependence of process outcomes on the nature of ligands of cobalt-compounds, catalyst components concentration and ratio, as well the influence of temperature on catalyst activity and selectivity. Developed catalytic systems permit in toluene solution an yield of 50.0–99.0% high molecular mass and highly crystallinity syndiotactic 1,2-polybutadiene with following characteristics: 1,2-content—94.0–99.0%; intrinsic viscosity—1.3–3.5 dL/g; crystallinity—63.0–98.0%; mp—175–212°C.

The optimal parameters of synthesis of high molecular mass and highly crystalline syndiotactic 1,2-polybutadiene: $[Co] = (1.0-2.0) \cdot 10^{-4}$ mol/L; $[M] = 3.0$ mol/L; Al:Co = (50–100): 1; T = 40–80°C was established.

Received data allowed explaining some proposed mechanism of formation of 1,4-cis-, 1,4-trans- and 1,2-structure in butadiene polymerization.

KEYWORDS

- butadiene
- cobalt alkyl xhanthogenate
- catalytic dithiosystems
- polymerization
- syndiotactic 1,2-polybutadiene

REFERENCES

1. Obata, Y.; Tosaki, C.; Homma, C. Bulk Properities of Syndiotactic 1,2-polybutadiene. III. Melting and Crystallization Properties. *Polym. J.* **1975,** *7*(3), 312–319.
2. Bertini, F.; Canetti, M.; Ricci, G. Crystallization and Melting Behavior of 1,2-syndiotactic Polybutadiene. *J. Appl. Polym. Sci.* **2004,** *92*(3), 1680–1687.
3. Chuvyrov, A. N. In *Physical and Mechanical Characteristics of 1,2-syndiotactic Polybutadiene Under Photoexcitation of Vinyl Bonds,* Seminar of the Polymer Council of Russian Academy of Sciences "Cutting Edge of the Polymer Science", Session 65, Feb 16, 2010.
4. Matsuda, Y.; Soeda, Y.; Zhang, X.; Tasaka, S. Lamellar Structure of a Reactive Blend of 1,2-Polybutadiene and Nylon 11. *J. Thermoplast. Compos. Mater.* **2009,** *22*(4), 353–354.
5. Jiali, C.; Qing, Yu.; Xuequan, Z.; Jiaping, L.; Liansheng, J. Control of Thermal Cross-linking Reactions and the Degree of Crystallinity of Syndiotactic 1,2-polybutadiene. *J. Polym. Sci. Part B* Polym. Phys. **2005,** *43*(20), 2885–2897.
6. Ashitaka, H.; Ishikawa, H.; Ueno, H.; Nagasaka, A. Syndiotactic 1,2-polybutadiene with Co-CS$_2$ Catalyst System. I. Preparation, Properties, and Application of Highly Crystalline Syndiotactic 1,2-polybutadiene. *J. Polym. Sci. Polym. Part A Chem. Ed.* **1983,** *21*, 1853–1860.
7. Ashitaka, H.; Inaishi, K.; Ueno, H.; Nagasaka, A. Syndiotactic 1,2-polybutadiene with Co-CS$_2$ Catalyst System. III. ^{1}H-and 13C-NMR Study of Highly Syndiotactic 1,2-Polybutadiene. *J. Polym. Sci. Part A Polym. Chem. Ed.* **1983,** *21*, 1973–1988.
8. Cheng-zhong, Z. Synthesis of Syndiotactic 1,2-polybutadiene with Silica Gel Supported CoCl$_2$-Al(I-Bu)$_3$-CS$_2$ Catalyst. *China Synth. Rubber Ind.* **1999,** 22(4), 243–246.
9. Chen, Y.; Yang, D.; Hu, Y. M.; Zhang, X. Q. Synthesis and Morphological Structure of Crystalline Syndiotactic 1,2-Polybutadiene. *Chem. J. Chin. Univ.* 2003, 24(11), 2119–2121.
10. Chen, Y.; Yang, D. C.; Hu, Y. M.; Zhang, X. Q. Single Crystal Structure of Crystalline Syndiotactic 1,2-Polybutadiene. *Chem. J. Chin. Univ.* **2003,** *24*(12), 2321–2323.
11. Hu, Y.; Dong, W.; Jiang, L.; Zhang, X.; Guo, Y.; Cao, L. Investigation on Synthesis of High Vinyl Polybutadiene with Iron-Based Catalysts. I. Effect of Triphenyl Phosphate. *Chin. J. Catal.* **2004,** *8*, 1219–1222.
12. Lan-guo, D. Cheng-zhong, Z.; Zhen, D.; Li-hong, N. Research Progress of Syndiotactic 1,2-Polybutadiene. *Chem. Propellants Polym. Mater.* **2005,** *4*, 302–305.
13. Luo, S. Polymerization Process for Preparing Syndiotactic 1,2-polybutadiene. Bridgestone Co., USA Patent 6720397 B2, April 13, 2004.
14. Hsu, W.-L. Halasa, A. F. Preparation of Syndiotactic Polybutadiene, Rubber Composition and Tire with Rubber Component. Goodyear Tire and Rubber Co, Patent USA 6956093 B1, October 18, 2005.
15. Jiang, L.; Zhang, X. Process for Manufacturing Vinyl-rich Polybutadiene Rubber, Changchun Institute of Applied Chemistry Chinese Academy of Science. Patent USA 7186785 B2, March 06, 2007.
16. Nasirov, F. A. Development of Scientific Foundation for Creation of Bifunctional Catalytic Dithiosystems for Reparation and Stabilization of Polybutadienes. Doctoral Dissertation (Chemistry), Baku: IPCP of Azerbaijan National Academy of Sciences, 2003, p 376 (Russian).

17. Nasirov, F. A. Bifunctional Nickel- or Cobalt Containing Catalyst-stabilizers for Poly-butadiene Production and Stabilization (Part I): Kinetic Study and Molecular Mass Stereoregularity Correlation. *Iran. Polym. J.* **2003,** *12,* 217–235.

18. Djanibekov, N. F. Dithiophosphates of Metals—Stabilizators of Polymer Materials. Doctoral Dissertation (Chemistry), IPCP, Azerbaijan Academy of Sciences: Baku, 1987, p 374(rus.).

19. Nasirov, F. A. Organic Dithioderivatives of Metals—Components and Modificators of Petrochemical Processes. *Petrochemistry* **2001,** *6,* 403–416 (rus.).

20. Rafikov, S. P.; Pavlova, S. A. Tvyordokhlebova I. I. *Methods of Determination Molec-ular Weight and Polydispersity of High Molecular Materials;* M.: Academy of Sciences of USSR: Russia, 1963; p 336.

21. Deyl, Z.; Macek, K.; Janak, J. *Liquid Column Chromatography;* Elsevier Scientific: Amsterdam, 1975; Vols. 1 and 2.

22. Rabek, J. F. *Experimental Methods in Polymer Chemistry: Physical Principles and Applications;* John Wiley & Sons: New York, 1982; p 365.

23. Zhang, H.-F.; Mo, Z.-S.; Gan, W.-J. X-Ray Crystallinity Determinations on the Syndio-tactic 1,2-Polybutadiene. *Acta Polym. Sin.* **1986,** *20*(3), 193–196.

24. Hohne, G. W. H.; Hemminger, W. F.; Flammersheim, H.-J. *Differential Scanning Calo-rimetry;* Springer: New York, 2003; p 356.

25. Nasirov, F. A.; Novruzova, F. M.; Azizov, A. G.; Djanibekov, N. F.; Golberg, I. P.; Guliev, N. A. Method of Producing Syndiotactic 1,2-polybutadiene. Patent 20010128, Azerbaijan, 1999 (azerb.).

26. Nasirov, F. A.; Guliev, N. A.; Novruzova, F. M.; Azizov, A. G.; Djanibekov, N. F. In *Pecularities of Butadiene Polymerization to Syndiotactic 1,2-polybutadiene in Toluene Solution,* Proceedings of IV Baku International Mamedaliev Petrochemistry Confer-ence, Baku, Sept, 2000, p 85.

27. Nasirov, F. A.; Azizov, A. H.; Novruzova, F. M.; Guliyev, N. A. In *Polymerization of Butadiene to Syndiotactic 1,2-polybutadiene in the Presence of Co-xantogenate + AlEt$_3$ Catalytic Systems,* Proceedings of Polychar—10 World Forum on Polymer Applications and Theory, Denton, USA, Dec, 2002; p 256.

CHAPTER 9

VINYLCYCLOPROPANES IN THE ADDITION AND POLYMERIZATION REACTIONS: A DETAILED REVIEW

ABASGULU GULIYEV[1,*], RITA SHAHNAZARLI[1], and GAFAR RAMAZANOV[2]

[1]*Institute of Polymer Materials of Azerbaijan, National Academy of Science, S. Vurgun Str., AZ 5004 Sumgait, Azerbaijan*

[2]*Department of Science, Sumgait State University, Badalbayli Street, AZ 5008 Sumgait, Azerbaijan*

*Corresponding author. E-mail: abasgulu@yandex.ru

CONTENTS

ABSTRACT

In this chapter, vinylcyclopropanes in addition and polymerization reactions is studied in detail.

9.1   BACKGROUND

In the field of chemistry of monomers and high-molecular weight compounds prepared from them, vinylcyclopropanes (VCP), a rapid development of which was observed recently, takes a special place. For some two to three decades it was opened a great number of new reactions with participation of VCP systems, to a number of which one can refer a specific addition of various addends, thermal and photochemical rearrangements proceeding with formation of substituted cyclopentanes, various isomerization reactions proceeding with intermediate formation of cyclopropylcarbinyl particles, etc.[1] Without going into details of the analysis of all these questions, let us consider a few aspects of chemistry of VCP, most important and relating to our investigations, including polymerization and copolymerization reactions.

A study of regularities and mechanism of the addition reactions to VCP has a great value both for theoretical organic chemistry and in practical plan. The compounds of a series of VCP are interesting with that the availability in their composition of two or more reactive centers gives these compounds sufficiently high reactivity for participation in the reactions with radical, electrophilic, and nucleophilic reagents.

A mutual influence of double bond and three-membered cycle makes these compounds relatively labile and that is why the compounds undergo various chemical reactions, including polymerization, for example, cyclopentenes,[2] diene systems,[3] and in some cases undergo the Cope rearrangement.[4]

The structure of cyclopropane ring supposes its participation in conjugation with neighboring unsaturated groups.[5] Consequently, an availability of conjugation between double bond and three-membered cycle makes these compounds more reactive in the various reactions. Considering its properties, they remind of the conjugated diene hydrocarbons.

In principle, every variety of these reactions, in which VCP can participate, is divided into the following groups: addition reactions, synchronous reactions, intramolecular rearrangements.

An interesting peculiarity of addition reaction proceeding on radical, electrophilic, and nucleophilic mechanisms is the fact that depending on nature of attacking addend and peculiarities of molecule of VCP, an addition proceeds either with participation of one of the reaction centers or with simultaneous participation of both reaction centers—double bond and cyclopropane ring.

The synchronous reactions usually proceed on agreed mechanism; these include cycloaddition reactions and various sigmatropic rearrangements (Cope rearrangement and retro-di-Pimentanova, etc).

The intramolecular rearrangements take place in thermal and photochemical processes and also in the presence of various catalysts in the system. The intermediates of such rearrangements are the unstable particles of types: biradical, zwitterions, and other similar radicals and ions.

The given classification of the reactions of substituted VCP in the various chemical reactions well reflects a variety of chemical properties of these compounds. For this reason, in the last two to three decades the attention of many researchers, both our country and foreign ones, has been given to the compounds of this class.

At first, we will consider the addition reactions of various addends to VCP and to its derivatives in the conditions of radical initiation.

9.2   REACTION OF VINYLCYCLOPROPANES WITH RADICALS

The first information on behavior of VCP in the addition reactions proceeding on radical mechanism appeared in the early 60s. Almost simultaneously, some groups of researchers have reported about possibility of radical polymerization of VCP and ability of these compounds to add such typical radical addends as mercaptanes and polyhalogen methanes.[5–9]

It has been established in these works that the VCP in the conditions of radical addition behave like conjugated diene hydrocarbons,[10] that is, an addition can proceed either with participation of only double bond or with inclusion of both reaction centers of double bond and three-membered cycle into reaction. In the first case, the products of 1,2-addition are formed, while in a case a participation of both functional groups in addition reaction leads to the formation of products of 1,5-addition.

For example, it has been shown in work[5,11,12] that bromotrichloromethane, carbon tetrachloride, thiophenol, and methyl mercaptanes are added to isopropenylcyclopropane with formation, depending on nature of addend, products having arranged (1,5-addition) and not rearranged (1,2-addition) structure:

$$\text{BrCCl}_3 \atop h\nu, 5°C$$ CCl_3 ⟶ Br   1,5-addition

$$\text{CCl}_4 \atop BPO, 80°C$$ CCl_3 ⟶ Cl   1,5-addition

$$\text{PhSH} \atop h\nu, 5°C$$ PhS   1,2-addition

$$\text{CH}_3\text{SH} \atop h\nu, 5°C$$ CH₃S $+$ CH₃S

Similarly, an addition of bromotrichloromethane to VCP proceeds.[6] As a result of photo-initiated reaction, it forms a mixture of *cis*- and *trans*-isomer products of 1,5-addition:

$$\text{BrCCl}_3 \atop h\nu, 16°C$$ CCl_3 Br $+$ CCl_3 Br

The addition of organostannanes to alkyl- and gem-dichloro-substituted [235, 236] VCP has been carried out in works,[7,8] where it has been established that the reaction of trimethyl- and tributyltin hydrides with these compounds leads to the formation of tin-organic products corresponding to structure of 1,5-addition. In its turn, these compounds as a result of protonolysis give the unsaturated hydrocarbons, formally corresponding to the process of selective hydrogenation of cyclopropane ring of the initial VCP:

R_3SnH ⟶ SnR_3 R_3SnH ⟶

The free radical addition of aliphatic and aromatic thiols to some polycyclic compounds having VCP fragment in its structure has been carried out.[9] It has been shown in these works that, as in a case of other VCP and radical addends, the reaction proceeds with formation mainly of products of 1,5-addition.

Now, it is considered that the reaction of radical addition proceeds on the following mechanism[5,10,13]: On the first stage of reaction as a result of addition of addend on double bond of VCP, it forms an adduct radical of cyclopropylcarbinyl type. Further, either forming radical or tearing radical is converted into product of 1,2-addition or is rearranged to thermodynamically more stable allylcarbinyl radical as a result of chain transfer which is then converted into product of 1,5-addition[14,15]:

$$R^{\bullet} + AX \longrightarrow A^{\bullet} + RX$$

adduct B

adduct A

In favor of this mechanism such experimental facts as a structure of forming adducts, kinetics and stereochemistry of addition process are indicated.

Indeed, a formation of adducts B, in which a residue of radical addend has been fixed in end methylene group is well agreed with known fact that a radical addition to mono- and gem-disubstituted olefins always proceeds against Markovnikov regulations.[16] A formation along with adducts B and also adducts A indicates to process behavior on type 1,5-addition. In other words, in the system, two types of radicals—cyclopropylcarbinyl and homoallyl—and no type "nonclassical radical" participate. Along with the proposed flow mechanism of the addition reaction of various addends to VCP, it has also proposed the other mechanism of reaction concluding that the intermediate compounds in this system are the nonclassical radicals of bicyclobutane type.[5]

The choice between two mechanisms has been made on the basis of analysis of structural peculiarities of adducts of reaction and study of kinetics and stereochemistry of process and also dependence of quantity of products on ratio of addend:VCP.

An existence of two radical particles in considered process is confirmed by the observation that the ratio of products 1,2- and 1,5-addition is mainly determined by rates of two competing processes:monomolecular arrangement (K) and bimolecular chain transfer (K_2). Consequence of this should be an existence of dependence of ratio of products 1,2- and 1,5-addition,[4,8,9,17-23] or exclusively rearranged adducts.[4] In this case, a ratio of products depends on concentration of addend,[13,17,23] its efficiency in chain transfer,[4,17] and on structure of peculiarities of initial VCP.[18–21] At the same time, the VCP containing substituents in cyclopropane ring, able to activate three-membered cycle and effectively to stabilize allylcarbinyl radical (CO_2R, CN, Rh, Cl, Br etc.) forming as a result of rearrangement, are reacting even with such high-efficient chain transmitter as thiophenol with formation of exclusively rearranged product.[17–24]

The kinetic data on homopolymerization of some VCP also confirm an existence of intermediate radical particles. Order lowering on monomer from theoretical, first order detected at polymerization of ethers of VCP–carboxylic acids and methoxymethyl-substituted VCP,[25,26] indicates to that during polymerization an additional stage takes place. This stage is the rearrangement of cyclopropylcarbinyl radical (CPCR) to allylcarbinyl.

In a number of works by means of EPR method, formation of CPCRs and their rearrangement to allylcarbinyl radicals (ACR) were directly observed.[27–30] It has been detected that the CPCRs arising in the system are in rapid equilibrium with ACR. Of course, equilibrium will be strongly displaced to the side of formation of ACR only in those cases when radical of cyclopropylcarbinyl type will be less stabilized in degree in comparison with radical of allylcarbinyl type forming from it:

CPCR-radicals   ACR-radicals

Let us note that stereoselectivity of addition reaction in literature has not been studied practically. There are a limiting number of experimental data on this question, besides having a contradictory character. It has been established in one of the works that the addition process proceeds nonstereoselectively and as a result of reaction it has formed a mixture of isomer products of 1,5-addition. For example, an addition of bromotrichloromethane to VCP[5] tributyl- and trimethyltin hydride to alkyl-substituted VCP[6] proceeds with formation of mixture of isomer 1,5-adducts. Similarly, it also proceeds homopolymerization of ethers of VCP–carboxylic acid, as a result of which it forms homopolymer containing both *trans*- and *cis*-isomer links in its composition.[8] In other works, the data indicating to high *trans*-stereoselectivity of free radical addition to VCP have been prepared. In particular, it has been shown in work[31] that formation of 1,5-adduct as a result of free radical addition of thiphenol to diethyl ether of cyclopropane-1,1-dicarboxylic acid has a *trans*-configuration. The similar results have been prepared in a case of addition of thiophenol to other disubstituted VCP. In most works,[6,8,31] a problem of stereoselectivity of process is not generally considered. The authors of work[5] consider that an addition of bromotrichloromethane to VCP proceeds stereoselectively with formation of either *trans*- or *cis*-product of 1,5-addition. The product formed as a result of reversible addition of bromine atom on internal double bond is isomerized in the process of reaction, as a result of which an equilibrium mixture of two isomer products is formed. In work,[22] the authors assume that cyclopropyl carbinyl radical arising on the first stage of reaction exists comparatively long, as a result of which a rotation has time to occur around a single bond before rearrangement to allylcarbinyl radical; therefore, the forming acyclic radical and also the products of its stabilization possess *trans*- and *cis*-configuration of internal double bond.

9.3   REACTION OF VINYLCYCLOPROPANES WITH CARBONIONS

A structure of three-membered cycle and its unsaturated character provide ability of cyclopropane compounds to take part in the reactions proceeding on mechanism of ion addition. Similar to olefins, the cyclopropanes having two electron-pullback substituents in geminal position are able to react with various nucleophilic reagents.[32]

In literature, there are a number of works devoted to the study of behavior of substituted VCP in the addition reactions proceeding on nucleophilic mechanism.[33-42]

In these reactions, mainly, the VCP of two types—with activated cyclopropane ring and with activated double bond—were studied.

The compounds of the first type are the derivatives of vinylcyclopropane-1,1-dicarboxylic acid. The compounds of the second type are the derivatives of cyclopropyl-substituted vinylidenmalonic acid:

where: $X = Y = CO_2R$ (a); $X = CO_2R$, $Y = CN$ (b); $X = CN$, $Y = CO_2R$ (c); $X = Y = CN$ (d).

In the reactions of nucleophilic addition along with compounds of the first type, some other VCP such as (2-vinyl-cyclopropyl)alkyl ketones,[43] phenyl VCP, and vinylologs of cyclopropyl ketone have been also studied.[42]

These compounds react with nucleophilic reagents such as aromatic or aliphatic thiols,[34,36,38] primary, secondary, and tertiary amines,[33–35,39,40] organic lithium cuprates,[40] etc. In this case depending on type of VCP on nature of nucleophile and reaction conditions either products of conjugated addition, proceeding with simultaneous participation both of cyclopropane ring and double bond or those and other products are simultaneously formed.

For compounds of the second type with activated double bond, it is mainly a characteristic formation of products of addition on one of the reaction centers. In this case, more often, a double bond is attacked. In particular, it has been shown in works[34,38,42,45,46] that an addition of butylmercaptane catalyzed alcoholates, dimethyl- and dibutyl lithium cuprates, and cyclohexene pyrrolidone proceeds mainly with formation of products of 2,1-addition. Along with these products, sometimes a formation of products of 5,1-addition is also observed:

In the presence of steric factors hindering nucleophilic attack of double bond, for example, introduction of methyl group in double bond, a flow of reaction with thiophenol and butylmercaptane toward formation of product of 5,3-addition is observed, that is, a process proceeds exclusively with opening of cyclopropane ring[37]:

Unlike compounds of the second type, the VCP with activated cyclopropane ring give a mixture of products of 1,5- and 3,5-addition in the nucleophilic addition reactions. Ratio of products, in this case, depends both on nature of nucleophile and initial VCP. With more strong nucleophiles such as lithium diallyl cuprates[41–43] and mercaptanes,[39–44] the diethyl and dimethyl ethers of 2-vinylcyclopropane acids react with formation mainly of products of 1,5-addition. However, with comparatively weak nucleophiles such as primary and secondary amines,[33,35,39] these compounds react with formation exclusively of products of 3,5-addition:

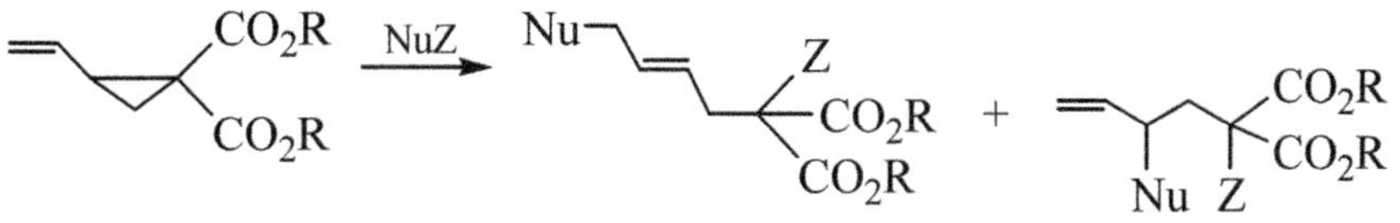

The characteristic peculiarity of the reactions with some amines is the formation along with basic side products. The formation of last ones has been connected with various conversions occurring in the process of reaction with primarily forming products. So, in work,[39] besides basic products of reaction—amine ethers—the other products forming as a result of its conversion have been also isolated:

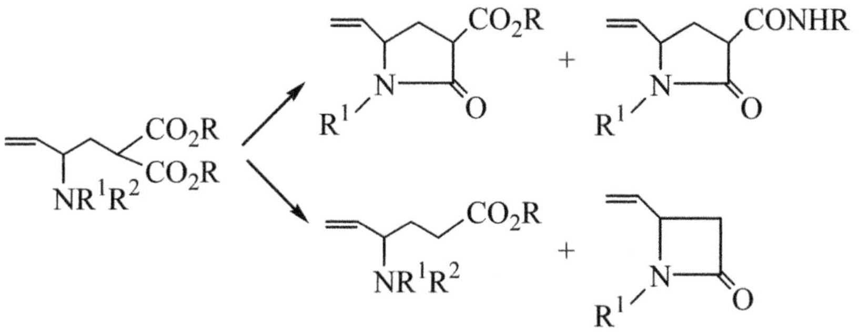

In attempt to add ethyl alcohol to compounds of type I in the presence of sodium ethylate instead of the basic reaction product, the condensation product has been prepared:

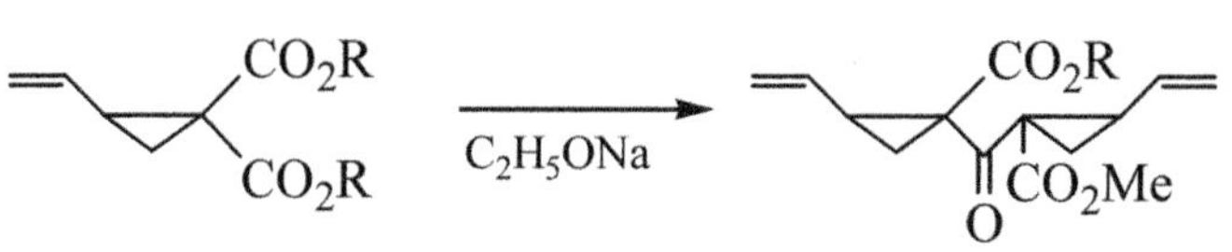

As it is shown in work,[44] a nature of substituents of VCP group plays an essential role in the reactions of nucleophilic addition. So, it has been established that an addition of thiophenol to gem-disubstituted VCP (compounds of type I) catalyzed by sodium ethylate proceeds, mainly, as 3,5-addition. In this case, a ratio of products of 1,5- and 3,5-additon is determined by nature of substituents in the cyclopropane ring:

In a case of compounds, where X, Y = Ph, it only formed a product of 3,5-addition, while in a case of compounds X, Y = CO_2Et and CN, both products of addition are formed (ratio of adducts 3,5- and 1,5-addition, respectively, is equal to 4 and 4,5). It is interesting that in the absence of catalyst with thiophenol they easily react and lead to the formation of adducts, a ratio of which corresponds to 2:1. The authors of this work assume that the formation of products takes place as a result of side reaction proceeding on radical mechanism.

For nucleophilic addition reactions to VCP,[34] the dipolar mechanism of opening of cyclopropane ring has been accepted.[35] Due to electron-pullback action of functional substituents, polarization of all molecules of VCP takes place. On carbon atom bearing electron-pullback groups, a partial negative charge is created and in remaining part of molecule, a partial positive charge is distributed. For compounds of the second group, an electrophilic center arises on cyclopropyl-substituted carbon atom; owing to conjugation of cyclopropyl group with this atom, a weaker electrophilic center is formed on carbon atoms of three-membered cycle. In a case of VCP of the first group, the electron-pullback substituents and electron-donor double bond form a system in which a negative charge of dipole is localized on carbon atom bearing electron-negative substituents, and a positive charge is delocalized in allyl system formed by double bond and cyclopropane carbon atom:

In the process of nucleophilic addition reaction, a nucleophile attacks one of the electrophilic centers of forming dipole, as a result of which an anion stabilized by substituents is formed. As a result of nucleophilic attack of environment, a carbanion is converted into products of reaction. It is natural that in the noted mechanism, the nucleophilic addition reaction to VCP should be extremely sensitive to nature of nucleophilic reagent, nature of functional substituent, steric factors, and conditions of carrying out of process, which in reality is observed when carrying out an experiment.

9.4 REACTION OF VINYLCYCLOPROPANES WITH CARBONIUM IONS

From addition reactions to VCP, the reactions of electrophilic type are most widely expanded as a double bond in these compounds is activated by electron-donor substituent—cyclopropane ring—and possesses higher inclination to addition of electrophilic reagents. Besides a cyclopropyl group, there is effectively stabilized group of intermediately forming carbonium-ions, which also increases an inclination of these compounds to be reacted on electrophilic mechanism.

In comparison with the reactions proceeding on radical and nucleophilic mechanisms, an electrophilic addition is characterized by considerably large variety. This has been connected with more wide spectrum of reagents able to be reacted on electrophilic mechanism with intermediate formation of cyclopropylcarbinyl cations and capacity of last ones to various intramolecular rearrangements leading, in particular, to the formation of cyclobutane compounds.[47]

Now there are sufficiently many works devoted to the study of conducting of VCP in the electrophilic addition reaction.[48–65] In these works, the mono-addition reactions of various addends to different alkyl- and aryl-substituted VCP,[48–56] polycyclic compounds containing VCP fragment in its structure,[57–61] to some functionally substituted VCP have been studied.[62,63]

The following electrophilic reagents were used in the addition reactions[48–64]: aromatic sulphenyl chlorides,[51,52,62] chlorosulfoisocyanates,[60,62] halogens,[48,51,62,63] inorganic, and organic acids,[48–50,56,59] strongly polarized cationoide reagents,[53–55,64] and other similar compounds.

A characteristic feature of the reactions proceeding on electophilic mechanism is the dependence of direction of reactions on nature of electrophilic reagent used, structural peculiarities of the initial VCP, and reaction conditions. The intermediates of these reactions, depending on peculiarities of reacting substances and reaction conditions, can be various cations on relative stability which are located to the following series[47]:

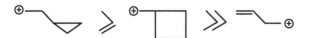

In the process of these reactions, such intermediate particles as bridge and opened carbonium ions can be formed, "close" and divided ionic pairs and so on.[65] In this connection, in the electrophilic addition reactions proceeding with participation of VCP depending on nature of transition compounds, a formation of various products is observed.

A basic factor determining a direction of the addition reaction to VCP is the nature of electrophilic reagent. In a case of weak electrophiles, an addition, mainly, proceeds without opening of cyclopropane ring and leads to the formation of products of 1,2-addition. In particular, it has been shown in works[51,52] that arylsulphenyl chlorides are added to VCP with formation of exclusively non-rearranged adducts:

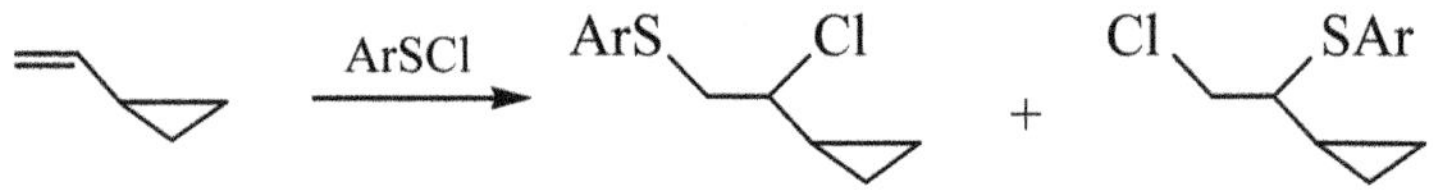

The formation of non-rearranged adducts was also observed in the reactions of more strong electrophiles. In particular, an addition of H, Cl, HCl, acidic-catalyzed addition of CH_3O and CH_3OH to bicyclo/3.0.1/hexene-2 proceeds, mainly, with formation of products of 1,2-addition.[57]

Similarly, without opening of cyclopropane ring there proceeds the acid-catalyzed hydration,[51] lactonization of VCP–carboxylic acids,[66] and migration of double bond in some bicyclic VCP.[61]

Generally, an addition of strong electrophiles to VCP proceeds with inclusion of cyclopropane ring into reaction; besides 1,2-adducts, the products of homoallyl rearrangement of intermediately forming cyclopropylcarbinyl cations are formed.

In particular, a bromination of VCP in acetic acid proceeds with formation of mixture of rearranged and non-rearranged products:

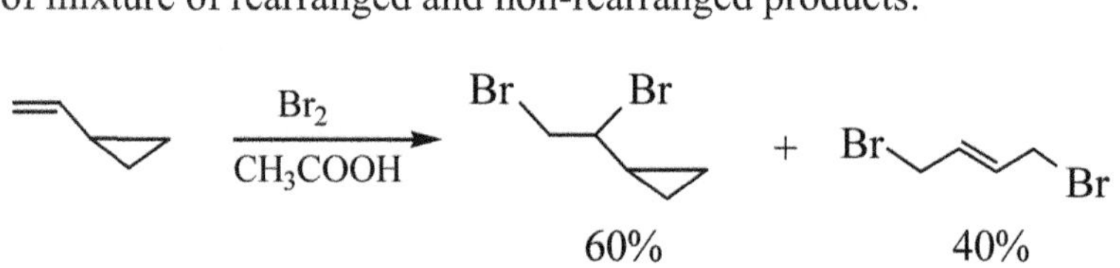

Similarly, with formation of products of 1,5-addition, the reactions of some polycyclic,[57,59,60] alkyl-, and aryl-substituted[48–50] VCP with organic and inorganic acids and other reagents proceed.

Strongly polarized cationoide reagents in the reactions with VCP lead to the formation of either acyclic unsaturated products of 1,5-addition[53] or compounds of cyclobutane series.[55,56] For example, it has been shown that an acylation of some VCP by various acylboron fluoride proceeds with formation of bicyclic compounds containing cyclobutane fragment:

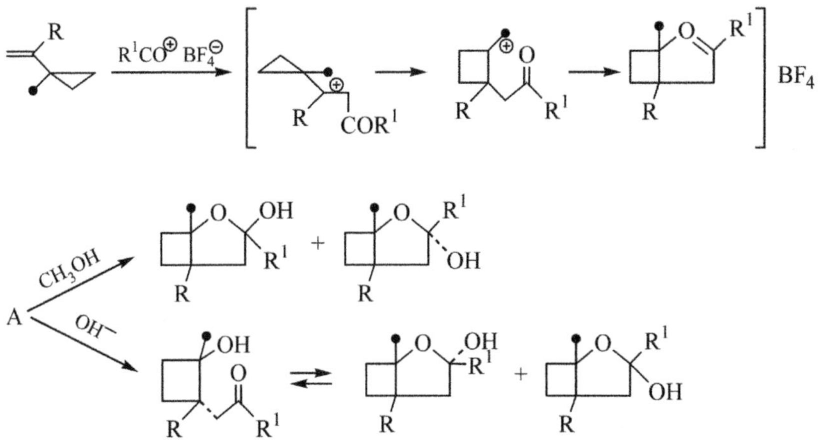

Let us note that a formation of cyclobutane derivatives in the conditions of electrophilic addition to VCPs was observed in the reactions with strongly polarized reagents and at cation polymerization of similar compounds.[53,55,56] In the rest cases, an addition proceeds with rearrangement of hydrocarbon skeleton.

The structural peculiarities of initial VCP and nature of substituents influence also on direction of addition reactions. In particular, it has been shown in work[52] that even an addition of such weak electrophilic reagent as 2,4-dinitrophenylsulfenyl chloride to substituted VCP proceeds with homoallyl rearrangement of intermediate particles and leads, mainly, to the products of 1,5-addition:

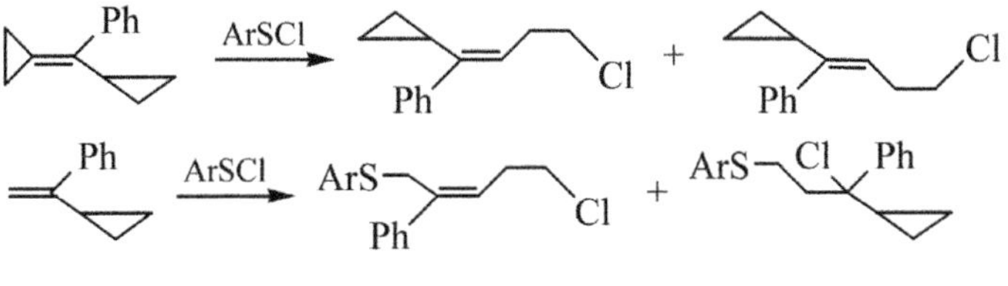

At the same time, an addition of strongly polarized cationoide reagent such as nitrinium boron fluoride to ethyl ethers of VCP–carboxylic acids proceeds without any rearrangement and leads to the formation of products of 1,2-addition[64]:

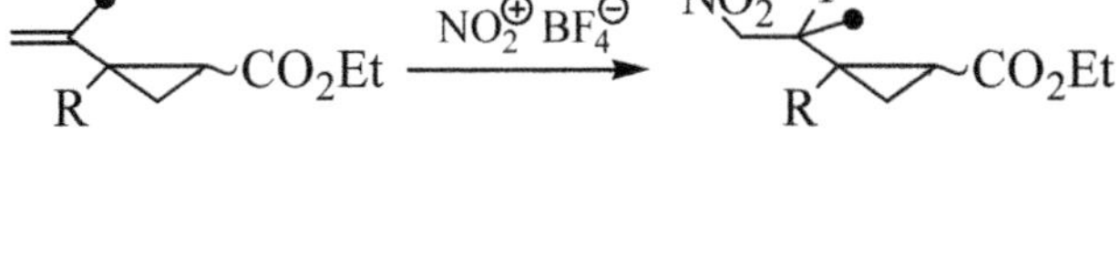

An influence of reaction conditions on flow of addition reaction can be demonstrated by the following example. It was previously mentioned that an addition of 2,4-dinitrophenylsulfenyl chloride to VCP proceeds with forma-tion of products of 1,2-addition. In this case, as it was noted,[52] an addition in nonpolar conditions (CCl_4) proceed with small conversion of the initial substances and in polar solvent (CH_3COOH) these products proceed almost with quantitative yield. In its turn, an introduction of catalytic quantities of lithium perchlorite cardinally changes a direction of this reaction. The domi-nating processes become homoallyl rearrangement of intermediately arising divided ionic pair and conjugated addition:

From all previously described literature data, follow three characteristic peculiarities of such investigations: first, investigations carried out in this field do not have systematic character; second, little attention is paid to stereo-chemical aspects of the studied reactions. The third peculiarity concluded in that is in the addition reactions proceeding on electrophile mechanism, the alkyl- and aryl-substituted VCP were mainly studied and functionally substituted VCP were practically not studied. The limited number of investi-gations,[63,64,67] in which as the object of investigation were used functionally substituted VCP, have an episodic character.

Thus, the literature information relatively mechanism of the addition reaction to VCP collected in this review show that up to the present this question cannot be considered as determined. Available conflicting conclu-sions require further checking and accurate specification. Apparently, it has also stimulated our investigations in this direction.

9.5 POLYMERIZATION OF VINYLCYCLOPROPANE

The cyclopropane and its alkenyl substituted and other derivatives in the conditions of radical initiation are not subjected to the polymerization. Only in the conditions of photoirradiation, some similar compounds form the oligomer products. Such compounds are usually polymerized in the ionic conditions. Therefore, here onwards, it will only be considered the radical polymerization of VCP.

An analysis of domestic and foreign literature shows that the VCP compounds easily undergo the polymerization and copolymerization with various monomers. In these processes, the VCP shows high reactivity, which has been undoubtedly connected with their electron structure and conformational disposition of substituents. Considering physical–chemical bases of formation process of polymers from VCP series, it is clarified that a conversion reaction of VCP to linear polymer is possible, if it proceeds with decrease of free energy, that is, if store of free energy of monomer VCP exceeds a store of free energy of polymer product. In other words, a polymer shows larger thermodynamic stability in the reaction conditions. It was known that a free energy ΔG, as it follows from conditions $\Delta G = \Delta H - T\Delta S$, is decreased with decrease of enthalpy ΔH (exothermal processes) and increase of entropy of system ΔS. In the process of monomers conversion into polymer, the new bonds and possibility of internal rotation around these bonds arise, that is, it increased a number of degrees of freedom. Thus, the conditions of polymerization are characterized by values of $\Delta H < 0$ and $\Delta S < 0$, which are characteristic for all vinyl monomers. In addition, a transition of strained cycle (owing to distortion of normal bond angles) to unstrained linear polymer also influences change of enthalpy. This polymerization process determines a field of thermodynamic stability of polymer. A field of temperatures, at which a polymer will be in thermodynamically stable state, can be determined from the following conditions:

$$\Delta G = \Delta H - T\Delta S \quad T > \frac{\Delta H}{\Delta S}$$

Literature data show that to prepare polymers from cyclopropanes in reality is very difficult if ΔG is negative.[68] However, it should be mentioned that a change in stabilization energy can play a considerable role in determination of polymerization heat of the investigated monomer. Due to not reaching special results in the field of polymerization and cyclopropane and its alkyl-substituted derivatives, many researchers have devoted the further works to the study of polymerization of vinyl-substituted cyclopropanes.

The first investigation in this field were patent works,[69,70] where were considered the possibilities of homopolymerization of vinyl-substituted cyclopropanes and their copolymerization with acrylonitrile, styrene, and diene hydrocarbons in the oxidative–reductive system. Unfortunately, the information about structure of the prepared polymers was absent.

If considering electronic structure, then it is easy to note that the cyclo-propane and its derivatives can possess higher reactivity. It is enough to note that in more favorable conformation conditions for conjugation of cyclopropane ring with adjacent groups, three-membered carbon cycle is able to undergo the conjugation and to give an influence of groups added to them. Thus, an overlapping of π-orbitals of substituent with orbital of cyclo-propane ring takes place.[71,72] The last one leads to increase of reactivity of general system. However, such systems are not too well polymerized, espe-cially in the presence of radical initiators. Some compounds with small cycle are polymerized in the presence of cation initiators or complex catalysts of Ziegler–Natt type. An introduction of alkenyl group into cyclopropane ring noticeably activates such monomers and they get tendency to be polymerized with sufficiently high rate in the presence of various catalysts. In this case, the VCP are mainly polymerized on type of 1,2- and 1,5-polymerization:

The polymerization reaction behavior on that or other mechanism depends on electron properties of substituents X, Y.

The first examples on polymerization of the VCP such as VCP,[73] 1,1-dichloro-2-vinylcyclopropane, 1,1-dibrom-2-vinylcyclopropane,[74] and ethyl ether of 2-propenylcyclopropane carboxylic acid,[75] carried out in the presence of radical initiators, lead to the formation of low-molecular weight oligomers, a structural link of which had a rearranged, acyclic structure, that is, met the polymerization process proceeding with simultaneous opening both of double bond and cyclopropane ring.

In the subsequent years, after these works, the intensive study of polym-erization of VCP began.

In works, particularly,[76–80] the homopolymerization of various VCP, in the presence of typical free radical initiators, was studied. In the polymerization,

very various VCP were involved—unsubstituted,[81–83] gem-dihalo-substi-
tuted, monohalo-substituted,[84] methoxymethyl-substituted and various
derivatives of VCP mono-[85,86] and dicarboxylic[87] acids.

The powdered polymer products have been prepared at polymerization
of 1,1-dichloro-2-vinylcyclopropane in the presence of benzoyl peroxide or
AIBN.[88] It has been established by investigation of structure of the prepared
polymers that the polymerization proceeds with simultaneous opening of
double bond and three-membered cycle:

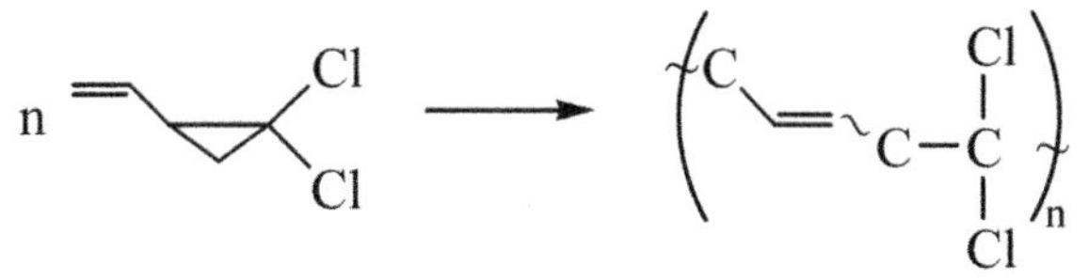

The VCP along with polymerization in the presence of ion catalysts gives
also a high-molecular weight product in radical initiation, and a structure
of forming polymers in this process is identical to structure prepared from
cyclopentene in the presence of complex catalysts[88,89]:

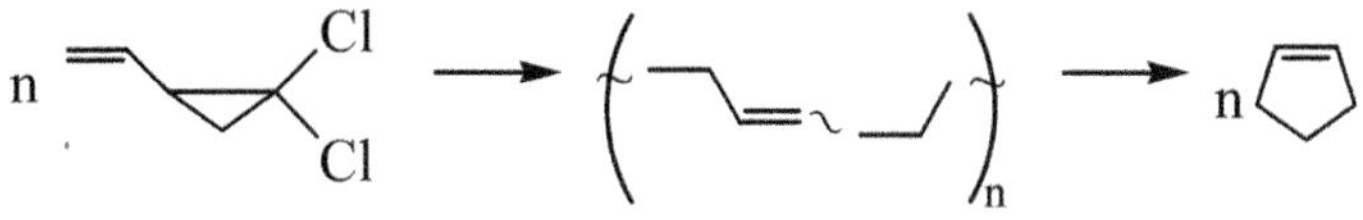

2-Carbethoxy-1-vinylcyclopropane as shown in work[90] is easily polymerized in the presence of radical initiators forming, in this case, sufficiently
high-molecular weight products having links, exclusively consisting of
1,5-structural units in the chain of macromolecules.

It has been established that the polymerization process begins from attack
on vinyl group with formation of intermediate cyclopropylcarbynyl radicals
which on the second stage are rearranged with opening of three-membered
cycle. The forming alkylcarbynyl radicals stabilized by ethoxycarbonyl
group again attack new molecule of VCP.

Comparatively, large stability of ethoxycarbonyl substituted radicals with
cyclopropylcarbynyl ones allows the rearrangement process. As a result,
the structure of polymer chain corresponds to pentenamer links, that is, the
polymerization process proceeds with simultaneous opening of double bond
and three-membered cycle:

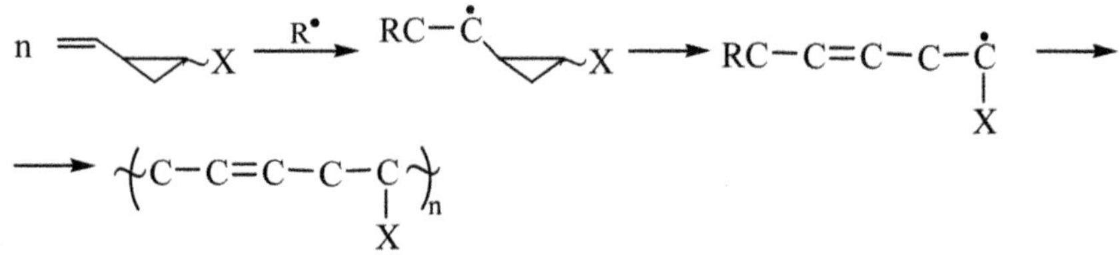

On analogous mechanism, the polymerization of 1-vinyl-2-carbethoxy-3-methyl-, 1-propenyl-2-carbethoxy-, and 1-isopropenyl-1-methyl-2-carbethoxycyclopropanes also proceeds.[91,92]

Some VCP containing alkyl or haloid substituents in ring have been also successfully polymerized. Although the polymerization products were the low-molecular weight compounds, nevertheless it has been shown by data of spectral analysis that the used compounds give 1,5-type of elementary links, that is, a process proceeds on the same two-stage mechanism.[70,76,81,93,94] The prepared results are explained by the authors on the basis of stability of intermediate-forming cyclopropylcarbynyl radicals.

Considering chain structure of homopolymers prepared at radical polymerization of 1,1-dicyclopropyl ethylene and comparing it with structure of polymer prepared from the same monomer in the cation conditions, the author's work[74,83,95,96] concluded that the most considerable difference in the IR-spectra is observed in the field of valence vibrations of double bond. In addition, availability of resonance signal at δ ~0.5 ppm, characterizing protons of cyclopropyl group, and considerably less intensive signal at δ ~5.7 ppm, indicating possibility of formation of two structures—rearranged and non-rearranged ones. In a case of radical polymerization, a rearrangement occurs in noticeable degree than in a case of cation process:

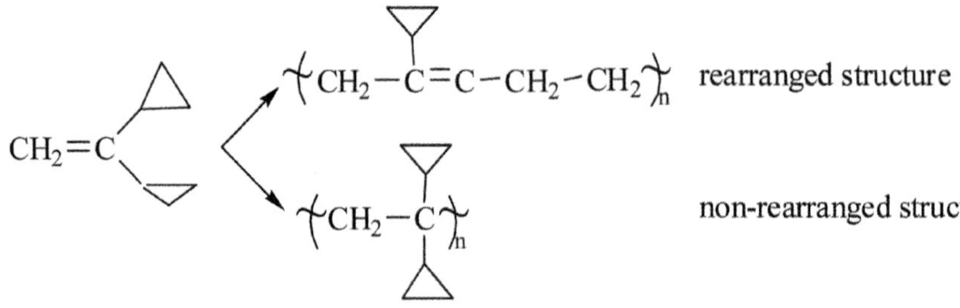

A quantitative determination of content of 1,5-type of structures on data of PMR-spectroscopy corresponds to >90%, whereas a polymer prepared in cation conditions contains only 25–30% 1,5-type of links.

In the investigations carried out, a formation of linear structures is explained by a level of stabilization of radicals forming in the process. In a case of availability of substituents able to stabilize allylcarbynyl radical forming in the process, the process completely proceeds as 1,5-polymerization.

In a case of introduction of substituent not containing π-electron system into VCP system, a polymerization behavior on mixed mechanism is observed, that is, a formation of statistic "copolymer" consisting of 1,2- and 1,5-type of structural units.[77] The conclusion made has been explained by the results of model reactions and determination of kinetic parameters of polymerization and also by establishment of structure of polymer chain by data of spectral and chemical analysis.

The VCP containing glycidyl substituent in cyclopropane group in the radical conditions has been converted into oligomer product having linear pentenamer links with side epoxy-groups.[97-99]

Korean researchers have shown that the gem-disubstituted VCP easily undergo the homopolymerization in the conditions of radical initiation.[31,100] Using 1,1,-dinitrile-, 1,1-diethoxycarbonyl-, 1-nitrile-1-ethoxy-carbonyl-, and 1,1-diphenyl-substituted VCP as the initial monomers, the authors have established that the copolymers with high molecular weights are prepared in a case by using ethoxycarbonyl- and nitrile-substituted VCP. Diphenyl substituted VCP is not polymerized. It has been revealed on the basis of data of the IR-spectroscopy that the double bonds, which arise as a result of polymerization on type 1,5- have *trans*-configuration. The results of model reactions confirmed the polymerization behavior on type of 1,5-addition. The authors explain the mechanism of the polymerization process upon levels of stabilization of intermediate and rearranged radicals forming in the system. An availability of two substituents with π-electronic system in radical center makes growing radicals more stabilized in comparison with intermediately forming cyclopropylcarbynyl radicals:

$$\text{(VCP structure)} \quad \xrightarrow{R^\cdot} \quad \{CH_2-CH=CH-CH_2-\underset{Y}{\overset{X}{C}}\}_n$$

X, Y = CO$_2$Et, CN, Ph

The order of reactivity found is as the following series: diphenyl- << diethoxycarbonyl- << ethoxycarbonyl nitrile- << dinitrile-vinylcyclopropane.

In literature, there are also investigations devoted to the polymerization of VCP in the other catalytic conditions. Certainly, this direction of investigation is of the large scientific and practical significance.

Let us first consider the polymerization of various VCP in the cation conditions.[101–104] In these works, the polymerization of alkyl-, aryl-, and functionally substituted VCP in the presence of various Lewis acids and Ziegler–Natt catalysts has been studied; it has been shown that the majority of these compounds are polymerized in the cation conditions, with formation of polymers having sufficiently high molecular weights.

A majority of the prepared polymers had the structural links of various types: nonarranged—corresponding to process of 1,2-polymerization; rearranged—corresponding to process of 1,5-polymerization. The relative content of these links changed depending on peculiarities of initial VCP. In particular, the VCPs in the presence of AlB_3 in a medium of C_2H_5Cl at $-78°C$ are polymerized with formation of three types of structural elements[82]:

$$-CH_2-\underset{\underset{\underset{CH_2}{\diagdown\diagup}}{\underset{CH_2-CH_2}{|\quad|}}}{CR}- \qquad\qquad -CH_2-CR-CH-CH_2-CH_2- \qquad \underset{\underset{CH_2-CH_2}{|\qquad|}}{CH_2-CR-CH}$$

$$\qquad\quad \text{I} \qquad\qquad\qquad\qquad\qquad \text{II} \qquad\qquad\qquad\qquad \text{III}$$

Occurrence of these structures has been stipulated by ability of cyclopropylcarbynyl cation to exist also as a nonclassic bicyclobutanium ion:

$$CH_2-CR-CH-CH_2 \xrightarrow{K^{\oplus}} KCH_2-\overset{\oplus}{CR}-CH-CH_2 \rightleftharpoons KCH_2-\overset{1}{CR}\overset{2}{=}CH$$

An attack of new molecule of monomer on C1, C2, and C4 leads to the formation of structural links of type I, II, and III.

T2 he polymerization of VCP also leads to the formation of polymer containing structural links of three types: nonarranged links (1,2)—60%, homoallyl links (1,5)—10%, and 30% links of the following structure[76]:

$$-CH_2-CH(CH_3)CH(CH_3)-$$

In its turn, isopropenylcyclopropane is polymerized with formation of polymer containing from 25% to 50% cyclopropane-containing links and

75–50% homoallyl structure having *cis*- and *trans*-configuration of internal double bond. At the same time, a cation polymerization of 1,1-dicyclopropyl ethylene proceeds with formation of polymer containing, mainly, 75% rearranged links.[101]

An interesting influence of substituents on process of cation polymerization during polymerization of aryl-substituted VCP has been revealed. It has been established in this work that depending on nature of substituents in aromatic nucleus of these VCP, the polymers with various content of links of rearranged and non-rearranged structure are formed:

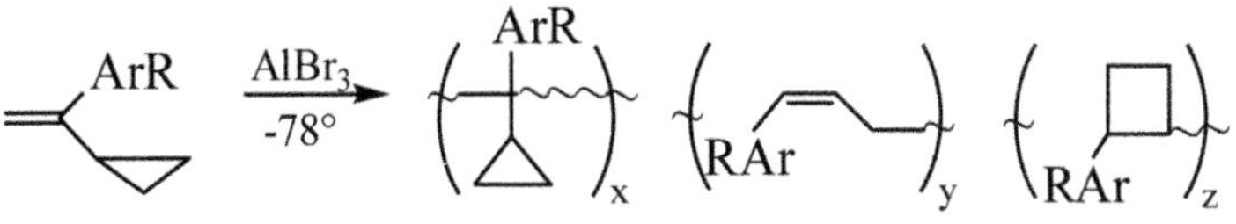

where : R= *o, p*-(NO$_2$)$_2$: x = 1,0
R= *i*-Pr : y = 0,7 x = 0,3
R= *p*-F : x = 0,2 y = 0,5 z = 0,3

The similar phenomenon has been also detected in a case of functionally substituted VCP. In particular, it was known that 1,1-dichloro-substituted VCP is not polymerized in the cation conditions.[83] At the same time, during polymerization of 1,1-dichloro-2-vinylcyclopropane, a low molecular polymer containing structural links of the following two types are formed:

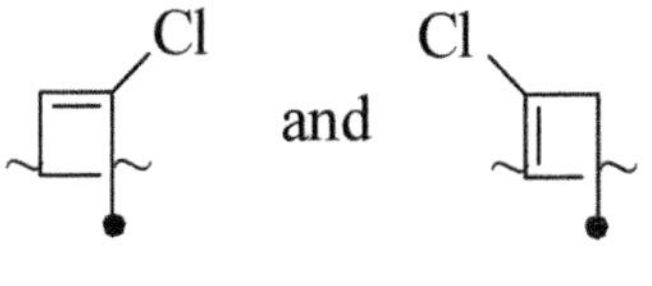

The important role of steric factors for period of cation polymerization has been also revealed in work.[102] It has been shown that a *cis*-isomer of monohalogen-substituted VCP is polymerized with formation of products having nonarranged structural links. At the same time, *trans*-isomer is not polymerized.

Now, let us consider two works devoted to the anionic polymerization of functionally substituted VCP.[45,105] The authors of these works have showed that the anion polymerization of a series of 1,1-disubstituted 2-vinylcyclopropanes proceeds with formation of polymers, a structural link of which has an acyclic structure with side double bonds, that is, the polymerization process corresponds to 3,5-addition and proceeds with conservation of double bonds of the initial VCP:

$$X = Y = CN; \ Y = CO_2Et, \ X = CN$$

It has been established in these works that in predominant majority of cases they are polymerized with formation of polypentenamers, that is, polymerization of these compounds proceeds as 1,5-addition. In this case, nature of substituent being in cyclopropane ring is the basic factor determining direction of polymerization (on 1,2- or 1,5-type) and molecular weight of forming polymers.

In a number of works,[106-108] the mechanism of polymerization was studied intensively. Considering the structural peculiarities of the initial monomer compounds, microstructures of polymers forming in the process, analyzing results of model reactions and kinetic investigations and also considering correlation between structure of VCP and reactivity of double bond, the authors have concluded that the VCP are polymerized in two-stage isomerization mechanism.

The polymerization of the VCP is characterized by that the radicals forming in the beginning of process are then rearranged to the radicals of other structure. Rearrangement occurs due to stabilization of radical center—electron system of substituent—because a monomolecular rearrangement rate constant in these processes predominates over growth rate constant on the primary reaction center. This circumstance is a necessary condition for polymerization of the VCP on 1,5-type.

Thus, the polymerization of derivatives of VCP proceeding on 1,5-type leads to the formation of polymers, a chain structure of which corresponds to polypentenamer links. Such structural links usually arise during polymerization of cycloolefin hydrocarbons in the presence of some complex catalysts on the basis of metals of type W, Mo, etc.[109]

As seen previously, once the polymerization of VCP systems in the presence of radical initiators is accompanied by simultaneous opening of double bond and cyclopropane ring, it leads to the formation of soluble polymers with pentenamer links in the main chain of macromolecules.

Now let us consider the copolymerization process proceeding with participation of derivatives of VCP. In the literature, there is a limited quantity of works on this theme. It should be first informed about two works of Japan researcher.[110,111] Analyzing results prepared on copolymerization of 1,1-dichloro-2-vinylcyclopropane with maleic anhydride, methyl methacrylate, and styrene, the author has concluded that in a case of copolymerization with maleic anhydride, where the last one undergoes the composition of copolymer in double quantity in its excess in the initial mixture an intramolecular cyclization with formation of six-membered cyclic structures takes place:

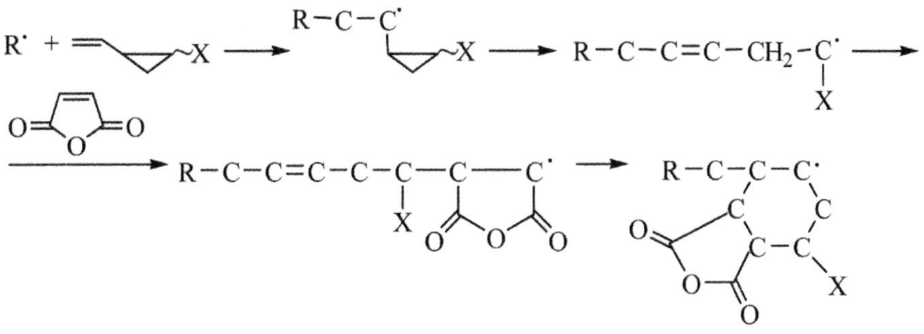

Such conclusion has been made on the basis of results of investigation of composition and structure of the prepared copolymers. It has been found that during carrying out copolymerization both in solution and in mass a dependence of composition of copolymer on composition of the initial mixture is observed. With increase of concentration of the VCP monomer, its quantity in copolymer grows to equimolar one. At its low concentrations, a formation of copolymer of composition is observed— maleic anhydride:VCP = 2:1. A kinetic scheme is considered, a mechanism of carrying out of process is proposed. In all rest cases of copolymerization, the statistic copolymers enriched by comonomers are formed. The constants determining relative reactivity of the investigated comonomers have been determined.

With the aim of establishment of influence of structure of the VCP monomer on its reactivity, the copolymerization of some ethers of VCP carboxylic acids and also their stereoisomes has been investigated.[112–115] Both in a case of structural and spatial isomers the formation of radicals of acrylate type in the copolymerization process very close radicals of acrylate type on reactivity has been established:

$$\left\{CH_2-CR^1=CR^2-CH_2-CH\right\}_n$$

The radicals forming in the system in enough degree are stabilized due to conjugation of unpaired electron with π-orbital of carbonyl group. Furthermore, a structural closeness makes them slightly differing on its reactivity. Therefore, the investigated systems are the simplified variant of the copolymerization process, where there are two different monomers competing with general radicals.

Some derivatives of VCP with maleic anhydride and styrene have been also successfully copolymerized.[116–121] The VCP and its methoxymethyl-, ethoxycarbonyl-, glycidoxymethyl, and carbglycidoxy-substituted derivatives in the presence of radical initiators undergo the copolymerization with simultaneous opening of double bond and three-membered cycle. In a majority of cases, a formation of copolymers with equimolar composition of the initial monomers is observed. In a case of copolymerization with styrene, the copolymers are formed, where links of separate comonomers are statistically distributed in copolymer.

For study of monomer compounds, the copolymerization constants and Q-e parameters characterizing specific activity of the initial monomers and polarity of radicals forming in the system have been determined. An availability of reactive functional groups in composition of VCP monomers gives chemical activity to form copolymers.

With the aim of preparation of self-cross-linking copolymers as the comonomers, vinylpyridine has been used.[122] It has been shown that the cross-linking rate depends mainly on ratio of monomer links in copolymer chain.

For copolymerization with participation of gem-disubstituted VCP monomers in work[31] the comonomers of three types, acrylonitrile, vinyl acetate, and styrene, have been chosen.

The spectral data about establishment of structure of the prepared copolymers and other experimental results showed that unlike data presented in work about intramolecular cyclization behavior,[110] in this case, the formation of six-membered cyclic structures in copolymer chain has not been detected.

Considering the possibilities of preparation of copolymers on the basis of vinyl alkyl ether and gem-dinitrile-substituted VCP, the authors of work[110] have observed that the structure of copolymer chain established on data of IR- and PMR-spectroscopy do not contain double bond. As far as in the spectra of copolymer samples, the resonance signals for protons of double bond are absent and in the IR-spectra there are also the absorption bands in the field of 2230 cm^{-1}, characteristic for nitrile group the authors have concluded that the VCP monomer undergoes the copolymerization on type 1,2 due to only vinyl group. The authors have called this process anomaly induced copolymerization, that is, the radicals of AK type arising in the system react first with electron-rich vinyl ether, forming thereby, the intermediate zwitterion radicals. The last circumstance, of course, promotes a formation of complexes of donor–acceptor type in the system. The presence of intermediate ion radicals is the most favorable condition for polymerization of vinyl alkyl ether:

Summarizing, one can conclude that many questions connected with polymerization of VCP remain still not studied. In particular, one can refer to them the investigations of kinetic regularities and mechanism of formation of homo- and copolymers, influence of method of preparation on their properties, interrelation between compositions, group, and physical–mechanical properties. Nevertheless, the intensive investigations and development of new types of polymers on the basis of substituted VCP, continuing now more and more, expand the fields of their application, open new perspective directions of their use.

Analyzing the information in review, it should be noted that the most perspective direction in the field of polymerization of VCP are the functionally substituted derivatives, that is, polymers prepared on their basis possess improved operational properties.

9.6 PECULIARITIES OF RING-OPENING POLYMERIZATION

The ring-opening polymerization phenomenon is very important and necessary for some branches of technique, as in number of cases of industrial application the monomers polymerizing without shrink or even in a number of cases with increasing volume are required. In particular, the products prepared as a result of ring-opening polymerization can be successfully used in the dental technology. The ring-opening polymerization is also interesting from the point of view of preparation of various classes of linear heterochain polymers with specific properties. For example, the preparation of block copolymers containing styrene, butadiene, and ε-caprolactone with defined structure is interesting.

In this connection, the ring-opening polymerization of cyclic monomers (cycloalkanes, cyclic ethers, acetals of esters, lactames, including functionally substituted) recently acquires a large practical value. Easiness of polymerization of such monomers depends on a series of factors, namely, on activity of functional group in cycle, on nature of used catalyst, on cycle size, etc. First, let us consider the polymerization process.

VCP suffer the simultaneous ring-opening polymerization and vinyl group as a result of which the corresponding polypentenamers are prepared. In this case, a polymerization ability of VCP monomers is increased by introduction of radical stabilizing such as alkoxycarbonyl, cyan, chlorine, etc. into cycle.[31,123-126] In VCP, a methyl group or chlorine atom in position 2 shows an essential influence on ring-opening polymerization rate. It has been established that the polymerization rate of monomers having substituents Me and Ph in position 2, are in ratio H:Me:Ph = $1:2:10^{127}$[127]:

$$n\ CH_2{=}CH{-}\underset{R}{\overset{R}{C}}{-}C\underset{CH_2}{\overset{Cl}{\diagdown}}Cl \longrightarrow \text{~}\!\!\left(\!CH_2{-}CH{=}\overset{R}{\underset{}{C}}{-}CH{-}CCl_2\!\right)_{\!n}\!\!\text{~}$$

R = H (Ia), CH$_3$ (Ib), Ph (Ic)

A mechanism of polymerization of VCP includes a process behavior in two stages: formation of intermediate CPCRs and their rearrangement to allylcarbinyl ones on which is continued a chain growth.[79] As a result of polymerization, the prepared polymers have 1,5-structural links, though a formation of linear structures has been connected with stabilization level of intermediately forming radicals:

$$n\ \overset{\triangleright}{\diagup}\!\!\!\text{~}X \xrightarrow{R^\bullet} RC{-}\overset{\bullet}{C}\overset{\triangleright}{\diagup}\!\!\!\text{~}X \longrightarrow RC{-}C{=}C{-}C{-}\underset{X}{\overset{|}{\overset{\bullet}{C}}} \longrightarrow$$

$$\longrightarrow \left(\!C{-}C{=}C{-}C{-}\underset{X}{\overset{|}{C}}\!\right)_{\!n}$$

It has been shown, based on results of model addition reactions of thio-phenol, that a limiting stage of polymerization is the first stage. In other words, a stage determining total polymerization rate is the addition of radical to monomer. Consequently, the polymerization rate is determined by reactivity of growing radical as a stabilizing effect on radical and spatial difficulties, which is created by substituent that grow in transition from H to Me and then to Ph.[26] A reactivity of the corresponding radicals is decreased and polymerization rate becomes slow:

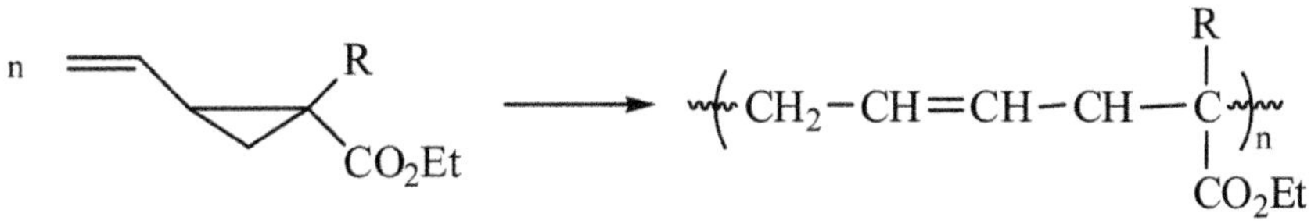

where R = H (Ia), Me (Ib), Ph (Ic)

VCP are subjected to the radical polymerization with formation of polymers containing links, preparing as a result of ring opening in posi-tion-1,5 and unidentified links.[123] These unidentified structures are probably formed as a result of intramolecular reaction and correspond to cyclobutane fragment:

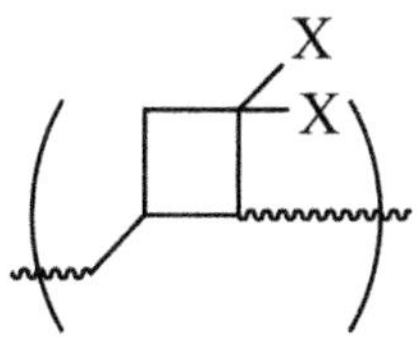

A comparison of data about contraction during polymerization of cyclic and vinyl monomers show the process behavior with small shrinkage. In this case, a linear dependence between volume shrinkage in the polymerization process and inverse value of MM has been established. As a result of polym-erization of cyclic monomers with ring opening the exact materials are prepared, in particular adhesives, as the process is accompanied by expan-sion of volume. Unlike polymerization of vinyl monomers, the ring-opening polymerization proceeds with less degree of change of volume.

Spirocyclic monomers consisting of VCP and ketene acetals during polymerization suffered double ring-opening, thereby, increasing a volume[31,124–126,128–133]:

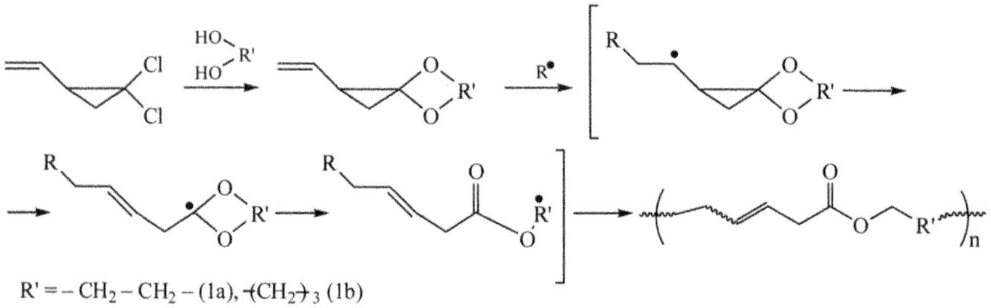

$R' = -CH_2-CH_2-$ (1a), $-(CH_2)_3$ (1b)

Apparently, the volume increase observed in replacement of the covalent bond by Van der Waals interaction, at break of the cycle, somewhat compensates the shrinkage characteristic for polymerization of vinyl monomers. Hence, it follows that during polymerization of monomers containing, at least, two breaking cycle, the process proceeds either even without shrinking or even with some volume increase. At such polymerization, an effect on size of ring appears, that is, monomer having five-membered acetal ring suffers polymerization only due to opening of cyclopropane fragment (acetal ring remains unaffected), while monomer having seven-membered acetal ring suffers polymerization due to double opening of both rings—VCP and acetal.[134–136]

On analogous scheme with formation of polymers containing unsaturated groups in the basic chain, proceeds a polymerization of vinylcyclobutane, heme-disubstituted VCP, and cation polymerization of methylene cyclobutane. 2-Methylene-1,3-dioxolanes in the presence of radical initiators are polymerized with ring-opening. It has been established according to data of spectral analysis that an opening proceeds quantitatively on C–O-bond[131]:

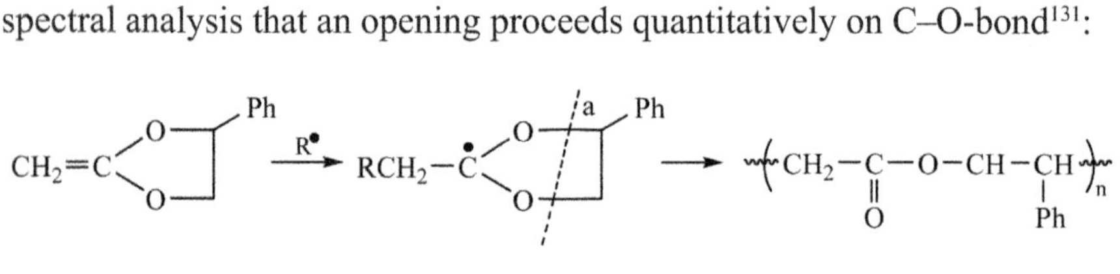

The polymerization of nitrogen-containing cyclic ketene acetals proceeds by 100% with C–O-bond rupture and formation of polyamides[134]:

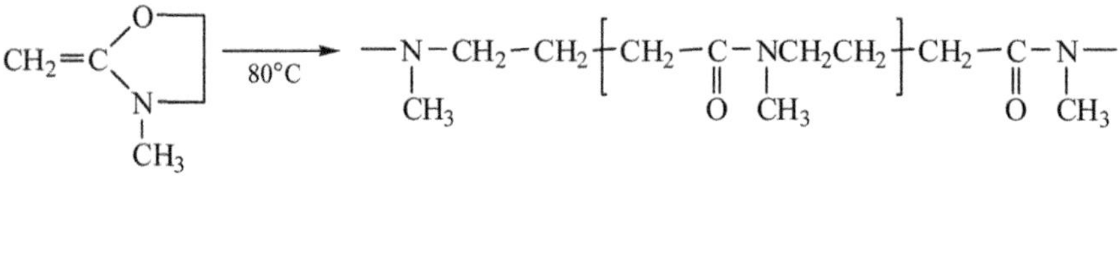

As a result of the ring-opening polymerization of six-membered azolatone, poly-N-acyl-β-peptides and poly-N-formyl-β-peptides are formed[135]:

$$n \underset{R}{\overset{R}{\underset{}{\bigotimes}}}\!\!\!\!\!\!\!\!\!\!\!\!\!\!\longrightarrow \left(\!\!-NCHCH_2C-\!\!\right)_n \quad R, R' = H, Me, Et, Pr, Ph$$

Sulfur-containing analogs of cyclic ketenacetals (thioethers) have higher energy than ethers and because of it, a degree of the ring-opening at their polymerization is decelerated. Even at 120°C, the ring-opening takes place only by 45%[136]:

$$CH_2=C\!\!\begin{array}{c}O\\ \\ S\end{array}\!\!\xrightarrow[\text{peroxide}]{120°C} \left(CH_2-\underset{O}{\overset{}{C}}-S-CH_2CH_2\right)_m\!\!\left(CH_2-C\right)_{n-m}$$

3-, 4-, and 5-membered cycles suffer practically the ring-opening polymerization of cyclic ethers: the cyclic ethers containing less 5 or more 6 members are polymerized relatively easily. 5-membered cyclic ethers are polymerized difficulty, substituted 5-membered cyclic ethers and acetals are usually inactive.[128,137,138]

The cyclic vinyl acetals such as 2-vinyl-1,3-dioxolane and 2-vinyl-1,3-dioxane possess small polymerizability owing to proceeding of degenerated chain transfer.[128,138] The polymers prepared from these monomers consisted of links forming as a result of ring-opening and characterized with very small MM. The polymers prepared from these monomers contain both vinyl links and links forming as a result of ring-opening. A degree of polymerization is in the range from 3 to 11:

$$\left(CH_2-CH\right)\!\!\left(CH-O-CH_2CH-O\right)\!\!\left(CH_2-CH_2-COCH_2CH\right)_n$$

The prepared polymer materials are not decomposed and are sufficiently stable, hydrolytically stable, and in the processing the various toxic products (e.g., HCl) are isolated. In addition, they are difficulty processed and in a case of availability in their composition metals (from catalyst) and other similar additions the process is more complicated. Besides all this, the ring-opening polymerization is interesting from some points of view.

First, such polymerization leads to the preparation of functional polymers a value of which for modern technique is simply invaluable (in terms of their solubility and reprocessing).

Second, a free-radical polymerization has an advantage over ionic one due to the absence of residue of metals in the composition of the prepared polymers.

Third, as a result of hydrolytic and thermal splitting of polymers one can prepare the bifunctional oligomers necessary for polycondensation processes.

KEYWORDS

- vinylcyclopropanes
- cyclopentanes
- isomerization
- cyclopropylcarbinyl
- polymerization

REFERENCES

1. Wong, H. N. C.; Hon, M.-Y.; Tse, C.-W.; Yip, Y.-C. Use of Cyclopropanes and Their Derivatives in Organic Synthesis. *Chem. Rev.* **1989**, *89*, 165–198.
2. Herman, G.; Richey, H. G.; Shull, D. W. The Large Effect of a 2-Dimethylamino Substituent on Thermal Rearrangement of Vinylcyclopropane. *Tetrahedron Lett.* **1976**, *8*, 575–576.
3. Berson, J. A.; Salomon, R. G. Thermal Sigmatropic Migration of the Carbomethoxy Group. *J. Am. Chem. Soc.* **1971**, *93*(18), 4620–4621.
4. Huyser, E. B.; Munson, L. R. Free-Radical Addition to 2-Cyclopropylpropene. *J. Org. Chem.* **1963**, *28*(12), 3442–3445.
5. Plemenkov, V. V. Electronic and Spatial Structure of Monofunctional Cyclopropanes. *J. Org. Chem.* **1997**, *33*(6), 849–859.

6. Ratier, M.; Pereyre, M. Organotin Route for Specific Reductive Ring-Opening of Vinyl-cyclopropanes. *Tetrahedron Lett.* **1976,** *26,* 2273–2276.

7. Seyferth, D.; Julia, T. F.; Deytouriz, H.; Pereyre, M. The Preparation of Group IV. Organometallic Compounds Containing Gem-Dihalocyclopropyl Groups. *J. Organomet. Chem.* **1968,** *11,* 63–76.

8. Bessmertnykh, A. G.; Blinov, K. A.; Grishin, Y. K.; Donskaya, N. A.; Beletskaya, I. P. Selective Hydrosilylation of Vinylcyclopropanes in the Presence of Chloroplatinic Acid. *J. Org. Chem.* **1995,** *31*(1), 49–53.

9. Cristol, S. J.; Kleeman, R. Bridged Polycyclic Compounds. LXX. Rearrangements Accompanying Free-Radical Addition of Thiophenol to 3-Methylenenortricyclene. *J. Org. Chem.* **1971,** *36*(14), 1866–1870.

10. Afanasyev, I. B.; Samokhvalov, G. I. Radical Addition Reactions and Telomerization of Conjugated Diende. *Usp. Khim.* **1969,** *38*(4), 687–710.

11. Ismailov, I. A.; Shahnazarli, R. Z.; Ramazanov, G. A.; Guliyev, A. M. Addition of Tribromtrichlormethane to Ethyl ethers of Chlorosubstituted Vinylcyclopropane Carboxylic Acids. *Chem. Petrochem.* **2005,** *1,* 29–33.

12. Ramazanov, G. A.; Shahnazarli, R. Z.; Nazaraliyev, K. G.; Guliyev, A. M. Bis-Adducts of Ethanedithiol with Dichlorosubstituted Vinylcyclopropanes—Decelerators of Prevulcanization of Rubber Mixtures. *Plastics* **2011,** *3,* 31–33.

13. Guliyev, A. M.; Lemeshev, A. N.; Gasanov, R. G.; Nefedov, O. M. EPR-Spectroscopic Investigation of Free-Radical Addition of Mercaptanes and Polyhalogen Methanes to Functionally Substituted Alkenyl Cyclopropanes by a Method of Spin Trap. *Izv. AN SSSR, Ser. Khim.* **1985,** *6,* 1341–1346.

14. Hiraguri, Y.; Endo, T. Synthesis and Radical Ring-Opening Polymerization of 1,2-Dicar-bomethoxy-3-Vinylcyclobutane. *J. Polym. Sci. Part C Polym. Lett.* **1989,** *27,* 333–337.

15. Endo, T.; Watanabe, M.; Suga, K.; Yokozawa, T. Radical Ring-Opening Polymerization Behavior of Vinylcyclopropane Derivatives Bearing an Ester Group. *J. Polym. Sci. Part A Polym. Chem.* **1987,** *25,* 3039–3048.

16. Devis, D.; Perret, M. *Free Radicals in Organic Synthesis*; Mir: Moscow, 1980; p 205.

17. Guliyev, A. M.; Kasimova, S. P.; Lemeshev, A. N.; Nefedov, O. M. Alkenylcyclo-propanes in the Addition Reactions. IV. Telomerization of Ethoxycarbonyl Substi-tuted Vinylcyclopropanes with Bromtrichloromethane. *J. Org. Chem.* **1990,** *26*(6), 1261–1270.

18. Guliyev, A. M.; Lemeshev, A. N.; Kasimova, S. P.; Lishanskiy, I. S. Alkenylcyclopro-panes in the Addition Reactions. I. Stereochemistry of Free-Radical Adition of Thio-phenol to Substituted Alkenyl Cyclopropanes. *J. Org. Chem.* **1983,** *19*(2), 346–353.

19. Guliyev, A. M.; Lemeshev, A. N.; Kasimova, S. P.; Nefedov, O. M. Alkenylcyclopro-panes in the Addition Reactions. II. Functionally Substituted Alkenylcyclopropanes in the Conditions of Asherah-Vovsi. *J. Org. Chem.* **1984,** *20*(11), 2333–2337.

20. Lishanskiy, I. S.; Zak, A. G.; Guliyev, A. M.; Fomina, O. S.; Khachaturov, A. S. Radical Addition Reaction of Thiols to Alkenylcyclopropanes. *DAN SSSR* **1966,** *170*(45), 1084–1087.

21. Lishanskiy, I. S.; Vinogradova, N. D.; Guliyev, A. M.; Zak, A. G.; Zvyagina, A. B.; Fomina, O. S.; Khachaturov, A. S. Isomerization of Alkenylcyclopropanes in Free-Radical Addition Reactions. *DAN SSSR* **1968,** *179*(4), 882–883.

22. Lishanskiy, I. S.; Zak, A. G.; Vinogradova, N. D.; Guliyev, A. M.; Fomina, O. S.; Khachaturov, A. S. Investigation of Reactivity of Alkenylcyclopropanes by Means of Model Reactions. *High-Mol. Compd.* **1968,** *10-A*(8), 1866–1877.

23. Lishanskiy, I. S.; Vinogradova, N. D.; Zak, A. G.; Zvyagina, A. B.; Guliyev, A. M.; Fomina, O. S. Influence of Structure on Reactivity of Some Unsaturated Derivatives of Cyclopropane. *J. Org. Chem.* **1974,** *10*(3), 493–503.

24. Shahnazarli, R. Z.; Ramazanov, G. A.; Guliyev, A. M. Biological Protection of Polymer Compositions with Application of Sulphur-Containing Adducts. In *Plastic Masses with Special Properties*; S-PB: St. Petersburg, 2011; pp 259–261.

25. Guliyev, A. M.; Ramazanov, G. A.; Guliyev, M. F.; Gasanova, S. S. Radical Polymerization of 1-Vinyl-2-Asetoxymethylcyclopropane. *High-Mol. Compd.* **1987,** *29*(8), 581–584.

26. Lishanskiy, I. S.; Zak, A. G.; Zherebetskaya, E. I.; Khachaturov, A. S. Poly-(3-Carboxyethoxypentenamer) at Radical Isomerization Polymerization of Ethyl Ether of 2-Vinylcyclopropane Carboxylic Acid. *High-Mol. Compd.* **1967,** *A-9*(9), 1895–1902.

27. Guliyev, A. M.; Guliyev, K. G.; Lishanskiy, I. S.; Safaraliyeva, G. M. Radical Addition of Thiophenol to Methoxymethyl Substituted Alkenylcyclopropane. *Azerb. Chem. J.* **1977,** *6*, 31–34.

28. Beckwith, A. L. J.; Serelis, A. K.; Affio, A.; Griller, D.; Ingold, K. U. Allulcarbinyl-Cyclopropylcarbinyl Rearrangement. *J. Am. Chem. Soc.* **1980,** *102*(5), 1734–1736.

29. Millard, B.; Forrest, D.; Ingold, K. U. Kinetic Application of Electron Paramagnetic Resonance Spectroscopy. 27. Isomerization of Cyclopropylcarbinyl to Allylcarbinyl. *J. Am. Chem. Soc.* **1976,** *98*(22), 7024–7026.

30. Krusee, P. R.; Mearkin, P.; Jesson, J. P. Electron Spin Resonance Studies of Conformations and Hindered Internal Rotation in Transient Free-Radicals. *J. Phys. Chem.* **1971,** *75*(22), 3438–3453.

31. Cho, I. K.; Ahn, D. Polymerization of Substituted Cyclopropanes. I. Radical Polymerization of 1,1-Disubstituted 2-Vinylcyclopropanes. *J. Polym. Sci. Polym. Chem. Ed.* **1979,** *17*(10), 3169–3182.

32. Danishefsky, S.; Rovnyak, G.; Cavanaugh, R. 1,7-Addition of Enamines to 1,1-Bisethoxycarbonyl-2-Vinylcyclopropane a Useful Method of Introduction a Six Carbon Fragments to Ketones. *J. Chem. Soc. D Chem. Commun.* **1969,** *12*, 636.

33. Danishefsky, S.; Rovnyak, G. Stereochemistry of the Ring-Opening of an Activated Vinylcyclopropane. *J. Chem. Soc. Chem. Commun.* **1972,** *14*, 821–822.

34. Danishefsky, S.; Rovnyak, G. Nucleophilic Additions to Diethylcyclopropylmethylidenemalonate. *J. Org. Chem.* **1974,** *34*(19), 2924–2925.

35. Danishefsky, S.; Rovnyak, G. Effects of Substituents on the Nucleophilic Ring Opening of Activated Cyclopropanes. *J. Org. Chem.* **1975,** *40*(1), 114–115.

36. Danishefsky, S.; Singh, R. Aspiroactivated Vinylcyclorpopane. *J. Org. Chem.* **1975,** *40*(25), 3807–3808.

37. Srewart, J. M.; Olsen, D. R. Participation of a Cyclopropane Ring in Extension of Conjugation. *J. Org. Chem.* **1968,** *33*(14), 4534–4536.

38. Reavers, W. A. Reaction of Vinylcyclopropanes with Butyl-Lithium. *Tetrahedron Lett.* **1979,** *19*, 1675–1678.

39. Stewart, J. M.; Pagenkof, G. K. Transmission of Conjugation by the Cyclopropane Ring. *J. Org. Chem.* **1969,** *34*(1), 7–10.

40. Westberg, H. H.; Stewart, J. M. Ring-Opening Addition Reactions of 1,1-Disubstituted Cyclopropanes with Amines. *Proc. Mont. Acad. Sci.* **1964,** *16*, 2461.

41. Marshall, D. A.; Ruden, R. A. The Addition of Lithium Dimethyl-Copper to Conjugated Cyclopropyl Enones. *J. Org. Chem.* **1972,** *37*(4), 659–664.

42. Grieco, P. A.; Finkelnor, R. Organocopper Chemistry Reaction of Lithium Dialkyl-Copper Reagents with Activated Vinylcyclopropanes an Instance of 1,7-Addition. *J. Org. Chem.* **1973,** *38*(11), 2100–2101.

43. Miyaura, N.; Itoh, M.; Sasaki, N.; Suzuki, A. The Reactions of Organoboranes and Lithium Dialkylcuprates with 1-Acyl-2-Vinylcyclopropanes. A Convenient New Route to γ,δ-Unsaturated Ketone Synthesis. *Synthesis* **1975,** *5*, 317–318.

44. Cho, I. K.; Ahn, D. Polymerization of Substituted Cyclopropanes. II. Anionic Polymerization of 1,1-Disubstituted 2-Vinylcyclopropanes. *J. Polym. Sci. Polym. Chem. Ed.* **1979,** *17*(10), 3183–3191.

45. Cho, I.; Kim, J.-B. Polymerization of Substituted Cyclopropanes. III. Anionic Polymerization of 2-Substituted Cyclopropane-1,1-Dicarbonitriles. *J. Polym. Sci. Polym. Chem. Ed.* **1980,** *18*, 3053–3057.

46. Berkowitz, W.; Greenetz, S. C. Cycloaddition of Enamine to an Activated Cyclopropane. *J. Org. Chem.* **1976,** *41*(1), 10–13.

47. Martin, R. A.; Landgrebe, J. A. Acetolysis of 1-Tosyloxy-2,2-dideuteriobicyclopropyl. *J. Org. Chem.* **1972,** *37*(12), 1996–1998.

48. Chmurny, A. B.; Gram, D. J. Studies in Stereochemistry. XLVI. Singlet Diradical Transition States in Epimerization Reactions of Substituted Cyclopropanes. *J. Am. Chem. Soc.* **1973,** *95*(13), 4237–4244.

49. Freeman, P. K.; Raymond, F. A.; Grostic, M. F. Reactive Intermediates in the Bicyclo[3,1,0]-hexyl- and Bicyclo[3,1,0]-hexylidene Synthesis. III. The Addition of Hydrogen Chloride and Deuterium Chloride to Bicyclo[3,1,0]-hexene-2. *H. Org. Chem.* **1967,** *32*(1), 24–28.

50. Hendrickson, J. B.; Boeckman, R. K. Cyclopropanes. II. An Electrophylic Addition with Nucleophile Reaction. *J. Am. Chem. Soc.* **1971,** *93*(18), 4491–4495.

51. Garrat, D. E.; Modro, A.; Oyama, K.; Schmidt, G. H.; Tidwell, T. T.; Jates, K. Effect of the Cyclopropyl Substituent on the Rates of Electrophylic Additions to Alkenes. *J. Am. Chem. Soc.* **1974,** *96*(16), 5295–5297.

52. Cercsus, T. R.; Girmadia, V. M.; Schmidt, G. M.; Tidwell, T. T. Reactivity of Cyclopropylalkenes with p-Chlorobenzenesulfenyl Chloride. *Can. J. Chem.* **1978,** *58*, 205–210.

53. Govel, G.; Feblentstein, S. S. AlCl₃-Induced Reactions with Rearrangement. *Tetrahedron Lett.* **1976,** *34*(7), 993–996.

54. Ibragimov, M. A.; Smith, V. A. Reaction of Arylthioalkylation of Trimethylsiloxy-alkenes. *Izv. AN SSSR, Ser. Khim.* **1982,** *9*, 2177–2178.

55. Pomytkin, I. A.; Balenkova, E. S.; Anfilova, S. N. Acylation of 1-Methyl-1-Vinyl-1-Methyl-1-Isopropenylcyclopropane by Acylborfluoride. *J. Org. Chem.* **1982,** *18*(3), 532.

56. Vasilyeva, A. A.; Balenkova, E. S. Acylation of Alkenes of Series Vinylcyclopropane with Acylborfluoride. *J. Org. Chem.* **1983,** *19*(2), 288–291.

57. Freeman, P. R.; Grostic, M. F.; Raymond, F. A. Reactive Intermediates in the Bicyclo[3,1,0]-Hexyl- and Bicyclo[3,1,0]-Hexylidene Systems. I. The Acid Catalyzed Additions of Methanol. *J. Org. Chem.* **1965,** *30*(3), 771–777.

58. Oth, J. F. M.; Merenyi, R.; Nielsen, J.; Schroder, G. Syntheses und Eigenschaften Einiger Monosubstituerten Bullvalene. *Chem. Ber.* **1965,** *10*, 3385–3400.

59. Daunben, W. G.; Freidrick, L. E.; Overnaun, S. P.; Aoyagi, E. J. Thujopsene Rearrangements the Cyclopropylcarbinyl Systems. *J. Org. Chem.* **1972,** *37*(1), 9–13.

60. Payuettle, L. A.; Kirshuer, S.; Malpass, J. R. CH₁₀CO₂-Interconversions. The Electrophilic Additions of Chlorosulfonyl Isocyanate to Bullvalene. *J. Am. Chem. Soc.* **1970,** *92*(14), 4330–4340.

61. Cooper, M. A.; Holden, C. M.; Loftus, P.; Whittaker, D. The Acid Catalyzed Hydration of Sabinene and α-Thujene. *J. Chem. Soc. Perkin Trans.* **1973**, *2*, 665–667.

62. Katz, T. J.; Nikolaou, K. C. Additions of Chlorosulfonyl Isocyanate and Sulfenyl Halides to Benzovolene. *J. Am. Chem. Soc.* **1974**, *96*(6), 1948–1949.

63. Tolstikov, G. A.; Lishtvanov, L. N.; Goryayev, M. I. Study of Substances Incoming in Composition of Ether Oils. VI. Hydration of Sabinene. *J. Org. Chem.* **1963**, *33*(2), 683.

64. Talybov, A. G.; Mursakulov, I. G.; Guliyev, A. M. Interaction of Nitronium Tetrafluorborate with Ethyl Ethers of Isopropenyl Substituted Cyclopropane Carboxylic Acid. *Azer. Chem. J.* **1982**, *1*, 40–42.

65. Delamar, P.; Bolton, R. *Electrophilic Addition to Saturated Systems*; Mir: Moscow, 1968; p 318.

66. Valter, R. E. Electron and Spatial Effects in Heterolytic Reactions of Intramolecular Cyclization. *Usp. Khim.* **1982**, *41*(5), 1374–1397.

67. Manakov, T. G.; Berdnikov, E. A.; Samuilov, Y. D. Interaction of Dichlorcarbene and Arensulfenyl Chlorides with Divinyl Oxide and Vinyl Ethyl Carbonate. *J. Org. Chem.* **2001**, *37*(3), 367–372.

68. Ibragimov, M. *Reactivity, Reaction Mechanism and Structure in the Chemistry of Polymers*; Dzhenkins, A., Ledvis, A., Eds.; Mir: Moscow, 1977; p 615.

69. Phillips Petroleum Co.; Jean, P. J. Polymerization of Vinylcyclopropane and Its Homologues and Derivatives. USA Patent 2540949 A., 1951.

70. Grace, W. R. & Co., Arthur, D. K. 1-Cyclopropyl-1-Phenyl Substituted Ethylene Polymers. Patent 3497483 A (USA), 1970.

71. Guliyev, A. M.; Shahnazarli, R. Z.; Ramazanov, G. A. Ring-Opening Polymerization of Vinylcyclopropanes. In *High-Performance Polymers for Engineering-Based Composites;* Apple Academic Press, Inc.: USA, 2015; pp 13–27.

72. Endo, T.; Watanabe, M.; Suga, K.; Yokozawa, T. Radical Ring-Opening Polymerization Behavior of Ethyl 1-Substituted 2-Vinylcyclopropanecarboxylate. *Macromol. Chem.* **1989**, *190*, 691–696.

73. Takahashi, T.; Yamashita, T. 1,5-Polymerization of Vinylcyclopropane. *J. Polym. Sci.* **1965**, *B-3*(4), 251–255.

74. Takahashi, T.; Yamashita, T. Polymerization of Vinylcyclopropane and its Derivatives. *J. Chem. Soc. Jpn. Ind. Chem. Soc.* **1965**, *68*(5), 869–872.

75. Guliyev, A. M. Investigation in the Field of Synthesis and Polymerization of Alkenyl Cyclopropanes. *Diss. Cand. Chem. Sci. L* **1970**, *4*, 250.

76. Takahashi, T.; Yamashita, I.; Miyakawa, T. The Polymerization of 1.1-Dichloro-2-Vinylcyclopropane. *Bull. Chem. Soc. Jpn.* **1964**, *37*(1), 131–132.

77. Guliyev, A. M.; Guliyev, K. G.; Lishanskiy, I. S. Free-Radical Polymerization of Methoxy-Substituted Alkenylcyclopropanes. *Azerb. Chem. J.* **1980**, *1*, 119–125.

78. Kolesov, S. V.; Vorobyeva, A. I.; Zlotskiy, S. S.; Khamidullina, A. R.; Muslukhov, R. R.; Spirikhin, L. V. Structure of Homopolymers of Vinyl-Gem-Dichlorocyclopropanes. *Dokl. Akademii Nauk.* **2008**, *418*(2), 203–204.

79. Kolesov, S. V.; Vorobyeva, A. I.; Zlotskiy, S. S.; Khamidullina, A. R.; Brusentsova, E. A.; Muslukhov, R. R.; Spirikhin, L. V.; Zaikov, G. E. Vinyl-Gem-Dichlorocyclopropanes in the Radical Polymerization Reactions. *J. Gen. Chem.* **2008**, *78*(5), 783–786.

80. Glazyrin, A. B.; Abdullin, M. I.; Dokichev, V. A.; Sultanova, R. M.; Muslukhov, R. R.; Yangirov, T. A. Synthesis and Properties of Cyclopropane Derivatives of Polybutadienes. *J. High-Mol. Compd. Ser. B* **2013**, *55*(12), 1510–1516.

81. Jones, J. P. Polymerization of Vinylcyclopropane and Its Homologues. USA Patent 2540949, 2540950, C.A., 1951, Volume 45, p 4086.

82. Takahashi, T. Polymerization of Vinylcyclopropane. II. *J. Polym. Sci.* **1968**, *A-1*(6), 403–414.

83. Kettley, A. D.; Berlin, A. J.; Fisher, L. P. Rearrangement of Propagating Chain End in the Cationic Polymerization of Vinylcyclopropane and Related Compounds. *J. Polym. Sci.* **1967**, *A-1*(5), 227–230.

84. Takahashi, T.; Yamashita, T.; Miyakawa, T. Polymerization of 1,1-Dichloro-Vinylcyclopropane. *Bull. Chem. Soc. Jpn.* **1964**, *37*, 131–132.

85. Mustafayeva, I. D.; Babakhanov, R. A.; Guliyev, K. G.; Guliyev, A. M. Isomerization Radical Polymerization of Alkenylcyclopropane Carboxylic Acids. *Azerb. Chem. J.* **1982**, *5*, 69–75.

86. Guliyev, A. M.; Lishanskiy, I. S. Stereoisomer Forms of Ethyl Ether of 2-Isopropenyl-2-Methylene Cyclopropane Carboxylic-1 Acid and Their Radical Reactivity. *Azerb. Chem. J.* **1965**, *1*, 51–54.

87. Cho, I. K.; Ahn, D. Facile Radical Isomerization Polymerization of Vinyl-Disubstituted Cyclopropane. *J. Polym. Sci. Polym. Lett. Ed.* **1977**, *15*(12), 751–753.

88. Overberger, C. G.; Borchert, A. E.; Katchman, A. Crystalline Polyvinyl-Cyclo-Propane, Polyvinylcyclopentane, and Polyvinylcyclohexane. *J. Polym. Sci.* **1960**, *44*, 491–504.

89. Natta, Y.; Dall'asta, G.; Mazanti, G. Stereospecific Homopolymerization of Cyclopentene. *Angew. Chem.* **1964**, *33*(11), 723–729.

90. Zak, A. G. Free-Radical Polymerization of Some Derivatives of Vinylcyclopropane. *Diss. Cand. Chem. Sci. L* **1968**, *6*, 180.

91. Guliyev, A. M.; Lishanskiy, I. S. Determination of Kinetic Parameters of Hompolymerization of Ethyl Ether of 2-Isopropenyl-2-Methyl Cyclopropane Carboxylic Acid. *Azerb. Chem. J.* **1972**, *2*, 92–100.

92. Guliyev, A. M.; Lishanskiy, I. S. Investigation in the Field of Free-Radical Polymerization of Ethers of Alkenylcyclopropane Carboxylic Acids. *Azerb. Chem. J.* **1973**, *4*, 68–72.

93. Lishanskiy, I. S.; Semenova, L. S. Polymerization of Some Derivatives of Vinylcyclopropane on the Complex Catalyst. *High-Mol. Compd.* **1971**, *13*(10), 2365–2369.

94. Takahashi, T.; Yamashita, T. A Review on Polymerization of Vinylcyclopropane. *Prep. Sci. Papers. Int. Symp. Macromol. Chem.* Tokyo, Kyoto, **1966**, *II-5.*

95. Guliyev, A. M.; Ramazanov, G. A.; Guliyev, M. F.; Gasanova, S. S. Radical Polymerization of 1-Vinyl-2-Acetoxymethylcyclopropane. *High-Mol. Compd.* **1987**, *29*(8), 581–584.

96. Lishanskiy, I. S.; Zak, A. G.; Zherebetskaya, E. I.; Khachaturov, A. S. Poly-(3-Carbethoxypentenamer) at Radical Isomerization Polymerization of Ethyl Ether of 2-Vinylcyclopropane Carboxylic Acid. *High-Mol. Compd.* **1967**, *A-9*(9), 1895–1902.

97. Guliyev, A. M.; Suleymanov, T. N.; Gadzhili, G. M.; Abbasova, A. G. Oligomer of 2-Glycidoxymethyl Vinylcyclopropane for Preparation of Thermostable Epoxide Resin Used as Binder of Glassplastics. A.C. 618373. Publ. in B.I. **1978**, *29.*

98. Guliyev, A. M.; Lishanskiy, I. S.; Suleymanov, T. N.; Gadzhili, T. M. Synthesis and Study of Properties of Unsaturated Epoxide Resins. In *Abstracts of Conf. Modification, Structure and Properties of Epoxide Polymers*; Kazan State University: Kazan, 1976; p 16.

99. Guliyev, A. M.; Lishanskiy, I. S.; Suleymanov, T. N.; Abbasova, A. G. Synthesis and Properties of Unsaturated Epoxide Resins. *Plastmassy* **1980**, *2*, 60–61.

100. Cho, I. K.; Lee, J.-J. Polymerization of Substituted Cyclopropanes. IV. Anamalous Radical Initiated Copolymerization of 2-Vinylcyclopropane-1,1-Dicarbonitrile with Alkyl Vinyl Ethers. *J. Polym. Sci. Polym. Lett. Ed.* **1980**, *18*, 639–642.

101. Kennedy, J. P.; Elliott, J. J.; Butler, H. E. Radical Isomerization of 1,1-Dicyclopropylethylene. *J. Macromol. Chem.* **1968**, *2*, 1415–1421.

102. Takahashi, T. Polymerization of Vinylcyclopropanes. III. Cationic Polymerization of Stereoisomers of 1-Halo-2-Vinylcyclopropanes. *J. Polym. Sci. A-1* **1968**, *6*(12), 3327–3331.

103. Shahnazarli, R. Z.; Aliyeva, S. G.; Guliyev, A. M. Synthesis and Cationic Polymerization of Some Substituted Vinylcyclopropyl Ethers. 1st International Turkic World Conference on Chemical Science and Technologies. ITWCCST Conference, Sarajevo. *CBU J. Sci.* **2015**, *11*(3), 349–352.

104. Li, L.; Zhang, G.; Kodama, K.; Yasutake, M.; Hirose, T. A Benign Initiating System for Cationic Polymerization of Isobutyl Vinyl Ether: Silver Salt/Aryl(alkyl) Halide/Lewis Base. *J. Polym. Sci. Part A Polym. Chem.* **2015**, *15*(7), 2050–2058.

105. Lishanskiy, I. S.; Vinogradova, A. D.; Guliyev, A. M.; Zak, A. G.; Zvyagina, A. B.; Fomina, O. S. Isomerization Mechanism of Free-Radical Reactions of Alkenylcyclopropanes. In *Abstracts of Symposium Structure, Reactivity and Mechanism of Conversions of Compounds with Multiple Bonds and Small Cycles*; Kazan State University, L, 1967; p 117.

106. Lishanskiy, I. S.; Vinogradova, A. D.; Guliyev, A. M.; Zak, A. G.; Zvyagina, A. B.; Koltsov, A. I.; Fomina, O. S.; Khachaturov, A. S. Free-Radical System, Isomerization Polymerization of Alkenylcyclopropanes. In *Coll. Synthesis, Sytructure and Properties of Polymers*; Nauka: Moscow, 1979; pp 35–42.

107. Lishanskiy, I. S.; Zak, A. G.; Fedorova, E. F.; Khachaturov, A. S. Polymerization of Derivatives of Vinylcyclopropane with Polar Substituents in Cycle. *High-Mol. Compd.* **1965**, *7*(6), 966–971.

108. Guliyev, A. M.; Ramazanov, G. A.; Gasanova, S. S.; Nefedov, O. M. Some Free-Radical Reactions of Allyl Ether of 2-Vinyl-2-Methylcyclopropane Carboxylic-1 Acid. *Izv. AN SSSR, Ser. Khim.* **1984**, *12*, 2729–2739.

109. Makovetskiy, K. D. Polymerization of Cycloolefins with Ring-Opening. Resume of Science and Technique. *Chem. Technol. High-Mol. Compd.* **1977**, *9*, 129–160.

110. Takahashi, T. Polymerization of Vinylcyclopropanes. IV. Radical Copolymerization of 1,1-Dichloro-2-Vinylcyclopropane with Maleic Anhydride. *J. Polym. Sci.* **1970**, *A-1*(8), 617–627.

111. Takahashi, T. Polymerization of Vinylcyclopropanes. V. Radical Copolymerization of 1,1-Dichloro-2-Vinylcyclopropane with Monosubstituted Ethylenes. *J. Polym. Sci.* **1970**, *A-1*(8), 739–749.

112. Guliyev, A. M.; Lishanskiy, I. S. Free-Radical Copolymerization of Some Ethers of Alkenylcyclopropane Carboxylic Acids. *Azerb. Chem. J.* **1973**, *3*, 58–62.

113. Lishanskiy, I. S.; Fomina, O. S. Investigation of Copolymerization in the System Two Monomer—One Radical With Use of Gas–Liquid Chromatography. *High-Mol. Compd.* **1969**, *A-11*(5), 1398–1402.

114. Shahnazarli, R. Z.; Guliyev, A. M. In *Reactivity of Double Bond of Vinylalkyl(cycloalkyl) Ethers in the Reactions with Ethoxycarbonyl Carbene*, Coll. Scient. Works of XXVI Intern. Scient. -Techn. Conf., Reagent-2012, Minsk, pp 76–84.

115. Ramazanov, G. A.; Shahnazarli, R. Z.; Guliyev, A. M. Synthesis and Properties of Functionally Substituted Unsaturated Polysulfones. *J. Appl. Chem.* **2005**, *78*(10), 1725–1728.

116. Guliyev, A. M. In *Free-Radical Copolymerization of Vinylcyclopropane with Maleic Anhydride*, Republic Conference on High-Molecular Compounds, Sumgait, 1971, p 45.

117. Guliyev, A. M.; Guliyev, K. G.; Lishanskiy, I. S. In *Radical Copolymerization of 1-Methoxy-2-Vinylcyclopropane with Maleic Anhydride*, V Intern. Symp. Polymer-75, Varna, 1975, pp 153–154.

118. Guliyev, A. M.; Gadzhili, G. M.; Lishanskiy, I. S. Investigation of Copolymerization of 1-Carbethoxy-2-Vinylcyclopropane with Maleic Anhydride. In *Coll. ICHOS AN Azerb. SSR.* 1977; 149–154.

119. Guliyev, A. M.; Ragimov, A. V.; Guliyev, K. G.; Suleymanov, S. S.; Safaraliyeva, G. M.; Mustafayeva, I. D.; Abasova, A. T. Method of Preparation of Linear Self-Cross-Linking Copolymers. A.C. 757545, 1980, Publ. In B.I. No. 31.

120. Guliyev, A. M.; Gadzhili, G. M.; Lishanskiy, I. S.; Abasova, A. G. Investigation of Copolymerization of Styrene with Glycidyl Ether of 2-Vinylcyclopropane Carboxylic-1 Acid. In *Coll. ICHOS AN Azerb. SSR*, 1977; pp 155–157.

121. Guliyev, A. M.; Lishanskiy, I. S.; Suleymanov, S. S. In *Radical Polymerization and Copolymerization of 1-Vinyl-2-Carboglycidoxycyclopropane*, International Symposium on Macromolecules, Rio (de) Janeiro, 1974.

122. Mustafayeva, I. D. In *Self-Crosslinking Copolymers of 2-Glycidoxymethylalkenylcyclo-propanes with 4-Vinyl Pyridine*, Mater. of Scien. Conf. of Post-graduate Students, AN Azerb. SSR, Baku, 1980; No.1, p 108.

123. Sanda, F.; Takata, T.; Endo, T. Radical Polymerization Behavior of 1,1-Disubstetuted 2-Vinylcyclopropanes. *Macromolecules* **1993**, *26*(8), 1818–1824.

124. Sanda, F.; Takata, T.; Endo, T. Vinylcyclopropane Cyclic Acetal: A Hybrid Monomer Undergoing Radical Double Ring-Opening Polymerization. *J. Polym. Sci. Polym. Chem.* **1993**, *31A*, 2659–2662.

125. Okazaki, T.; Sanda, F.; Endo, T. Radical Ring-Opening Polyaddition of a Bifunctional Vinylcyclopropane Bearing a Spiroacetal Moiety with Dithiols. *J. Polym. Sci. Part A Polym. Chem.* **1997**, *35*, 2487–2492.

126. Sanda, F.; Takata, T.; Endo, T. Syntheses and Radical Ring-Opening Polymerization of a Vinylcyclopropan Bearing a Cyclic Carbonate Moiety, 1-Vinyl-5,7-Dioxaspiro[2,5] octan-6-One. *Macromolecules* **1994**, *27*, 3986–3991.

127. Endo, T.; Suga, K. Radical Ring-Opening Polymerization Behavior of Vinylcyclopropanes Derived from Dienes and Chloroform. *J. Polym. Sci. Polym. Chem. Ed.* **1989**, *27A*, 1831–1842.

128. Bailey, W. J.; Ni, Z.; Wu, S. R. Synthesis of Poly-E-Caprolactone Via and Free-Radical Mechanism. Free-Radical Ring-Opening Polymerization of 2-Methylene-1,3-Dioxolane. *J. Polym. Chem.* **1982**, *20*(11), 3021–3030.

129. Ueda, M.; Takashi, M. Radical-Initiated Homo- and Copolymerization of α-Methylene-γ-Butyrolactone. *J. Polym. Chem. Ed.* **1982**, *20*, 2819–2828.

130. Han, Y. K.; Chol, S. K. Synthesis and Polymerization of 8,9-Benzo-2-Methylene-1,4,6-trioxaspiro[4,4]nonane (BMTN). *J. Polym. Sci. Polym. Chem. Ed.* **1983**, *21*(2), 353–364.

131. Kobayashi, S. Radical Polymerization of 2-Methylene-1,3-Dithiane. *Macromol. Rep.* **1991**, *A28*(Suppl. 1), 1–5.

132. Moszner, N.; Völkel, T.; Rheinberger, V. Polymerization of Cyclic Monomers, 3. Synthesis, Radical and Cationic Polymerization of Bicyclic 2-Methylene-1,3-Dioxepanes. *Macromol. Chem Phys.* **1997**, *198*(2), 749–762.

133. Jin, S.; Gonsalves, K. E. Synthesis and Characterization of Functionalized Poly(ε-Caprolactone) Copolymers by Free-Radical Polymerization. *Macromolecules* **1998,** *31*(4), 1010–1015.

134. Bailey, W. J.; Arfaei, A.; Chen, P. Y., et al. Proc. IUPAC 28th Macromolecular Symposium, Amherst, Massachusetts, 1982, p 214.

135. Kobayachi, S.; Tsukamoto, Y.; Saegusa, T. Ring-Opening Polymerization of 5,6-Dihydro-4H-1,3-Oxazin-6-Ones, Six-Membered "Azlactones" to Poly (N-Acyl-β-Peptide)s. *Macromolecules* **1990,** *23*(10), 2609–2612.

136. Sidney, L.; Shaffer, S. E.; Bailey, W. J. *J. Am. Chem. Sci. Div. Polym. Chem. Prepr.* **1981,** *22*(2), 373.

137. *Polymerization of Vinyl Monomers*; Khem, D., Ed.; Khimiya: Moscow, 1973; pp 250–302. (Trans. from Eng.).

138. Yamashita, N.; Nishii, Y.; Maeshima, T. Anionic Polymerizabilities of 2-Vinyl-1,3-Dioxolane and 2-Vinyl-1,3-Dioxane. *J. Polym. Sci. Polym. Lett. Ed.* **1979,** *17*(8), 521–526.

PART II
Bioorganic and Biological Chemistry

CHAPTER 10

HYALURONAN: BIOLOGICAL FUNCTION AND MEDICAL APPLICATIONS

MAYSA M. SABET[1,2], TAMER M. TAMER[1,2,*], and AHMED M. OMER[1]

[1]*Polymer Materials Research Department, Advanced Technologies and New Materials Research Institute (ATNMRI), City of Scientific Research and Technological Applications (SRTA-City), New Borg El-Arab City, P.O. Box 21934 Alexandria, Egypt*

[2]*Laboratory of Bioorganic Chemistry of Drugs, Institute of Experimental Pharmacology and Toxicology, SK-81404 Bratislava, Slovakia*

Corresponding author. E-mail: ttamer85@gmail.com

CONTENTS

ABSTRACT

In this chapter, biological function and medical applications of hyaluronan is reviewed in detail.

10.1 INTRODUCTION

Hyaluronan (HA) is a unique biopolymer with repeating disaccharide construction units that are formed from D-glucuronic acid in β-(1-3)-linkage to N-acetyl-D-glucosamine, which in turn, is in β-1,4 linkage to the next D-glucuronic acid (Fig. 10.1). Both sugars are spatially linked to glucose, which in β-configuration involves all of its large groups (the hydroxyls, the carboxylate moiety, and the anomeric carbon of the adjacent sugar) to be in sterically favorable equatorial positions, while all of the small hydrogen atoms occupy the less sterically favorable axial positions. Thus, the structure of the disaccharide is energetically very stable. Each glucuronate unit possesses an anionic charge at physiological pH, and there are negative charges associated with its carboxylate group which are balanced by mobile cations such as Na^+. HA is also novel in its size, reaching up to several million daltons and is synthesized at the plasma membrane rather than in the Golgi, where sulfated glycosaminoglycans are added to protein cores.

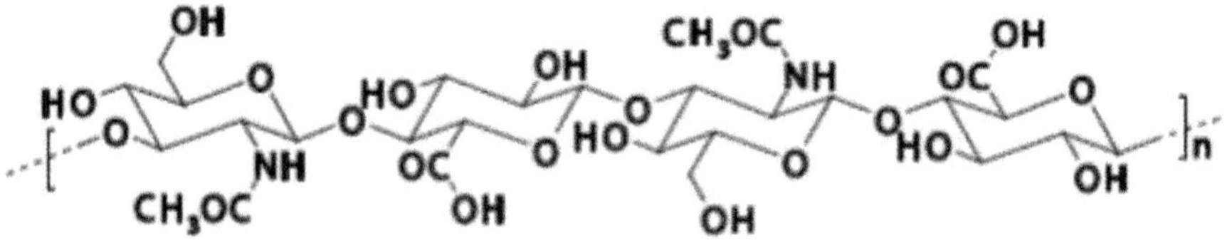

FIGURE 10.1   Chemical structure of HA.

Hyaluronic acid (HA) was discovered and isolated from bovine vitreous humor by Meyer and Palmer.[40] It is most frequently referred to HA because it exists in vivo as a polyanion and not in the protonated acid form. Survey on the Scopus database was done for the studies reporting about HA. Criteria for inclusion were articles addressed to HA and publication years between 1934 and 2015 (Figs. 10.2 and 10.3). Dramatic increases in publications indicate a growing interest in HA over the recent years.

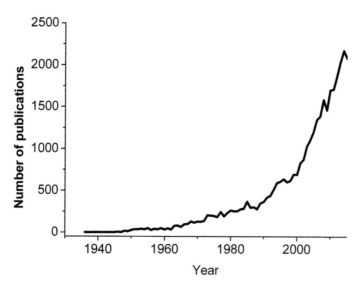

FIGURE 10.2 Annual number of articles indexed in Scopus over the 1934–2015 period by HA.

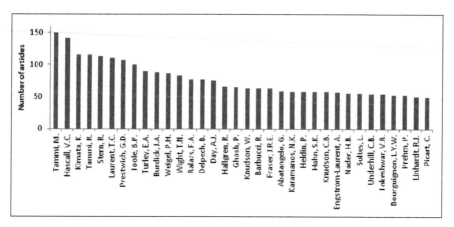

FIGURE 10.3 Top indexed scientists contribute on HA publications according to Scopus over the 1934–2015 period.

10.2 PROPERTIES OF HA

HA is employed in medical applications for its unique physicochemical properties. Properties of HA include: (1) it is very hydrophilic; its viscous solutions have the most unusual rheological properties and are exceedingly lubricious, (2) biodegradability, biocompatibility, and bioresorbability, (3) HA is a major intracellular component of connective tissues, where it plays a significant role in lubrication, cell differentiation, and cell growth, (4) HA contains functional groups (carboxylic acids and alcohols) along its backbone that can be utilized to chemical modification or structure crosslinking.

10.2.1   HA NETWORKS

The physicochemical characteristics of HA were studied in detail from 1950 onwards.[7,8,13] The molecules perform in the solution as highly hydrated randomly kinked coils. The complexity point can be seen both by sedimentation analysis and viscosity.[32,42] Scott et al. (1991)[56] has given evidence that the chains when entangling also interact with each other and form stretches of double helices so that the network becomes mechanically more firm.

10.2.2   RHEOLOGICAL PROPERTIES

HA gives a highly viscous solution, viscoelastic, and the viscosity is particularly shearing dependent.[42] Beyond the entanglement point, the viscosity rises rapidly and exponentially with strength $(\sim c^{3.3})$[42] and a solution of 10 g/L may have a viscosity at a low shear of $\sim 10^6$ times the viscosity of the solvent. At high shear, the viscosity may decline as much as $\sim 10^3$ times. The elasticity of the system improves with expanding molecular weight and concentration of HA as supposed for a molecular network. The rheological properties of HA have been combined with lubrication of joints and tissues, and HA is regularly found in the body separating surfaces that move toward each other, for example, cartilage surfaces and muscle bundles.[11]

10.2.3   WATER HOMEOSTASIS

In the solution, the polymer takes up a stiffened helical configuration, attributed to hydrogen bonding between the hydroxyl groups along the chain, forming a coil structure that traps approximately 1000 times its weight in water. A fixed polysaccharide network offers a great defense to bulk flow of solvent.[13] It was reported that hyaluronidase treatment removes a high interference to water flow through a fascia. Thus, HA and other polysaccharides limit excessive fluid fluxes through tissue compartments. Moreover, the osmotic pressure of the HA solution is nonideal and grows exponentially with the strength. In spite of the high molecular weight of the polymer, the osmotic pressure of a 10 g/L HA solution is of the similar order as the 10 g/L albumin solution. The exponential link performs HA and other polysaccharides excellent osmotic buffering substances—moderate changes in concentration lead to significant changes in osmotic pressure. Flow resistance together with osmotic buffering causes HA an ideal regulator of the water homeostasis in the body.

10.3 BIOLOGICAL FUNCTION OF HA

Naturally, HA has fundamental roles in body functions according to organ type in which it is distributed.[31]

10.3.1 SPACE FILLER

The particular duties of HA in joints are yet actually unknown. The easiest description for its presence would be that a flow of HA through the joint is needed to keep the joint cavity open and provide extended movements of the joint. HA is regularly secreted into the joint and displaced by the synovium. The total mass of HA in the joint cavity is resolved by these two processes. The half-life of the polysaccharide at steady-state is in the order of 0.5–1 day in rabbit and sheep.[12,18] The volume of the cavity is determined by the pressure conditions (hydrostatic and osmotic) in the cavity and its enclosing. HA could, by its osmotic contributions and its formation of flow barriers in the limiting layers, be a regulator of the pressure and flow rate.[39] It is interesting that in fetal development the formation of joint cavities is parallel with a local increase in HA.[16]

10.3.2 LUBRICATION

HA has been observed as an excellent lubricant in the joints due to its shear-dependent viscosity[45] but its function in lubrication has been opposed by others.[46] However, there are now ideas to consider that the role of HA is to form a layer separating the cartilage surfaces. The pressure on the joints may push out water and low-molecular weight solutions from the HA layer into the cartilage matrix. As a result, the concentration of HA increases and a gel structure of micrometric thickness is formed, which preserves the cartilage surfaces from frictional damage.[20] This mechanism to form a protecting cover is much less useful in arthritis, when the synovial HA concentration decreases and molecular weight of HA becomes lower than normal. Another difference in the arthritic joint is the protein formation of the synovial fluid. Fraser et al.[17] presented that addition of various serum proteins to HA substantially improved the viscosity, and this has drawn a renewed attention to discovered hyaladherins. TSG-6 and inter-α-trypsin inhibitor and other acute phase reactants, such as haptoglobin, are concentrated to arthritic synovial fluid.[26] It is not identified to what extent they are affecting the rheology and lubricating properties.

10.3.3   SCAVENGER FUNCTIONS

HA has also been appointed to function as a scavenger in the joints. It has been known since the 1940s that various oxidizing systems and ionizing irradiation degrade HA, and we are aware today that the popular denominator is a chain cleavage induced by free radicals, typically hydroxy radicals.[43] Within this reaction, HA serves as a very active scavenger of free radicals. Whether this fact has any physiological significance in preserving the joint against free radicals is unknown. The fast turnover of HA in the joints has led to the proposal that it also functions as a scavenger of cellular debris.[33] The cellular material could be detected in the HA network and eliminated at the same rate as the polysaccharide.[51,83] As discussed above, more recently proposed roles of HA are based on its special interactions with hyaladherins. One exciting aspect is the fact that HA influences angiogenesis but the impact is varied depending on its concentration and molecular weight.[54] High concentrations and high molecular weight of the polymer inhibit the generation of capillaries, while oligosaccharides can produce angiogenesis. There are also statements of HA receptors on vascular endothelial cells, whereby HA could perform on the cells. The vascularity of the joint cavity could be a result of HA inhibition of angiogenesis. Another interaction of some concern in the joint is the adhesive of HA to cell surface proteins. Lymphocytes and other cells may find their way to joints through this interaction. Injection of high doses of HA intraarticularly could attract cells expressing these proteins. Cells can also change their expression in HA-binding proteins in states of disease, whereby HA may affect immunological reactions and cellular traffic in the path of physiological processes in cells.[15] The research often described that intraarticular injections of HA alleviate pain in joint diseases and may show a direct or indirect interaction with pain receptors.[1]

10.3.4 HA AND SYNOVIAL FLUID

In a healthy joint, the synovial fluid, which consists of an ultrafiltrate of blood plasma and glycoproteins contains HA macromolecules of molar mass varying between 6×10^6 and 10×10^6 Da.[50] SF also serves as a lubricating and crash absorbing barrier layer between moving parts of synovial joints. SF overcomes friction and damage and tear of the synovial joint sporting thus a vital role in the lubrication and stability of the joint tissues from destruction through motion.[47] As SF of healthy humans displays no activity of hyaluronidase, it has been suggested that oxygen-derived free radicals are involved in a self-perpetuating process of HA catabolism within the joint.[19,53,69] This radical-mediated process is considered to account for ca. 12-h half-life of native HA macromolecules in SF.

10.4 APPLICATIONS

10.4.1 DRUG DELIVERY

Biological and chemical properties of HA fit this macromolecule as a prospective carrier of drugs particularly for local application and/or targeting to the lymphatic system. It is immunologically inactive and harmlessly demoted in lysosomes of many cells. Thus, it can be applied in a wide range of molecular sizes. It also implements suitable chemical groups for spacer-drug side chain attachment. Although, several points call for experimental checking via the effect of a substituent on the immunology, receptor affinity, and degradation, pharmacokinetics should be also observed after different ways of administration. Conjugation of HA with antitumor drugs is pursued to gain several advantages, including a prolonged plasma half-life, an improved solubility, and improved pharmacokinetic features of the resulting conjugate. In addition, all these polymer-anticancer drug conjugates take advantage of the so called "enhanced permeability and retention" effect.[35]

10.4.2 HA WOUND HEALING

The recent development of wound healing and wound management guide scientists to find and modify new materials, not only to isolate or cover wounds but also to help and accelerate wound healing. Due to antioxidative activity, nontoxicity, biocompatibility, and biodegradability of HA, high-molecular-weight HA appears as an ideal material for wound dressing and tissue regeneration applications.

The combination of HA with some other biomaterials for certain wound healing was studied by several researchers in the last few years. HA blends with gelatin or collagen in the presence or absence of growth factors (i.e., epidermal growth factor or basic fibroblast growth factor, which showed promising results with laboratory animals.[28,37,38,41,55,99,100]

Antimicrobial wound dressing based on HA was done by combining HA with antibiotics, nanosilver, or blending with antimicrobial polymers such as chitosan.[14,27,29,34,36,48]

10.4.3 EVALUATION OF DRUGS

Sensitivity of HA to free radicals made it an appropriate sensor for evaluation of antioxidants and inflammation drugs. Several years ago Soltes and his group established a methodology concerning to degradation of HA in SF under oxidation stress to evaluate anti-inflammatory drugs.[2,5,9,10,57,59,62,65-67,74-77,79,80,90-92,94-96]

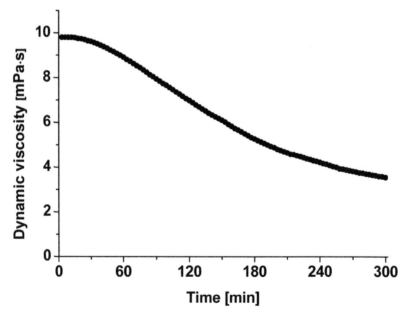

FIGURE 10.4 High-molar-mass HA degradation in vitro induced by Weissberger's biogenic oxidative system (WBOS).

Acceleration of degradation of high-molecular-weight HA occurring under inflammation and/or oxidative stress is accompanied by impairment and loss of its viscoelastic properties (Fig. 10.4).[23,24,30,49,62,70,85,87–89,97] Low molecular weight HA was found to exert different biological activities compared to the native high-molecular-weight biopolymer. HA chains of 25–50 disaccharide units are inflammatory, immunestimulatory, and highly angiogenic. HA fragments of this size appear to function as endogenous danger signals, reflecting tissues under stress.[44,58,63,72,98] Figure 10.5 describes the fragmentation mechanism of HA under free radical stress.

1. Initiation phase: the intact HA macromolecule entering the reaction with the HO$^{\bullet}$ radical formed via the Fenton-like reaction:

$$Cu^+ + H_2O_2 \rightarrow Cu^{2+} + HO + OH \qquad (10.1)$$

   H_2O_2 has its origin in the oxidative action of WBOS (see Fig. 10.5)

2. Formation of an alkyl radical (C-centered HA macro radical) initiated by the HO$^{\bullet}$ radical attack.
3. Propagation phase: formation of a peroxy-type C-macro radical of HA in a process of oxygenation after entrapping a molecule of O_2.
4. Formation of a HA-derived hydroperoxide via the reaction with another HA macromolecule.
5. Formation of a highly unstable alkoxy-type C-macro radical of HA on undergoing a redox reaction with a transition metal ion in a reduced state.
6. Termination phase: rapid formation of alkoxy-type C fragments and the fragments with a terminal C=O group due to the glycosidic bond scission of HA. Alkoxy-type C fragments may continue the propagation phase of the free-radical degradation of HA. Both fragments are represented by reduced molar masses.[4,22,53,64,73,85,86]

Soltes and his coworkers examined the effect of drugs and essential product compounds on inhibition of the degradation kinetics of a high-molecular-weight HA in vitro. High-molecular-weight HA samples were exposed to free-radical chain degradation reactions induced by ascorbate in the presence of Cu(II) ions, the so-called WBOS. The concentrations of both reactants [ascorbate, Cu(II)] were comparable to those that may occur during an early stage of the acute phase of joint inflammation (see Fig. 10.6).[3,6,21,25,51,52,57,60,61,63,71,81–85,87–89,93]

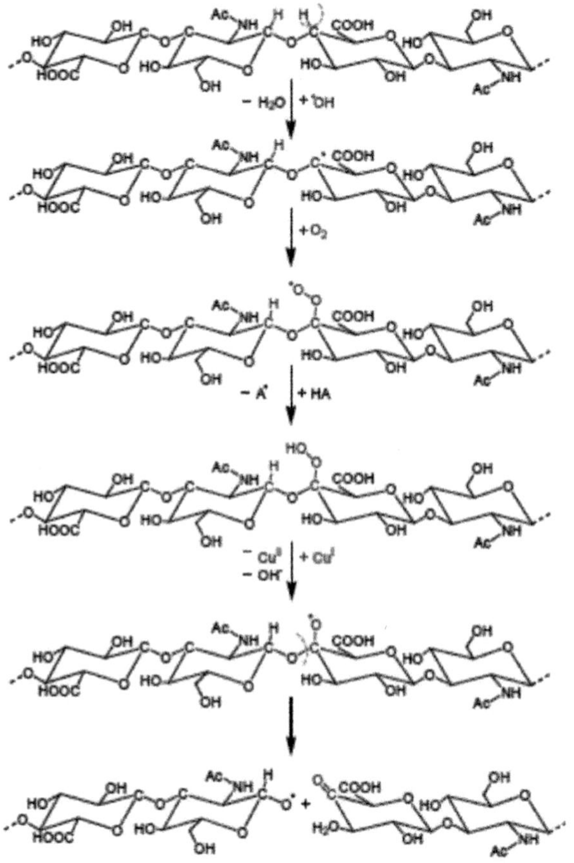

FIGURE 10.5 Schematic degradation of HA under free radical stress.[22]

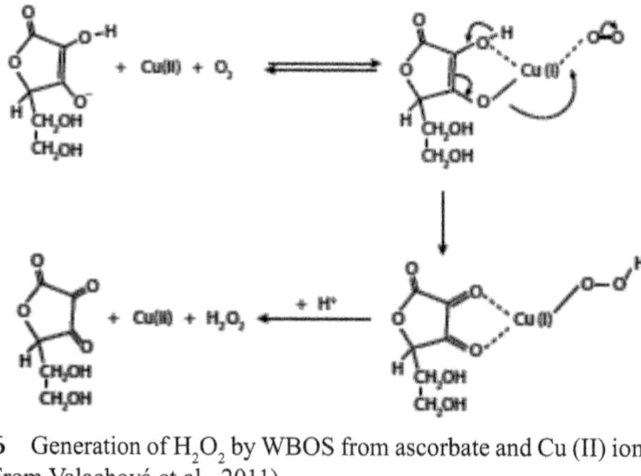

FIGURE 10.6 Generation of H_2O_2 by WBOS from ascorbate and Cu (II) ions under aerobic conditions. (From Valachová et al., 2011)

ACKNOWLEDGMENT

The authors would like to thank the Institute of Experimental Pharmacology and Toxicology for inviting them and orienting them in the field of medical research. They would also like to thank Slovak Academic Information Agency for funding them during their work at the institute.

KEYWORDS

- hyaluronan
- hydrogen atoms
- disaccharide
- biopolymer
- Golgi

REFERENCES

1. Adams, M. E. Viseosupplementation: A Treatment for Osteoarthritis. *J. Rheumatol.* **1993**, *20*(39), 1–24.
2. Balkose, D.; Horák, D.; Šoltés, L. *Key Engineering Materials Volume I Current State-of-the-Art on Novel Materials;* Apple Academic Press/Taylor & Francis Group: Canada/USA, 2014; pp 45–60.
3. Baňasová, M.; Valachová, K.; Hrabárová, E.; Priesolová, E.; Nagy, M.; Juránek, I.; Šoltés, L. In *Early Stage of the Acute Phase of Joint Inflammation. In Vitro Testing of Bucillamine and its Oxidized Metabolite SA981 in the Function of Antioxidants*, 16th Interdisciplinary Czech-Slovak Toxicological Conference in Prague. Interdisciplinary Toxicology, *4*(2), 22, 2011a.
4. Baňasová, M.; Valachová, K.; Juránek, I.; Šoltés, L. Effect of Thiol Compounds on Oxidative Degradation of High Molar Hyaluronan In Vitro. *Interdiscip. Toxicol.* **2012**, *5*(1), 25–26.
5. Baňasová, M.; Valachová, K.; Juránek, I.; Šoltés, L. Aloe Vera and Methylsulfonyl-methane as Dietary Supplements: Their Potential Benefit to Arthritic Patients with Diabetic Complications. *J. Inf. Intell. Knowl.* **2013**, *5*, 51–68.
6. Baňasová, M.; Valachová, K.; Rychlý, J.; Priesolová, E.; Nagy, M.; Juránek, I.; Šoltés, L. Scavenging and Chain Breaking Activity of Bucillamine on Free-Radical-Mediated Degradation of High Molar Mass Hyaluronan. *ChemZi* **2011b**, *7*, 205–206.
7. Baňasová, M.; Kerner, L.; Juráek, I.; Putala, M.; Valachová, K.; Šoltés, L. Radical Scavenging Capacity of N-(2-Mercapto-2-Methylpropionyl)-L-Cysteine. Design and Synthesis of its Derivative with Enhanced Potential to Scavenge Hypochlorite. *J. Inf., Intell. Knowl.* **2014c**, *6*, 453–470.

8. Pearce, E. M.; Howell, B. A.; Pethrick, R. A.; Zaikov, G. E. In *Physical Chemistry Research for Engineering and Applied Sciences Volume 1: Principles and Technological Implications*; Pearce, E. M., Howell, B. A., Pethrick, R. A., Zaikov, G. E., Eds.; Apple Academic Press, Inc.: New Jersey, USA, 2015; pp 71–92.

9. Baňasová, M.; Valachová, K.; Juráek, I.; Šoltés, L. Dithiols as More Effective Than Monothiols in Protecting Biomacromolecules from Free-Radical-Mediated Damage: In Vitro Oxidative Degradation of High-Molar-Mass Hyaluronan. *Chem. Pap.* **2014a**, *68*, 428–1434.

10. Baňasová, M.; Valachová, K.; Rychlý, J.; Janigová, I.; Csomorová, K.; Mendichi, R.; Mislovičová, D.; Juránek, I.; Šoltés, L. Free-Radical-Mediated Degradation of High-Molar-Mass Hyaluronan. Action of an Anti immflamatory Drug. *Polymers* **2014b**, *6*, 2625–2644.

11. Bothner, H.; Wik, O. Rheology of Hyaluronate. *Acta Otolaryngol* **1987**, *442*, 25–30.

12. Brown, M. B.; Jones, S. A. Hyaluronic Acid: A Unique Topical Vehicle for the Localized Delivery of Drugs to the Skin. *J. Eur. Acad. Dermatol. Venereol.* **2005**, *19*, 308–318.

13. Comper, W. D.; Laurent, T. C. Physiological Function of Connective Tissue Polysaccharides. *Physiol. Rev.* **1978**, *58*, 255–315.

14. Correia, C. R.; Moreira-Teixeira, L. S.; Moroni, L.; Reis, R. L.; van Blitterswijk, C. A.; Karperien, M.; Mano, J. F. Chitosan Scaffolds Containing Hyaluronic Acid for Cartilage Tissue Engineering. *Tissue Eng. Part C Methods* **2011**, *7*, 717–730.

15. Edwards, J. C. W. Consensus Statement. Second International Meeting on Synovium. Cell Biology, Physiology and Pathology. *Ann. Rheum. Dis.* **1995**, *54*, 389–391.

16. Edwards, J. C. W.; Wilkinson, L. S.; Jones, H. M. The Formation of Human Synovial Cavities: A Possible Role for Hyaluronan and CD44 in Altered Interzone Cohesion. *J. Anat.* **1994**, *185*, 355–367.

17. Fraser, J. R. E.; Foo, W. K.; Maritz, J. S. Viscous Interactions of Hyaluronic Acid with Some Proteins and Neutral Saccharides. *Ann. Rheum. Dis.* **1972**, *31*, 513–520.

18. Fraser, J. R. E.; Kimpton, W. G.; Pierscionek, B. K.; Cahill, R. N. P. The Kinetics of Hyaluronan in Normal and Acutely Inflamed Synovial Joints—Observations with Experimental Arthritis in Sheep. *Semin. Arthritis Rheum.* **1993**, *22*, 9–17.

19. Grootveld, M.; Henderson, E .B.; Farrell, A.; Blake, D. R.; Parkes, H. G.; Haycock, P. Oxidative Damage to Hyaluronate and Glucose in Synovial Fluid During Exercise of the Inflamed Rheumatoid Joint. Detection of Abnormal Low-Molecular-Mass Metabolites by Proton-N.M.R. Spectroscopy. *Biochem. J.* **1991**, *273*, 459–467.

20. Hlaváček, M. The Role of Synovial Fluid Filtration by Cartilage in Lubrication of Synovial Joints. *J. Biomech.* **1993**, *26*(10), 1145–1150.

21. Hrabárová, E.; Valachová, K.; Juránek, I.; Šoltés, L. Free-Radical Degradation of High-Molar-Mass Hyaluronan Induced by Ascorbate Plus Cupric Ions. Antioxidative Properties of the Piešťany-spa Curative Waters From Healing Peloid and Maturation Pool. In *Kinetics, Catalysis and Mechanism of Chemical Reactions;* Zaikov, G. E. Nova Science Publishers: New York, 2011; pp 29–36.

22. Hrabárová, E.; Valachová, K.; Juránek, I.; Šoltés, L. Free-Radical Degradation of High-Molar-Mass Hyaluronan Induced by Ascorbate Plus Cupric Ions: Evaluation of Antioxidative Effect of Cysteine-Derived Compounds. *Chem. Biodivers.* **2012**, *9*, 309–317.

23. Hrabárová, E.; Gemeiner, P.; Šoltés, L. Peroxynitrite: In Vivo and In Vitro Synthesis and Oxidant Degradative Action on Biological Systems Regarding Biomolecular Injury and Inflammatory Processes. *Chem. Pap.* **2007**, *61*, 417–437.

24. Hrabárová, E.; Valachová, K.; Rapta, P.; Šoltés, L. An Alternative Standard for Trolox-Equivalent Antioxidant-Capacity Estimation Based on Thiol Antioxidants. Comparative 2, 2'-Azinobis [3-Ethylbenzothiazoline-6-Sulfonic acid] Decolorization and Rotational Viscometry Study Regarding Hyaluronan Degradation. *Chem. Biodivers.* **2010,** *7*(9), 2191–2200.

25. Hrabárová, E.; Valachová, K.; Rychlý, J.; Rapta, P.; Sasinková, V.; Gemeiner, P.; Šoltés, L. High-Molar-Mass Hyaluronan Degradation by the Weissberger's System: Pro- and Anti-oxidative Effects of Some Thiol Compounds. *Polym. Degrad. Stab.* **2009,** *94*, 1867–1875.

26. Hutadilok, N.; Ghosh, P.; Brooks, P. M. Binding of Haptoglobin. Inter-α-Trypsin Inhibitor, and 1 proteinase Inhibitor to Synovial Fluid Hyaluronateand the Influence of These Proteins on its Degradation by Oxygen Derived Free Radicals. *Ann. Rheum. Dis.* **1988,** *47*, 377–385.

27. Kemp, M. M.; Kumar, A.; Clement, D.; Ajayan, P.; Mousa, S.; Linhardt, R. J. Hyaluronan- and Heparin-Reduced Silver Nanoparticles with Antimicrobial Properties. *Nanomedicine* **2009,** *4*(4), 421–429.

28. Kondo, S.; Kuroyanagi, Y. Development of a Wound Dressing Composed of Hyaluronic Acid and Collagen Sponge with Epidermal Growth Factor. *J. Biomater. Sci. Polym. Ed.* **2012,** *23*(5), 629–643.

29. Kontoes, P. P.; Vrettou, C. P.; Loupatatzi, A. N.; Marayiannis, K. V.; Foukas, P. G.; Vlachos, S. P. Wound Healing After Laser Skin Resurfacing: The Effect of a Silver Sulfadiazine-Hyaluronic Acid-Containing Cream Under an Occlusive Dressing. *J. Cosmet. Laser Ther.* **2010,** *1*, 10–13.

30. Lath, D.; Csomorová, K.; Kollárikóvá, G.; Stankovská, M.; Šoltés, L. Molar Mass Intrinsic Viscosity Relationship of High-Molar-Mass Myaluronans: Involvement of Shear Rate. *Chem. Pap.* **2005,** *59*, 291–293.

31. Laurent, T. C.; Laurent, U. B. G.; Fraser, J. R. E. The Structure and Function of Hyaluronan: An Overview. *Immunol. Cell. Biol.* **1996,** *74*, 1–7.

32. Laurent, T. C.; Ryan, M.; Pictruszkiewicz, A. Fractionation of Hyaluronic acid. The Polydispersity of Hyaluronic Acid From the Vitreous Body. *Biochim. Biophys. Acta* **1960,** *42*, 476–485.

33. Laurent, T. C.; Laurent, U. B. G.; Fraser, J. R. E. Functions of Hyaluronan. *Ann. Rheum. Dis.* **1995,** *54*, 429–432.

34. Lu, H. D.; Zhao, H. Q.; Wang, K. Novel Hyaluronic Acid-Chitosan Nanoparticles as Non-Viral Gene Delivery Vectors Targeting Osteoarthritis. *Int. J. Pharm.* **2011,** *420*(2), 358–365.

35. Maeda, H.; Nakamura, H.; Fang, J. EPR Effect for Macromolecular Drug Delivery to Solid Tumors: Improvement of Tumor Uptake, Lowering of Systemic Toxicity, and Distinct Tumor Imaging In Vivo. *Adv. Drug Deliv. Rev.* **2013,** *65*, 71–79.

36. Mathews, S.; Bhonde, R.; Gupta, P. K.; Totey, S. A Novel Tripolymer Coating Demonstrating the Synergistic Effect of Chitosan, Collagen Type 1 and Hyaluronic Acid on Osteogenic differentiation of Human Bone Marrow Derived Mesenchymal Stem Cells. *Biochem. Biophys. Res. Commun.* **2011,** *414*(1), 270–276.

37. Matsumoto, Y.; Arai, K.; Momose, H.; Kuroyanagi, Y. Development of a Wound Dressing Composed of a Hyaluronic Acid Sponge Containing Arginine. *J. Biomater. Sci. Polym. Ed.* **2009,** *20*(7–8), 993–1004.

38. Matsumoto, Y.; Kuroyanagi, Y. Development of a Wound Dressing Composed of Hyaluronic Acid Sponge Containing Arginine and Epidermal Growth Factor. *J. Biomater. Sci. Polym. Ed.* **2010,** *21*(6), 715–726.

39. McDonald, J. N.; Leviek, J. R. Effect of Intra-Articular Hyaluronan on Pressure-Flow Relation Across Synovium in Anaesthetized Rabbits. *J. Physiol.* **1995**, *485*, 179–193.

40. Meyer, K.; Palmer, J. W. The Polysaccharide of the Vitreous Humor. *J. Biol. Chem.* **1934**, *107*, 629–634.

41. Mineo, A.; Suzuki, R.; Kuroyanagi, Y. Development of an Artificial Dermis Composed of Hyaluronic Acid and Collagen. *J. Biomater. Sci. Polym. Ed.* **2013**, *24*(6), 726–740.

42. Morris, E. R.; Rees, D. A.; Welsh, E. J. Conformation and Dynamic Interactions in Hyaluronate Solutions. *J. Mol. Biol.* **1980**, *138*, 383–400.

43. Myint, P. The Reactivity of Various Free Radicals with Hyaluronic Acid Steady-State and Pulse Radiolysis Studies. *Biochim. Biophys. Acta* **1987**, *925*, 194–202.

44. Noble, P. W. Hyaluronan and Its Catabolic Products in Tissue Injury and Repair. *Matrix Biol.* **2002**, *21*, 25–29.

45. Ogston, A. G.; Stanier, J. E. The Physiological Function of Hyaluronic Acid in Synovial Fluid Viscous, Elastic and Lubricant Properties. *J. Physiol.* **1953**, *199*, 244–252.

46. Radin, E. L.; Swann, D. A.; Weisser, P. A. Separation of a Hyaluronate-Free Lubricating Fraction from Synovial Fluid. *Nature* **1970**, *228*, 377–378.

47. Oates, K. M. N.; Krause, W. E.; Colby, R. H. Using Rheology to Probe the Mechanism of Joint Lubrication: Polyelectrolyte/Protein Interactions in Synovial Fluid. *Mater. Res. Soc. Syrnp. Proc.* **2002**, *711*, 53–58.

48. Park, H.; Choi, B.; Hu, J.; Lee, M. Injectable Chitosan Hyaluronic Acid Hydrogels for Cartilage Tissue Engineering. *Acta Biomater.* **2013**, *9*(1), 4779–4786.

49. Parsons, B. J.; Al-Assaf, S.; Navaratnam, S.; Phillips, G. O. Comparison of The Reactivity of Different Oxidative Species (ROS) Towards Hyaluronan. In *Hyaluronan: Chemical, Biochemical and Biological Aspects*; Kennedy, J. F., Phillips, G. O., Williams, P. A., Hascall, V. C., Eds.; Woodhead, Publishing Ltd: Cambridge, MA, Massachusetts, 2002, pp 141–150.

50. Praest, B. M.; Greiling, H.; Kock, R. Effects of Oxygen-Derived Free Radicals on the Molecular Weight and the Polydispersity of Hyaluronan Solutions. *Carbohydr. Res.* **1997**, *303*, 153–157.

51. Rapta, P.; Valachová, K.; Gemeiner, P.; Šoltés, L. High-Molar-Mass Hyaluronan Behavior During Testing Its Radical Scavenging Capacity in Organic and Aqueous Media: Effects of the Presence of Manganese (II) Ions. *Chem. Biodivers.* **2009**, *6*, 162–169.

52. Rapta, P.; Valachová, K.; Zalibera, M.; Šnirc, V.; Šoltés, L. Hyaluronan Degradation by Reactive Oxygen Species: Scavenging Effect of the Hexapyridoindole Stobadine and Two of its Derivatives. In *Monomers, Oligomers, Polymers, Composites, and Nanocomposites*; Pethrick, R. A., Petkov, P., Zlatarov, A., Zaikov, G. E., Rakovsky, S. K. Nova Science Publishers: New York, 2010; pp 113–126 (Chapter 7).

53. Rychlý, J.; Šoltés, L.; Stankovská, M.; Janigová, I.; Csomorová, K.; Sasinková, V.; Kogan, G.; Gemeiner, P. Unexplored Capabilities of Chemiluminescence and Thermoanalytical Methods in Characterization of Intact and Degraded Hyaluronans. *Polym. Degrad. Stab.* **2006**, *91*(12), 3174–3184.

54. Sattar, A.; Kumar, S.; West, D. C. Does Hyaluronan Have a Role in Endothelial Cell Proliferation of the Synovium. *Semin. Arthritis Rheum.* **1992**, *22*, 37–43.

55. Sawa, M.; Kuroyanagi, Y. Potential of a Cryopreserved Cultured Dermal Substitute Composed of Hyaluronic Acid and Collagen to Release Angiogenic Cytokine. *J. Biomater. Sci. Polym. Ed.* **2013**, *24*(2), 224–238.

56. Scott, J. E.; Cummings, C.; Brass, A.; Chen, Y. Secondary and Tertiary Structures of Hyaluronan in Aqueous Solution, Investigated by Rotary Shadowing Electron Microscopy and Computer Simulation. *Biochem. J.* **1991**, *274*, 600–705.

57. Šoltés, L.; Brezová, V.; Stankovská, M.; Kogan, G.; Gemeiner, P. Degradation of High-Molecular-Weight Hyaluronan by Hydrogen Peroxide in the Presence of Cupric Ions. *Carbohydr. Res.* **2006a**, *341*, 639–644.

58. Šoltés, L.; Kogan, G. Impact of Transition Metals in the Free-Radical Degradation of Hyaluronan Biopolymer. In *Kinetics and Thermodynamics for Chemistry and Biochemistry:* Pearce, E. M., Zaikov, G. E., Kirshenbaum, G., Eds.; Nova Science Publishers: New York, 2009; Vol. 2, pp 181–199.

59. Šoltés, L.; Mendichi, R.; Kogan, G.; Mach, M. Associating Hyaluronan Derivatives: A Novel Horizon in Viscosupplementation of Osteoarthritic Joints. *Chem. Biodivers.* **2004**, *1*, 468–472.

60. Šoltés, L.; Mendichi, R.; Kogan, G.; Schiller, J.; Stankovska, M.; Arnhold, J. Degradative Action of Reactive Oxygen Species on Hyaluronan. *Biomacromolecules* **2006b**, *7*, 659–668.

61. Šoltés, L.; Stankovská, M.; Brezová, V.; Schiller, J.; Arnhold, J.; Kogan, G.; Gemeiner, P. Hyaluronan Degradation by Copper (II) Chloride and Ascorbate: Rotational Viscometric, EPR Spin-Trapping, and MALDI-TOF Mass Spectrometric Investigations. *Carbohydr. Res.* **2006c**, *341*, 2826–2834.

62. Šoltés, L.; Stankovská, M.; Kogan, G.; Germeiner, P.; Stern, R. Contribution of Oxidative Reductive Reactions to High Molecular Weight Hyaluronan Catabolism. *Chem. Biodivers.* **2005**, *2*, 1242–1245.

63. Šoltés, L.; Valachová, K.; Mendichi, R.; Kogan, G.; Arnhold, J.; Gemeiner, P. Solution Properties of High-Molar-Mass Hyaluronans: The Biopolymer Degradation by Ascorbate. *Carbohydr. Res.* **2007**, *342*, 1071–1077.

64. Šoltés, L. Hyaluronan—A High-Molar-Mass Messenger Reporting on the Status of Synovial Joints: Part II. Pathophysiological Status In *New Steps in Chemical and Biochemical Physics. Pure and Applied Science;* Pearce, E. M., Kirshenbaum, G., Zaikov, G. E., Eds.; Nova Science Publishers: New York, 2010; pp 137–152.

65. Šoltés, L.; Kogan, G. Hyaluronan: A Harbinger of the Status and Functionality of the Joint. In *Engineering of Polymers and Chemicals Complexity Volume II: New Approaches, Limitations and Control;* Focke, W., Radusch, H.-J., Eds.; Apple Academic Press/Taylor & Francis Group: Canada/United States, 2014; pp 259–286.

66. Šoltés, L.; Kogan, G. Hyaluronan: A Harbinger of the Status and Functionality of the Joint. *Polym. Res. J.* **2014**, *8*, 49–73.

67. Šoltés, L.; Kogan, G. Hyaluronan: A Harbinger of the Status and Functionality of the Joint. In *News in Chemistry, Biochemistry and Biotechnology State of the Art and Prospects of Development*; Zaikov, G. E., Nyszko, G., Krylova, L. P., Varfolomeev, S. D., Eds.; Nova Science Publishers: New York, 2014; pp 1–26.

68. Stankovská, M.; Arnholz, J.; Rychly, J.; Spalteholz, H.; Gemeiner, P.; Šoltés, L. In Vitro Screening of the Action of Non-Steroidal Anti-Inflammatory Drugs on Hypochlorous Acid-Induced Hyaluronan Degradation. *Polym. Degrad. Stab.* **2007**, *92*, 644–652.

69. Stankovská, M.; Hrabarová, E.; Valachová, K.; Molnárová, M.; Gemeiner, P.; Šoltés, L. The Degradative Action of Peroxynitrite on High-Molecular-Weight Hyaluronan. *Neuroendocrinol. Lett.* **2006**, *27*(2), 31–34.

70. Stankovská, M.; Šoltés, L.; Vikartovska, A.; Gemeiner, P.; Kogan, G.; Bakos, D. Degradation of High-Molecular-Weight Hyaluronan: A Rotational Viscometry Study. *Biologia* **2005,** *60*(17), 149–152.

71. Stankovská, M.; Šoltés, L.; Vikartovská, A.; Mendichi, R.; Lath, D.; Molnárová, M.; Gemeiner, P. Study of Hyaluronan Degradation by Means of Rotational Viscometry: Contribution of the Material of Viscometer. *Chem. Pap.* **2004,** *58*, 348–352.

72. Stern, R.; Kogan, G.; Jedrzejas, M. J.; Šoltés, L. The Many Ways to Cleave Hyaluronan. *Biotechnol. Adv.* **2007,** *25*, 537–557.

73. Surovčíková, L.; Valachová, K.; Baňasová, M.; Šnirc, V.; Priesolova, E.; Nagy, M.; Juránek, I.; Šoltés, L. Free-Radical Degradation of High-Molar-Mass Hyaluronan Induced by Ascorbate Plus Cupric Ions: Testing of Stobadine and Its Two Derivatives in Function as Antioxidants. *General Physiol. Biophys.* **2012,** *31*, 57–64.

74. Tamer, M. A.; Valachová, K.; Mohamed, S. M.; Šoltés, L. Free Radical Scavenger Activity of Cinnamyl Chitosan Schiff Base. *J. Appl. Pharmaceut. Sci.* **2016a,** *6*, 130–136.

75. Tamer, M. T.; Valachová, K.; Mohyeldin, M. S.; Šoltés, L. Free Radical Scavenger Activity of Chitosan and its Aminated Derivative. *J. App. Pharmaceut. Sci.* **2016b,** *6*, 195–201.

76. Tamer, T. M. Hyaluronan and Synovial Joint: Function, Distribution and Healing. *Interdisc. Toxicol.* **2013,** *6*, 111–125.

77. Tamer, T. M.; Valachová, K.; Šoltés, L. Inhibition of Free Radical Degradation in Medical Grade Hyaluronic Acid. In *Hylauronic Acid for Biomedical and Pharmaceutical Applications*; Collins, M., Ed.; Smithers Rapra Technology Ltd.: UK, 2014; pp 103–117.

78. Topoľská, D.; Valachová, K.; Hrabárova, E.; Rapta, P.; Baňasová, M.; Juránek, I.; Šoltés, L. Determination of Protective Properties of Bardejovske Kupele Spa Curative Waters by Rotational Viscometry and Abts Assay. *Balneo. Res. J.* **2014a,** *5*, 3–15.

79. Topoľská, D.; Valachová, K.; Nagy, M.; Šoltés, L. Determination of Antioxidative Properties of Herbal Extracts: *Agrimonia herba*, *Cynare folium*, and *Ligustri folium*. *Neuroendocrinol. Lett.* **2014b,** *35*, 192–196.

80. Topoľská, D.; Valachová, K.; Rapta, P.; Šilhár, S.; Panghyová, E.; Horváth, A.; Šoltés, L. Antioxidative Properties of *Sambucus nigra* Extracts. *Chem. Pap.* **2015,** *69*, 1202–1210.

81. Valachová, K.; Hrabárová, E.; Gemeiner, P.; Šoltés, L. Study of Pro- and Anti-Oxidative Properties of D-Penicillamine in a System Comprising High Molar-Mass Hyaluronan, Ascorbate, and Cupric Ions. *Neuroendocrinol. Lett.* **2008a,** *29*, 697–701.

82. Valachová, K.; Kogan, G.; Gemeiner, P.; Šoltés, L. Hyaluronan Degradation by Ascorbate: Protective Effects of Manganese (II). Cellulose. *Chem. Technol.* **2008b,** *42*(9–10), 473–483.

83. Valachová, K.; Rapta, P.; Kogan, G.; Hrabárová, E.; Gemeiner, P.; Šoltés, L. Degradation of High-Molar-Mass Hyaluronan by Ascorbate Plus Cupric Ions: Effects of D-Penicillamine Addition. *Chem. Biodivers.* **2009a,** *6*, 389–395.

84. Valachová, K.; Kogan, G.; Gemeiner, P.; Šoltés, L. Hyaluronan Degradation by Ascorbate: Protective Effects of Manganese (II) Chloride. In *Progress in Chemistry and Biochemistry. Kinetics, Thermodynamics, Synthesis, Properties and Application;* Nova Science Publishers: New York, 2009b; pp 201–215 (Chapter 20).

85. Valachová, K.; Šoltés, L. Effects of Biogenic Transition Metal Ions Zn(II) and Mn(II) on Hyaluronan Degradation by Action of Ascorbate Plus Cu(II) Ions. In *New Steps in Chemical and Biochemical Physics. Pure and Applied Science, Nova Science;* Pearce,

E. M., Kirshenbaum, G., Zaikov, G. E., Eds.; Nova Science Publishers: NewYork, 2010; pp 153–160 (Chapter 10).

86. Valachová, K.; Mendichi, R.; Šoltés, L. Effect of L-Glutathione on High-Molar-Mass Hyaluronan Degradation by Oxidative System Cu(II) Plus Ascorbate. In *Monomers, Oligomers, Polymers, Composites, and Nanocomposites;* Pethrick, R. A., Petkov, P., Zlatarov, A., Zaikov, G. E., Rakovsky, S. K., Eds.; Nova Science Publishers: New York, 2010; pp 101–111 (Chapter 6).

87. Valachová, K. Hrabárová, E.; Priesolová, E.; Nagy, M.; Baňasová, M.; Juránek, I.; Šoltés, L. Free-Radical Degradation of High-Molecular-Weight Hyaluronan Induced by Ascorbate Plus Cupric Ions. Testing of Bucillamine and Its SA981-Metabolite as Antioxidants. *J. Pharm. Biomed. Anal.* **2011a,** *56,* 664–670.

88. Valachová, K.; Vargová, A.; Rapta, P.; Hrabárová, E.; Dráfi, F.; Bauerová, K.; Juránek, I.; Šoltés, L. Aurothiomalate in Function of Preventive and Chain Breaking Antioxidant at Radical Degradation of High-Molar-Mass Hyaluronan. *Chem. Biodivers.* **2011b,** *8,* 1274–1283.

89. Valachová, K.; Hrabárová, E.; Juránek, I.; Šoltés, L. Radical Degradation of High-Molar-Mass Hyaluronan Induced by Weissberger Oxidative System. Testing of Thiol Compounds in the Function of Antioxidants. 16th Interdisciplinary Slovak-Czech Toxicological Conference in Prague. *Interdiscip. Toxicol.* **2011c,** *4*(2), 65.

90. Valachová, K.; Rapta, P.; Slováková, M.; Priesolová, E.; Nagy, M.; Mislovičová, D.; Dráfi, F.; Bauerová, K.; Šoltés, L. Radical Degradation of High-Molar-Mass Hyaluronan induced by Ascorbate Plus Cupric Ions. Testing of Arbutin in the Function of Antioxidant. In *Advances in Kinetics and Mechanism of Chemical Reactions*; Zaikov, G. E., Valente, A. J. M., Iordanskii, A. L., Eds.; Apple Academic Press, Waretown, NJ, USA, 2013a; pp 1–19.

91. Valachová, K.; Baňasová, M.; Machová, L.; Juránek, I.; Bezek, Š.; Šoltés, L. Testing Various Hexahydropyridoindoles to Act as Antioxidants. In *Pharmaceutical and Medical Biotechnology. New Perspectives*; Orlicki, R., Cienciala, C., Krylova, L. P., Pielichowski, J., Zaikov, G. E., Eds.; Nova Science Publishers: New York, 2013b; pp 93–110.

92. Valachová, K.; Baňasová, M.; Machová, L.; Juránek, I.; Bezek, Š.; Šoltés, L. Practical Hints on Testing Various Hexahydropyridoindoles to Act as Antioxidants. In *Chemistry and Physics of Modern Materials*; Aneli, J. N., Jimenez, A., Kubica, S., Eds.; Apple Academic Press/Taylor & Francis Group: Toronto/USA, 2013c; pp 137–158.

93. Valachová, K.; Topoľská, D.; Nagy, M.; Gaidau, C.; Niculescu, M.; Matyašovský, J.; Jurkovič, P.; Šoltés, L. Radical Scavenging Activity of *Caesalpinia spinosa. Neuroendocrinol. Lett.* **2014,** *35*(2), 197–200.

94. Valachová, K.; Tamer, M. T.; Šoltés, L. Comparison of Free-Radical Scavenging Properties of Glutathione Under Neutral and Acidic Conditions. In *Chemistry and Chemical Biology, Methodologies and Applications*; Joswik, R., Dalinkevich, A. A., Eds.; Apple Academic Press/Taylor & Francis Group: New Jersey/USA, 2015a; pp 227–245.

95. Valachová, K.; Baňasová, M.; Topoľská, D.; Sasinková, V.; Juránek, I.; Collins, M. N.; Šoltés, L. Influence of Tiopronin, Captopril and Levamisole Therapeutics on the Oxidative Degradation of Hyaluronan. *Carbohydr. Polym.* **2015b,** *134,* 516–523.

96. Valachová, K.; Tamer, M. T.; Mohy, E. M.; Šoltés, L. Radical-Scavenging Activity of Glutathione, Chitin Derivatives and Their Combination. *Chem. Pap.* **2016a,** *70,* 820–827.

97. Valachová, K.; Topoľská, D.; Mendichi, R.; Collins, M. N.; Sasinková, V.; Šoltés, L. Hydrogen Peroxide Generation by the Weissberger Biogenic Oxidative System During Hyaluronan Degradation. *Carbohydr. Polym.* **2016b,** 148, 189–193.

98. West, D. C.; Hampson, I. N.; Arnold, F.; Kumar, S. Angiogenesis Induced by Degradation Products of Hyaluronic Acid. *Science* **1985,** *228,* 1324–1326.

99. Yamamoto, A.; Shimizu, N.; Kuroyanagi, Y. Potential of Wound Dressing Composed of
Hyaluronic Acid Containing Epidermal Growth Factor to Enhance Cytokine Production
by Fibroblasts. *J. Artif. Organs* **2013,** *16,* 489–494.

100. Yu, A.; Niiyama, H.; Kondo, S.; Yamamoto, A.; Suzuki, R.; Kuroyanagi, Y. Wound
Dressing Composed of Hyaluronic Acid and Collagen Containing EGF or bFGF:
Comparative Culture Study. *J. Biomater. Sci. Polym.* **2013,** *24*(8), 1015–1026.

CHAPTER 11

PROTECTIVE EFFECTS OF CATALASE INDICATE HYDROGEN PEROXIDE INVOLVEMENT IN HYALURONAN DEGRADATION INITIATED BY CU(II) IONS PLUS ASCORBATE IN SITU

KATARÍNA VALACHOVÁ[1,*], DOMINIKA TOPOL'SKÁ[1], RANIERO MENDICHI[2], JOZEF RYCHLÝ[3], and LADISLAV ŠOLTÉS[1]

[1]Institute of Experimental Pharmacology and Toxicology of SAS, Slovak Academy of Sciences, Bratislava, Slovakia

[2]Istituto per lo Studio delle Macromolecole, Consiglio Nazionale delle Ricerche, Milano, Italy

[3]Polymer Institute, Slovak Academy of Sciences, Slovakia

*Corresponding author. E-mail: katarina.valachova@savba.sk

CONTENTS

ABSTRACT

High-molar-mass hyaluronan was exposed to oxidative degradation induced by Cu(II) ions and ascorbate (the Weissberger biogenic oxidative system; WBOS), which resulted in generation of hydrogen peroxide, followed by production of hydroxyl radicals. Catalase was examined to prove the involvement of hydrogen peroxide in the hyaluronan and WBOS reaction mixture. Complementary methods, such as size-exclusion chromatography-multiangle light scattering (SEC-MALS) and nonisothermal chemiluminometry, were used to analyze the recovered hyaluronan samples. Catalase was effective in preventing hyaluronan degradation induced by both hydroxyl and peroxy-type hyaluronan macroradicals as showed by the results of rotational viscometry. Results of SEC-MALS showed that catalase was protective against hyaluronan degradation by WBOS. Results of nonisothermal chemiluminometry showed several populations of hydroperoxides in both native and recovered hyaluronan samples.

11.1   INTRODUCTION

Hyaluronan is a nonsulfated glycoaminoglycan, composed of N-acetyl-D-glucosamine and D-glucuronic acid linked by β-(1→4) and β-(1→3) linkages, present in all vertebrates. In mammals, hyaluronan, which is an important component of the extracellular matrix involved in the structure of connective tissues, can modulate a variety of cellular and tissue functions.[9,11,15,17–19,34,36,45–48,52,53]

Differently sized hyaluronans trigger different signal transduction pathways. For example, tetrasaccharidic hyaluronan is antiapoptotic and induces heat shock proteins. Hyaluronan oligomers with 8–16 repeating units, stimulated angiogenesis *in vivo*, and endothelial proliferation *in vitro*; smaller hyaluronans (1000 kDa) induce proinflammatory responses in macrophages. Larger hyaluronan (1000–5000 kDa) suppress angiogenesis, immune responses, and inflammation.[14]

High-molar-mass hyaluronan is readily degraded into small molecules after tissue injury, primarily by increased levels of hyaluronidases (HYALs), reactive oxygen species, and reactive nitrogen species.[59]

The mechanism of hyaluronan removal or turnover facilitated by HYALs, which, in mammals, consists of a family of enzymes including HYALs: HYAL1, HYAL2, HYAL3, HYAL4, PH20, and HYALP1. HYALs as a class are highly homologous endoglycosidases, and in terms of hyaluronan catabolism, they specifically hydrolyze β-1→4 linkages of the hyaluronan macromolecule. The acidic environment necessary for HYAL2 activity is provided by Na^+–H^+ exchanger, facilitating the generation of hyaluronan polymers of approximately 20 kDa (or hyaluronan macromolecules with 50 disaccharide units). Then, 20 kDa hyaluronan fragments are internalized and transported first to endosome and then to lysosomes, where lysosomal HYAL1 further degrades hyaluronan into short tetrasaccharide units.[26]

Reactive oxygen and nitrogen species degrading hyaluronan are generated during the inflammatory response in sepsis, tissue inflammation, and ischemia–reperfusion injury. The most direct evidence for this has been accumulated in synovial fluid, where inflammatory oxidation leads to degradation of native high-molar-mass hyaluronan with resulting decrease in synovial fluid viscosity and cartilage degeneration, and in the airways, where reactive oxygen species can degrade luminal epithelial hyaluronan.[4,10,12,31,42,49,54]

Our studies are focused on hyaluronan degradation induced by Cu(II) ions and ascorbate, which results in the production of hydrogen peroxide, followed by production of hydroxyl radicals (Fig. 11.1).[13,29,37,43,50,55]

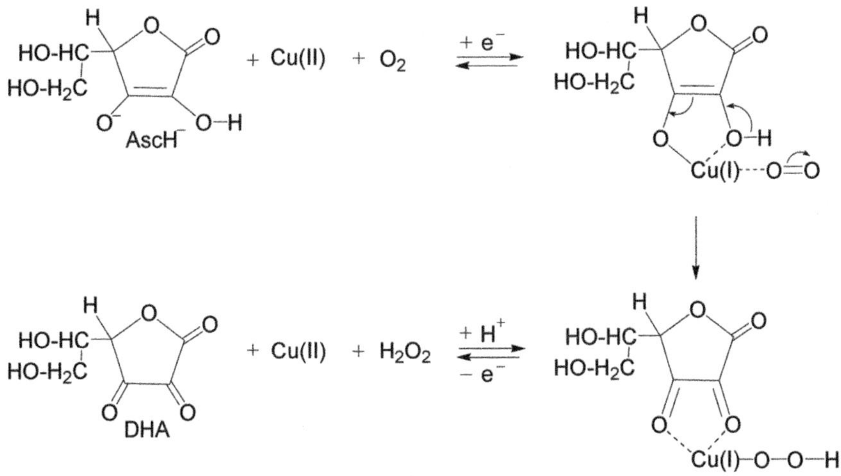

$$H_2O_2 + Cu(I)\text{---complex} \rightarrow {}^{\bullet}OH + Cu(II) + OH^-$$

FIGURE 11.1 Chemistry of Weissberger's biogenic oxidative system: hydrogen peroxide is generated by oxidation of ascorbate under catalytic action of Cu(II) ions (adapted from Hrabárová[7]).

Hydrogen peroxide is a cytotoxic compound, whose amount is necessary to be eliminated by using antioxidative enzymes. In the absence of transition metals, hydrogen peroxide is slightly reactive.[60] Usually, hydrogen peroxide molecules are formed by dismutation of superoxide anion radicals, which are primary radicals formed during inflammation.[61,62] Concentration of hydrogen peroxide *in vivo* is about 10^{-7}–10^{-8} mol/L.[63] Its concentrations ≥ 50 µmol/L are toxic for many animals, plants, and bacteria. Hydrogen peroxide molecules serve as a mild oxidizing and reducing agent. High amounts of the hydrogen peroxide reagent are present in drinks, for example, in tea or coffee. In a site of inflammation, hydrogen peroxide produced by activated fagocytes, functions the way that it controls proliferation of cells, their apoptosis, and aggregation of platelets.[6] Hydrogen peroxide decomposes aggregates of proteoglycans in human articular cartilage *in vitro*.[64,68] Unlike more reactive oxygen species, biological membranes are permeable for hydrogen peroxide. Production of hydrogen peroxide in cells is catalyzed by Mn-superoxide dismutase, which is present in mitochondria and CuZn-superoxide dismutase, which is present in cytoplasm.[65] Bacteria contain also Fe-superoxide dismutase.[66] Hydrogen peroxide itself does not degrade hyaluronan, however, trace amounts of transition metal ions in hyaluronan preparations can decompose hydrogen peroxide, which can lead to production of hydroxyl radicals.[38,40]

Šoltés et al.[35,39] and Rychlý et al.[30] studied hyaluronan degradation induced by Cu(II) ions and hydrogen peroxide. The molecules of hydrogen peroxide are *in vivo* eliminated by peroxidases, especially by glutathione peroxidase, compounds containing thioredoxin, and catalase. Reaction products of hydrogen peroxide, including lipid hydroperoxides, are also biologically active.[5,6,16,20] Catalase is a common enzyme found in nearly all living organisms exposed to oxygen. It consists of four identical subunits, each containing a single heme for hydrogen peroxide decomposition to water molecules and molecular oxygen. The enzyme catalase is present in cytoplasma and peroxizomes.[44,67]

Hydroxyl radical has the highest reductive potential, that is, 2.31 V. The OH radicals react with all compounds containing −CH groups with the second rate order 10^9–10^{10} $M^{-1} \cdot s^{-1}$. Hydroxyl radical is a primary radical degrading hyaluronan molecule in the way that an atom of hydrogen is primarily abstracted from D-glucuronic acid or *N*-acetylglucosamine, followed by the formation of C-centered macroradicals. No radicals are formed on C-2 molecule of hyaluronan.[18] These macroradicals undergo immediately to reaction with molecular oxygen yielding by that way AOO˙—the peroxyl-type macroradicals and/or AO˙—alkoxyl-type macroradicals of hyaluronan, and on the basis of the cleavage of glycosidic bond, the macromolecule of hyaluronan has been fragmented.[8,17,18,30,32,69]

The aim of the study was:

1. To confirm the inhibitory action of catalase added to the reaction mixture consisting of hyaluronan, ascorbate, and cupric ions before hyaluronan degradation begins
2. To confirm changes in molar mass parameters of hyaluronan samples
3. To confirm the inhibitory action of catalase added to the hyaluronan reaction mixture 1 h after hyaluronan degradation begins
4. To demonstrate the presence of hydroperoxides in the formed hyaluronan fragments by using nonisothermal chemiluminometry

11.2 MATERIALS AND METHODS

11.2.1 CHEMICALS

Hyaluronan sample (sodium salt) of molar mass Mw = 709.3 kDa coded P0207–1A was from Lifecore Biomedical, Chaska, MN, USA. Hyaluronan coded F1750762 of molar mass 1.38 MDa was obtained from Thomas Laboratories, Tolleson, AZ, USA. Analytical purity grade NaCl and $CuCl_2.2H_2O$ were purchased from Slavus, Bratislava, Slovakia. Ascorbic acid, K_2HPO_4, and KH_2PO_4 were the products of Sigma Aldrich, Bratislava, Slovakia. Catalase from bovine liver was purchased from Sigma Aldrich, Bratislava, Slovakia. Redistilled deionized high quality grade water, with conductivity of < 0.055 $\mu S \cdot cm^{-1}$, was produced using the TKA water purification system from Water Purification Systems, Niederelbert, Germany.

11.2.2 PREPARATION OF STOCK AND WORKING SOLUTIONS

The hyaluronan samples (24 mg) were dissolved in 0.15 mol/L of aqueous NaCl solution for 24 h in the dark. The hyaluronan sample solutions were prepared in two steps: first, 4.0 mL and after 6 h: 3.85, 3.65, 3.4, or 2.9 mL of 0.15 mol/L NaCl were added when working in the absence or presence of the catalase. Solutions of ascorbate (16 mmol/L) and cupric chloride (320 μmol/L) were prepared also in 0.15 mol/L aqueous NaCl. Catalase (1 mg/mL) was prepared in potassium phosphate buffer (50 mmol/L).

11.2.3 UNINHIBITED/INHIBITED HYALURONAN DEGRADATION

First, hyaluronan degradation was induced by the oxidative system comprising $CuCl_2$ (2.0 $\mu mol/L$) and ascorbic acid (100 $\mu mol/L$). The procedure was as follows: a volume of 50 μL of 320 $\mu mol/L$ $CuCl_2$ solution was added to the hyaluronan solution (6.9 mL), and the mixture was left to stand for 7 min 30 s at room temperature after a 30 s stirring. Then, 1 mL of 50 mmol/L phosphate buffer was added and stirred again for 30 s. Finally, 50 μL of ascorbic acid solution (16 mmol/L) was added to the solution and stirred for 30 s. The solution was then immediately transferred into the viscometer Teflon® cup reservoir.

The procedures to investigate the influence of catalase on oxidatively degraded hyaluronan are as follows:

1. A volume of 50 μL of 160 $\mu mol/L$ $CuCl_2$ solution was added to the hyaluronan solution (7.85, 7.65, 7.4, or 6.9 mL), and the mixture, after 30 s stirring, was allowed to stand for 7 min 30 s at room temperature. Then, 50, 250, 500, or 1000 μL of catalase were added to the solution, followed by stirring again for 30 s. Finally, 50 μL of ascorbic acid solution (16 mmol/L) was added to the solution, and the mixture was stirred for 30 s. The solution was then immediately transferred into the viscometer Teflon cup reservoir.

2. In the second experimental setting, a procedure similar to that described in (1) was applied, however, after standing for 7 min 30 s at room temperature, 50 μL of ascorbic acid solution (16 mmol/L) was added to the mixture and a 30 s stirring followed. After 1 h, finally 50, 250, 500, or 1000 μL of catalase were added to the solution, followed by 30 s stirring. The solution mixture was then immediately transferred into the viscometer Teflon cup reservoir.

11.2.4 ROTATIONAL VISCOMETRY

Dynamic viscosity of the reaction mixture (8 mL in 0.15 mol/L aqueous NaCl) containing hyaluronan (3 mg/mL), phosphate buffer (6.25 mmol/L), and ascorbate (100 $\mu mol/L$) plus Cu(II) ions (2 $\mu mol/L$) in the absence and presence of catalase (100–250, 500–1250, 1000–2500, and 2000–5000 U) was monitored by a Brookfield LVDV-II+PRO digital rotational viscometer (Brookfield Engineering Labs., Middleboro, MA, USA) at 25.0 ± 0.1°C and at a shear rate of 237.6 s^{-1} for 4 h in the Teflon cup reservoir.[1,21,30,33,40,41,51,55,56]

After finishing experiments, the hyaluronan solutions were precipitated in 96% ethanol for 24 h, centrifuged for 5 min at 3000 rpm and dried in a desiccator.

11.2.5   SEC-MALS ANALYSES

The molar mass distribution of samples was determined by a modular multidetector size-exclusion chromatographic (SEC) system. This system consisted of an Alliance 2695 separation module from Waters company (Milford, MA, USA) equipped with two on-line detectors: A multiangle light scattering photometer (MALS Dawn DSP-F) from Wyatt company (Santa Barbara, CA, USA) and a 2414 differential refractometer (DRI) from Waters company (Milford, MA, USA); the latter was used in the function as a polymer concentration detector. The setup of this multidetector chromatographic system was serial in the following order: Alliance–MALS–DRI. The wavelength of the MALS laser was 632.8 nm. The light scattering signal was detected simultaneously at 15 scattering angles ranging from 14.5° to 151.3°. The calibration constant was calculated using toluene as standard, assuming a Rayleigh factor of 1.406×10^{-5} cm^{-1}. The angular normalization was performed by measuring the scattering intensity of a concentrated solution of bovine serum albumin (BSA) globular protein in the mobile phase, assumed to act as an isotropic scatterer. Considering the high-molar-mass of the samples, a relatively low flow rate, 0.5 mL/min, was used to avoid shear-degradation of the polymer in the chromatographic columns. The experimental methodology for a reliable use of the SEC-MALS chromatographic system was described in detail.[23–25,57,58] The experimental conditions of the SEC-MALS system were the following: two Shodex (KB806 and KB805) chromatographic columns from Tosoh Bioscience (Stuttgart, Germany); mobile phase: 0.20 mol/L aqueous NaCl; temperature: 35°C; injection volume: 150 µL; polymer concentration: 0.4–1.0 mg/mL depending on the molar mass of the polymer sample. The refractive index increment, dn/dc, with respect to the 0.20 mol/L aqueous NaCl solvent was measured by a KMX-16 DRI from LDC Milton Roy (Riviera Beach, FL, USA). The dn/dc value determined was 0.150 mL/g.

11.2.6   NONISOTHERMAL CHEMILUMINOMETRY

Chemiluminescence measurements were performed with a photon-counting instrument Lumipol 3, manufactured at the Polymer Institute of the Slovak

Academy of Sciences (Bratislava, Slovakia). The sample was placed on an aluminum pan in the sample compartment. The gas flow (pure nitrogen) through the sample cell was 3.0 l/h. The temperature in the sample compartment of the apparatus was raised from 40°C to 250°C, with a linear gradient of 2.5°C/min. The signal from the photocathode was recorded at 10-s data collection intervals. The amount of samples used for each measurement was in the range of 1.5–3.2 mg.

11.3   RESULTS AND DISCUSSION

At first, high-molar-mass hyaluronan was exposed to oxidative degradation by Cu(II) ions (2.0 μmol/L) in the presence of ascorbate (100 μmol/L) and potassium phosphate buffer (6.25 mmol/L), observing the decrease in dynamic viscosity (η) of the hyaluronan solution by 5.1 mPa·s (Fig. 11.2, curve 1). Curve 1 served as a reference.

WBOS used in our experiments is a source of hydrogen peroxide, so we examined the ability of catalase to decompose hydrogen peroxide. Figure 11.2, left panel represents the addition of different concentrations of catalase into the hyaluronan reaction system. The catalase was demonstrated to suppress hyaluronan degradation dose-dependently, which can result in decomposition of hydrogen peroxide and thereby in prevention of the production of hydroxyl radicals, which can degrade hyaluronan molecules to fragments. In the presence of catalase at the highest concentration

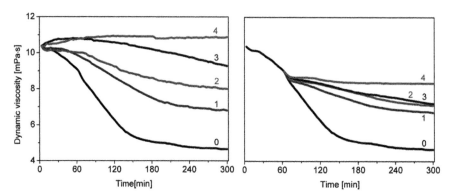

FIGURE 11.2   Hyaluronan degradation induced by the oxidative system Cu(II) ions (2.0 μmol/L) and ascorbate (100 μmol/L) with addition of potassium phosphate buffer (1 mL, 50 mmol/L) (curve 0). Addition of catalase (100–250 U, curve 1), (500–1250 U, curve 2), (1000–2500 U, curve 3), and (2000–5000 U, curve 4) into the reactive system before hyaluronan degradation begins (left panel) or 1 h later (right panel).

no decrease in dynamic viscosity of the hyaluronan solution was observed (see Fig. 11.2, left panel, curve 4).

However, catalase decreased the rate of hyaluronan degradation in part also in the reactive system, in which the production of peroxy-type hyaluronan macroradicals predominates (Fig. 11.2, right panel). Dynamic viscosity of the hyaluronan solution with the addition of catalase at the highest concentration was observed to decrease by 0.68 mPa·s (Fig. 11.2, right panel, curve 4).

Lovstad[22] found an inhibitory effect of catalase on copper catalyzed oxidation of ascorbate, which, as he considered, originates probably from the binding of copper ions to the enzyme macromolecule; however, our results do not support such a statement.

Figure 11.3 displays the results of comparison of differential molar mass distribution of six hyaluronan samples. The molar mass averages of the native hyaluronic sample and of each polymeric fraction determined by the on-line MALS detector of the SEC analyses are absolute, not relative to some particular calibration standards, and the final molar mass distribution of the hyaluronan samples were obtained directly without any (relative) calibration. As seen, molar masses of hyaluronan resulted from the experiments

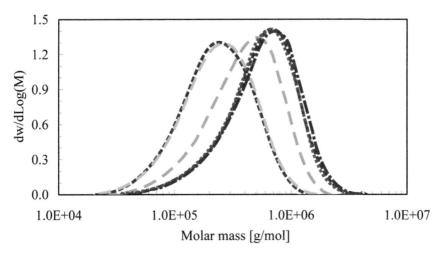

FIGURE 11.3 Comparison of differential molar mass distribution of native hyaluronan P0207–1A (red), precipitated native hyaluronan P0207–1A (blue), precipitated hyaluronan samples prepared by treating the native hyaluronan with WBOS: in the presence of phosphate buffer added before hyaluronan degradation begins (cyan) and 1 h later (brown) or in the presence of catalase (1000–2500 U, dissolved in phosphate buffer) added before hyaluronan degradation begins (violet) and 1 h later (green).

at which the catalase has been involved (added before hyaluronan degradation begins, violet curve in Fig. 11.3) and the precipitated native hyaluronan (blue curve) were a bit lower compared with the native hyaluronan (red curve in Fig. 11.3). In contrast, molar masses of hyaluronan samples containing phosphate buffer was twofold lower. Selected parameters of hyaluronan samples are summarized in Table 11.1. Results of SEC-MALS are in well accordance with the results of rotational viscometry.

TABLE 11.1   SEC-MALS Results Obtained with the Native Hyaluronan Sample, Native Hyaluronan After Precipitation, and the Fragmented Polymers.

Sample	M_n (kg/mol)	M_w (kg/mol)	M_z (kg/mol)	M_w/M_n	M_z/M_w	Rg_z (nm)
1	412.5	709.3	999.6	1.72	1.41	83.4
2	368.5	662.4	951.8	1.80	1.44	80.8
3	372.1	649.4	917.2	1.75	1.41	80.5
4	291.9	486.1	686.9	1.66	1.41	65.9
5	175.9	286.3	416.8	1.63	1.46	45.0
6	176.3	281.0	407.1	1.59	1.45	46.2

M_n, number average of molar masses; M_w, weight average of molar masses; M_z, z-average of molar masses; M_w/M_n and M_z/M_w, polymer dispersities; Rg_z, z-average of radiuses of gyration.

Samples: 1, native hyaluronan; 2, precipitated native hyaluronan; 3, addition of catalase (1000–2500 U) to hyaluronan reaction system before hyaluronan degradation begins; 4, addition of catalase (1000–2500 U) to hyaluronan reaction system 1 h after hyaluronan degradation begins; 5, addition of phosphate buffer (50 mmol/L) to hyaluronan reaction system 1 h after hyaluronan degradation begins; 6, addition of phosphate buffer (50 mmol/L) to hyaluronan reaction system 1 h after hyaluronan degradation begins.

To complete the results from rotational viscometry, we have evaluated the momentaneous state of hyaluronan by chemiluminescence tests. The chemiluminescence signal at medium temperatures reflects the bimolecular decomposition of hydroperoxides and thus the chemiluminescence intensity is directly proportional to their concentration. The light emission comes from exothermic disproportionation reactions of peroxyl-type radical in the polymeric macromolecule being formed from hydroperoxides or hydrogen peroxide, the reactions that provide the heat necessary for the excitation of potential emitters[2] such as carbonyl groups or singlet oxygen. As seen in Figure 11.4, nontreated hyaluronan samples rather than hyaluronan samples treated by catalase give much more intensive signal when exposed to the temperature from 40°C to 250°C. We examined two different hyaluronan precipitates (P0207–1A and F1750762) to be sure that the results may be considered as more general. As evident in Figure 11.4, hyaluronan samples treated with catalase demonstrate not only reduced intensity of chemilumi-nescence observed at lower temperatures but also the treated samples contain some portion of more stable peroxides (compare the maximum of the peaks at 160°C and 200°C). However, their concentration is also reduced when treated with catalase longer time (3 h).

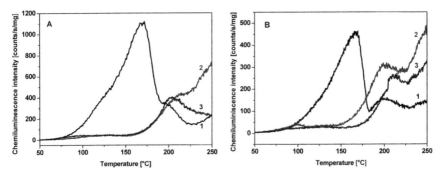

FIGURE 11.4 Nonisothermal chemiluminometry of hyaluronan samples carried out in nitrogen. Native hyaluronan samples: P0207–1A (A1) and F1750762 (B1), recovered polymers prepared by treating native hyaluronan sample with addition of catalase (1000–2500 U) after 1 h (A,B 2) and after 4 h (A,B 3).

Campanella et al. [3] studied the catalytic mechanism of catalase on organic hydroperoxides in organic solvents. Pigeolet et al.[27] studied the effect of hydroperoxides on catalase and they observed that the hydro-peroxides did not inactivate catalase. Radi et al.[28] studied catalase on hydroperoxide-dependent lipid peroxidation in rat heart and kidney mitochondria.

ACKNOWLEDGMENT

The study was supported by VEGA grant 2/0065/15.

KEYWORDS

- dynamic viscosity
- enzymes
- glycosaminoglycans
- radical degradation
- rotational viscometry

REFERENCES

1. Baňasová, M.; Valachová, K.; Juránek, I.; Šoltés, L. *Aloe vera* and Methylsulfonyl-methane as Dietary Supplements: Their Potential Benefits for Arthritic Patients with Diabetic Complications. *J. Inf. Intelligence Knowl.* **2013**, *5*, 51–68.
2. Baňasová, M.; Valachová, K.; Rychlý, J.; Janigová, I.; Csomorová, K.; Mendichi, R.; Mislovičová, D.; Juránek, I.; Šoltés, L. Effect of Bucillamine on Free-Radical-Mediated Degradation of High-Molar-Mass Hyaluronan Induced In Vitro by Ascorbic Acid and Cu(II) Ions. *Polymers* **2014**, *6*, 2625–2644.
3. Campanella, L.; Spuri Capesciotti, G.; Russo, M. V.; Tomassetti, M. Study of the Cata-lytic Mechanism of the Enzyme Catalase on Organic Hydroperoxides in Non-Polar Organic Solvent. *Curr. Enzym. Inhib.* **2008**, *4*, 86–92.
4. Cyphert, J. M.; Trempus, C. S.; Garantziotis, S. Size Matters: Molecular Weight Speci-ficity of Hyaluronan Effects in Cell Biology. *Int. J. Cell Biol.* **2015**, *2015*, 563818.
5. Griendling, K. K.; Harrison, D. G. Dual Role of Reactive Oxygen Species in Vascular Growth. *Circ. Res.* **1999**, *85*, 562–563.
6. Halliwell, B.; Clement, M. V.; Long, L. H. Hydrogen Peroxide in the Human Body. *FEBS Lett.* 2000, *486*, 10–13.
7. Hrabárová, E. Free-Radical Degradation of High-Molar-Mass Hyaluronan by Oxygen Free Radicals. Evaluation of Antioxidant Properties of Endogenic Andexogenic Compounds with Thiol Groups in Their Structure. Ph.D. Thesis, Faculty of Chemical and Food Technology, Bratislava, Slovak. 2012c.
8. Hrabárová, E.; Rychlý, J.; Sasinková, V.; Valachová, K.; Janigová, I.; Csomorová, K.; Juránek, I.; Šoltés, L. Structural Characterisation of Thiol-Modified Hyaluronans. *Cellulose* **2012**, *19*, 2093–2104.
9. Hrabárová, E.; Valachová, K.; Juránek, I.; Šoltés, L. Free-Radical Degradation of High-Molar-Mass Hyaluronan Induced by Ascorbate Plus Cupric Ions. Anti-Oxidative Properties of the Piešťany-Spa Curative Waters from Healing Peloid and Maturation Pool. **In:** *Kinetics, Catalysis and Mechanism of Chemical Reactions: From Pure to Applied Science. Volume 2—Tomorrow and Perspectives*; Islamova, R. M.; Kolesov, S. V.; Zaikov, G. E., Eds.; Nova Science Publishers: New York, 2012; Vol. 2, pp 29–36.
10. Hrabárová, E.; Valachová, K.; Juránek, I.; Šoltés, L. Free-Radical Degradation of High-Molar-Mass Hyaluronan Induced by Ascorbate Plus Cupric Ions: Evaluation of

Anti-Oxidative Effect of Cysteine-Derived Compounds. *Chem. Biodivers.* **2012**, *9*, 309–317.

11. Hrabárová, E.; Šoltés, L. Peroxynitrite: A Potent Endogenous Pro-Oxidant Agent Playing an Important Role in Degradation of Hyaluronan Biopolymer. In: *Monomers, Oligomers, Polymers, Composites, and Nanocomposites* (Polymer Yearbook); Pethrick, R. A.; Petkov, P.; Zlatarov, A.; Zaikov, G. E.; Rakovsky, S. K., Eds.; Nova Science Publishers: New York, 2010; Vol. 23, pp 127–138.

12. Hrabárová, E.; Juránek, I.; Šoltés, L. Pro-Oxidative Effect of Peroxynitrite Regarding Biological Systems: A Special Focus on High-Molar-Mass Hyaluronan Degradation. *Gen. Physiol. Biophys.* **2011**, *30*, 223–238.

13. Hrabárová E., Valachová K., Rychlý J., Rapta P., Sasinková V., Malíková M., Šoltés L. High-Molar-Mass Hyaluronan Degradation by Weissberger's System: Pro- and Anti-Oxidative Effects of Some Thiol Compounds. *Polym. Degrad. Stab.* **2009**, *94*, 1867–1875.

14. Duan, J.; Kasper, D. L. Oxidative Depolymerization of Polysaccharides by Reactive Oxygen/Nitrogen Species. *Glycobiology* **2011**, *21*, 401–409.

15. Juránek I., Šoltés L. Reactive Oxygen Species in Joint Physiology: Possible Mechanism of Maintaining Hypoxia to Protect Chondrocytes from Oxygen Excess via Synovial Fluid Hyaluronan Peroxidation. In: *Kinetics, Catalysis and Mechanism of Chemical Reactions: From Pure to Applied Science. Volume 2—Tomorrow and Perspectives;* Islamova, R. M.; Kolesov, S. V.; Zaikov, G. E., Eds.; Nova Science Publishers: New York, 2012; Vol. 2, pp 1–10.

16. Kalpakcioglu, B.; Şenel, K. The Interrelation of Glutathione Reductase, Catalase, Glutathione Peroxidase, Superoxide Dismutase, and Glucose-6-Phosphate in the Pathogenesis of Rheumatoid Arthritis. *Clin. Rheumatol.* **2008**, *27*, 141–145.

17. Kogan, G.; Šoltés, L.; Stern, R.; Gemeiner, P. Hyaluronic Acid: A Natural Biopolymer with a Broad Range of Biomedical and Industrial Applications. *Biotechnol. Lett.* **2007a**, *29*, 17–25.

18. Kogan, G.; Šoltés, L.; Stern, R.; Mendichi, R. Hyaluronic Acid: A Biopolymer with Versatile Physico-Chemical and Biological Properties. In: *Handbook of Polymer Research: Monomers, Oligomers, Polymers and Composites*; Pethrick, R. A.; Ballada, A.; Zaikov, G. E., Eds., Nova Science Publishers: New York, 2007b; pp 393–439.

19. Kogan, G.; Šoltés, L.; Stern, R.; Schiller, J.; Mendichi, R. Hyaluronic Acid: Its Function and Degradation in In Vivo Systems. In: *Studies in Natural Products Chemistry.* Atta-ur-Rahman, Ed., (Bioactive Natural Products, Part D), Elsevier: Amsterdam, 2008; Vol. 34, pp 789–882.

20. Kunsch, C.; Sikorski, J. A.; Sundell, C. L. Oxidative Stress and the Use of Antioxidants for the Treatment of Rheumatoid Arthritis. *Immunol. Endocr. Metab. Agents Med. Chem.* **2005**, *5*, 249–258.

21. Lath, D.; Csomorová, K.; Kolláriková, G.; Stankovská, M.; Šoltés, L. Molar Mass-Intrinsic Viscosity Relationship of High-Molar-Mass Hyaluronans: Involvement of Shear Rate. *Chemical Papers.* **2005**, *59*, 291–293.

22. Løvstad, R. A. Copper Catalyzed Oxidation of Ascorbate (Vitamin C). Inhibitory Effect of Catalase, Superoxide Dismutase, Serum Proteins (Ceruloplasmin, Albumin, Apotransferrin) and Amino Acids. *Int. J. Biochem.* **1987**, *19*, 309–313.

23. Mendichi, R.; Schieroni, A. G.; Grassi, C.; Re, A. Characterization of Ultra-High Molar Mass Hyaluronan: 1. Off-Line Static Methods. *Polymer* **1998**, *39*, 6611–6620.

24. Mendichi, R.; Schieroni, A. G. Fractionation and Characterization of Ultra-High Molar Mass Hyaluronan: 2. On-Line Size Exclusion Chromatography Methods. *Polymer* **2002**, *43*, 6115–6121.

25. Mendichi, R.; Šoltés, L.; Schieroni, A. G. Evaluation of Radius of Gyration and Intrinsic Viscosity Molar Mass Dependence and Stiffness of Hyaluronan. *Biomacromolecules* **2003**, *4*, 1805–1810.

26. Natowicz, M. R.; Short, M. P.; Wang, Y.; Dickersin, G. R.; Gebhardt, M. C.; Rosenthal, D. I.; Sims, K. B.; Rosenberg, A. E. Clinical and Biochemical Manifestations of Hyaluronidase Deficiency. *N. Engl. J. Med.* **1996**, *335*, 1029–1033.

27. Pigeolet, E.; Corbisier, P.; Houbion, A.; Lambert, D.; Michiels, C.; Raes, M.; Zachary, M. D.; Remacle, J. Glutathione Peroxidase, Superoxide Dismutase, and Catalase Inactivation by Peroxides and Oxygen Derived Free Radicals. *Mech. Ageing Dev.* **1990**, *51*, 283–297.

28. Radi, R.; Sims, S.; Cassina, A.; Turrens, J. F. Roles of Catalase and Cytochrome C in Hydroperoxide Lipid Peroxidation and Chemiluminiscence in Rat Heart and Kidney Mitochondria. *Free Radic. Biol. Med.* **1993**, *15*, 653–659.

29. Rapta, P.; Valachová, K.; Zalibera, M.; Šnirc, V.; Šoltés, L. Hyaluronan Degradation by Reactive Oxygen Species: Scavenging Effect of the Hexahydropyridoindole Stobadine and Two of its Derivatives. In: *Monomers, Oligomers, Polymers, Composites, and Nanocomposites* (Polymer Yearbook); Pethrick, R. A.; Petkov, P.; Zlatarov, A.; Zaikov, G. E.; Rakovsky, S. K., Eds.; Nova Science Publishers: New York, 2010: Vol. 23, pp 113–126.

30. Rychlý, J.; Šoltés, L.; Stankovská, M.; Janigová, I.; Csomorová, K.; Sasinková, V.; Kogan, G.; Gemeiner, P. Unexplored Capabilities of Chemiluminescence and Thermoanalytical Methods in Characterization of Intact and Degraded Hyaluronans. *Poly. Degrad. Stab.* **2006**, *91*, 3174–3184.

31. Schiller, J.; Volpi, N.; Hrabárova, E.; Šoltés, L. Hyaluronic Acid: A Natural Biopolymer. In: *Handbook of Biopolymers and Their Applications*; Kalia, S.; Averous, L., Eds.; Wiley & Scrivener Publishing, 2011: pp 3–34.

32. Stankovská, M.; Arnhold, J.; Rychlý, J.; Spalteholz, H.; Gemeiner, P.; Šoltés, L. In Vitro Screening of the Action of Non-Steroidal Anti-Inflammatory Drugs on Hypochlorous Acid-Induced Hyaluronan Degradation. *Poly. Degrad. Stab.* **2007**, *92*, 644–652.

33. Stankovská M., Šoltés L., Vikartovská A., Gemeiner P., Kogan G., Bakoš D. Degradation of High-Molecular-Weight Hyaluronan: A Rotational Viscometry Study. *Biologia.* **2005**, *60*, (Suppl. 17), 149–152.

34. Surovčíková, L.; Valachová, K.; Baňasová, M.; Šnirc, V.; Priesolová, E.; Nagy, M.; Juránek, I.; Šoltés, L. Free-Radical Degradation of High-Molar-Mass Hyaluronan Induced by Ascorbate Plus Cupric Ions: Testing of Stobadine and Its two Derivatives in Function as Antioxidants. *Gen. Physiol. Biophys.* **2012**, *31*, 57–64.

35. Šoltés, L.; Brezová, V.; Stankovská, M.; Kogan, G.; Gemeiner, P. Degradation of High-Molecular-Weight Hyaluronan by Hydrogen Peroxide in the Presence of Cupric Ions. *Carbohydr. Res.* **2006**, *341*, 639–644.

36. Šoltés, L.; Kogan, G. Hyaluronan: An Information Rich Messenger Reporting on the Physiological and Pathological Status of Synovial Joints. *Polym. Res. J.* **2014**, *8*, 49–73.

37. Šoltés, L.; Kogan, G. Impact of Transition Metals in the Free-Radical Degradation of Hyaluronan Biopolymer. In: *Kinetics & Thermodynamics for Chemistry & Biochemistry Volume 2;* Pearce, E. M.; Zaikov, G. E.; Kirshenbaum, G., Eds.; Nova Science Publishers: New York, 2009; Vol. 2, pp 181–199.

38. Šoltés, L.; Kogan, G.; Stankovská, M.; Mendichi, O. R.; Rychlý, J.; Schiller, J.; Gemeiner, P. Degradation of High-Molar-Mass Hyaluronan and Characterization of Fragments. *Biomacromolecules* **2007**, *8*, 2697–2705.
39. Šoltés, L.; Stankovská, M.; Brezová, V.; Schiller, J.; Arnhold, J.; Kogan, G.; Gemeiner, P. Hyaluronan Degradation By Copper(II) Chloride and Ascorbate: Rotational Viscometric, EPR Spin-Trapping, and MALDI-TOF Mass Spectrometric Investigations. *Carbohydr. Res.* **2006b**, *341*, 2826–2834.
40. Šoltés, L.; Valachová, K.; Mendichi, R.; Kogan, G.; Arnhold, J.; Gemeiner, P. Solution Properties of High-Molar-Mass Hyaluronans: The Biopolymer Degradation by Ascorbate. *Carbohydr. Res.* **2007**, *342*, 1071–1077.
41. Topoľská, D.; Valachová, K.; Hrabárová, E.; Rapta, P.; Baňasová, M.; Juránek, I.; Šoltés, L. Determination of Protective Properties of Bardejovske Kupele Spa Curative Waters by Rotational Viscometry and ABTS Assay. *Balneo Res. J.* **2014**, *5*, 3–15.
42. Topoľská, D.; Valachová, K.; Rapta, P.; Šilhár, S.; Panghyová, E.; Horváth, A.; Šoltés, L. Antioxidative Properties of *Sambucus nigra* Extracts. *Chem. Papers.* **2015**, *69*, 1202–1210.
43. Valachová, K.; Kogan, G.; Gemeiner, P.; Šoltés, L. Protective Effects of Manganese(Ii) Chloride on Hyaluronan Degradation by Oxidative System Ascorbate Plus Cupric Chloride. *Interdiscip. Toxicol.* **2010**, *3*, 26–34.
44. Yano, S.; Arroyo, N.; Yano, N. Catalase Binds GRB2 in Tumor Cells When Stimulated with Serum or Ligands for Integrin Receptors. *Free Radic. Biol. Med.* **2004**, *36*, 1542–1554.
45. Valachová, K.; Rapta, P.; Slováková, M.; Priesolova, E.; Nagy, M.; Mislovičová, D.; Dráfi, F.; Bauerová, K.; Šoltés, L. Radical Degradation of High Molar Mass Hyaluronan Induced by Ascorbate Plus Cupric Ions Testing of Arbutin in the Function of Antioxidant. In: *Advances in Kinetics and Mechanism of Chemical Reactions*; Zaikov, G. E.; Valente, A. J. M.; Iordanskii, A. L., Eds.; Apple Academic Press: Waretown, NJ, 2013a; pp 1–19.
46. Valachová, K.; Baňasová, M.; Machová, L.; Juránek, I.; Bezek, S.; Šoltés, L. Antioxidant Activity of Various Hexahydropyridoindoles. *J. Inf. Intelligence Knowl.* **2013b**, *5*, 15–32.
47. Valachová, K.; Baňasová, M.; Machová, Ľ.; Juránek, I.; Bezek, S.; Šoltés, L. Practical Hints on Testing Various Hexahydropyridoindoles to Act as Antioxidants. In: *Chemistry and Physics of Modern Materials*; Aneli, J. N.; Jimenez, A.; Kubica, S., Eds.; Apple Academic Press, Taylor & Francis Group: Toronto, 2013c; pp 137–158.
48. Valachová, K.; Baňasová, M.; Topoľská, D.; Sasinková, V.; Juránek, I.; Collins, M. N.; Šoltés, L. Influence of Tiopronin, Captopril and Levamisole Therapeutics on the Oxidative Degradation of Hyaluronan. *Carbohydr. Polym.* **2015b**, *134*, 516–523.
49. Valachová, K.; Hrabárová, E.; Priesolová, E.; Nagy, M.; Baňasová, M.; Juránek, I.; Šoltés, L. Free-Radical Degradation of High-Molecular-Weight Hyaluronan Induced by Ascorbate Plus Cupric Ions. Testing of Bucillamine and Its Sa981-Metabolite as Antioxidants. *J. Pharm. Biomed. Anal.* **2011**, *56*, 664–670.
50. Valachová, K.; Kogan, G.; Gemeiner, P.; Šoltés, L. Hyaluronan Degradation by Ascorbate: Protective Effects of Manganese(II) chloride. In: *Kinetics & Thermodynamics for Chemistry & Biochemistry*; Pearce, E. M.; Zaikov, G. E.; Kirshenbaum, G., Eds.; Nova Science Publishers: New York, 2009; Vol. 2, pp 201–215.
51. Valachová, K.; Mendichi, R.; Šoltés, L. Effect of L-glutathione on High-Molar-Mass Hyaluronan Degradation By Oxidative System Cu(II) Plus Ascorbate. In: *Monomers,*

Oligomers, Polymers, Composites, and Nanocomposites (Polymer Yearbook); Pethrick, R. A.; Petkov, P.; Zlatarov, A.; Zaikov, G. E.; Rakovsky, S. K., Eds.; Nova Science Publishers: New York, 2010a; Vol. 23, pp 101–111.

52. Valachová, K.; Rapta, P.; Slováková, M.; Priesolová, E.; Nagy, M.; Mislovičová, D.; Dráfi, F.; Bauerová, K.; Šoltés, L. Radical Degradation of High-Molar-Mass Hyaluronan Induced by Ascorbate Plus Cupric Ions. Testing of Arbutin in the Function of Antioxidant. In: *Kinetics, Catalysis and Mechanism of Chemical Reactions: From Pure to Applied Science. Volume 2 - Tomorrow and Perspectives*; Islamova, R. M.; Kolesov, S. V.; Zaikov, G. E., Eds.; Nova Science Publishers: New York, 2012; Vol. 2, pp 11–28.

53. Valachová K., Tamer M. T., Šoltés L. Comparison of Free-Radical Scavenging Properties of Glutathione under Neutral and Acidic Conditions. In.: *Chemistry and Chemical Biology, Methodologies and Applications*; Joswik, R.; Dalinkevich, A. A., Eds.; Apple Academic Press, Taylor & Francis Group: New Jersey, 2015a; pp 227–245.

54. Valachová, K.; Vargová, A.; Rapta, P.; Hrabárová, E.; Dráfi, F.; Bauerová, K.; Juránek, I.; Šoltés, L. Aurothiomalate as Preventive and Chain-Breaking Antioxidant in Radical Degradation of High-Molar-Mass Hyaluronan. *Chem. Biodiver.* **2011**, *8*, 1274–1283.

55. Valachová, K.; Rapta, P.; Kogan, G.; Hrabárová, E.; Gemeiner, P.; Šoltés, L. Degradation of High-Molar-Mass Hyaluronan by Ascorbate Plus Cupric Ions: Effects of D-Penicillamine Addition. *Chem. Biodiver.* **2009**, *6*, 389–395.

56. Valachová, K.; Šoltés, L. Effects of Biogenic Transition Metal Ions Zn(II) and Mn(II) on Hyaluronan Degradation by Action of Ascorbate Plus Cu(II) ions. In: *New Steps in Chemical and Biochemical Physics. Pure and Applied Science*; Pearce, E. M.; Kirshenbaum, G.; Zaikov G. E., Eds.; Nova Science Publishers: New York, 2010b; pp 153–160.

57. Weissberger, A.; LuValle, J. E.; Jr. Thomas, D. S. Oxidation Processes. XVI. The Autoxidation of Ascorbic Acid. *J. Am. Chem. Soc.* **1943**, *65*, 1934–1939.

58. Wyatt, P. J. Light Scattering and the Absolute Characterization of Macromolecules. *Analytica Chimica Acta.* **1993**, *272*, 1–40.

59. Xing, G.; Ren, M.; Verma, A. Divergent Temporal Expression of Hyaluronan Metabolizing Enzymes and Receptors with Craniotomy vs. Controlled-Cortical Impact Injury in Rat Brain: A Pilot Study. *Front. Neurol.* **2014**, *5*, 173.

60. Duarte, T. L.; Jones, G. D. Vitamin C Modulation of H_2O_2-induced Damage and Iron Homeostasis in Human Cells. *Free Radic. Biol. Med.* 2007, *43*(80), 1165–1175.

61. Shukla, R.; Barve, V.; Padhye, S.; Bhonde, R. Reduction of Oxidative Stress Induced Vanadium Toxicity by Complexing with a Flavonoid, Quercetin: A Pragmatic Therapeutic Approach for Diabetes. *Biometals* 2006, *19*(6), 685–693.

62. Al-Assaf, S; Navaratnam, S; Parsons, B. J.; Phillips, G. O. Chain Scission of Hyaluronan by Peroxynitrite. *Arch. Biochem. Biophys.* **2003**, *411*(1), 73–82.

63. Carter, D. E. Oxidation-Reduction Reactions of Metal Ions. *Environ. Health Perspect.* **1995**, *103*(suppl 1), 17–19.

64. Henrotin, Y. E.; Bruckner, P.; Pujol, J. P. The Role of Reactive Oxygen Species in Homeostasis and Degradation of Cartilage. *Osteoarthritis Cartilage* **2003**, *11*(10), 747–755.

65. Afonso, V.; Champy, R.; Mitrovic, D.; Collin, P.; Lomri, A. Reactive Oxygen Species and Superoxide Dismutases: Role in Joint Diseases. *Joint Bone Spine* 2007, *77*(4), 324–329.

66. Whittaker, J. W. The Irony of Manganese Superoxide Dismutase. *Biochem. Soc. Trans.* **2003**, *31*(6), 1318–1321.

67. Fisher, A. E. O.; Naughton, D. P. Vitamin C Contributes to Inflammation via Radical Generating Mechanisms: A Cautionary Note. *Med. Hypotheses* 2003, *61*(5–6), 657–660.

68. Tsuji, G.; Koshiba, M.; Nakamujra, H.; Kosak, H.; Hatachi, S.; Kurimoto, C.; Kurosaka, Y.; Hayashi, Y.; Yodoi, J.; Kamagai, S. Thioredoxin Protects Against Joint Destruction in a Murine Arthritis Model. *Free Radic. Biol. Med.* **2006**, *40*, 1721–1731.

69. Aruoma, O. I. Nutrition and Health Aspects of Free Radicals and Antioxidants. *Food Chem. Toxicol.* **1994**, *32*(7), 671–683.

CHAPTER 12

PROTEIN STRUCTURE PREDICTION

RAJEEV SINGH[1] and ANAMIKA SINGH[2,*]

[1]*Department of Environmental Studies, Satyawati College, University of Delhi, New Delhi, India*

[2]*Department of Botany, Maitreyi College, University of Delhi, New Delhi, India*

Corresponding author. E-mail: arjumika@gmail.com

CONTENTS

ABSTRACT

In this chapter, we show how protein modeling can be used to identify information about the protein of interest and to observe important properties.

12.1 INTRODUCTION

Proteins play a key role in almost all biological processes. Proteins are important for maintaining the structural integrity of cell, transport, storage, regulation, signaling, immunity, and act as a catalyst. Proteins are made up of small peptides and these peptides are actually chains of amino acids. Proteins fold themselves in different conformations and these conformations are responsible for functions of protein. An accurate structural characterization of proteins is provided by X-ray but there are limitations of these methods. In spite of that, these methods are time-taking methods and labor intensive. Nowadays, protein structures are created computationally and deposited in Protein Data Bank (PDB) (www.rcsb.org). On the basis of structures, proteins are divided into four major groups.

12.1.1 PRIMARY STRUCTURE

It is the linear sequence of amino acids, which is chemically a polypeptide chain, joined by peptide bonds.[1,2] Peptide bonds are covalent in nature and many peptide bonds form polypeptide chain. These peptide bonds are formed during translation (protein synthesis). There are two terminals in any polypeptide chain C-terminus (carboxyl group) and the N-terminus (amino group). Counting of residues always starts at the N-terminal end (NH_2 group), which is the end where the amino group is not involved in a peptide bond. The primary structure of a protein is determined by the gene corresponding to the protein. Sequence of nucleotide, that is, DNA forms mRNA which is known as transcription and when mRNAs are synthesizing proteins, it is known as translation and the whole process is known as central dogma. Each protein has its specific amino acid sequence.[3,4] These specific orders of amino acids are unique for every protein and they determine the structure and function of the proteins. Order of amino acids can be determined by Edman degradation method r mass spectrometry. There are approximately 10,000 different proteins in any human being and all are composed from just 20 different types of amino acids residues.[5] Protein formation is followed by post-translational modifications which include disulphide bond formation, addition of glycosyl molecules (glycosylation), and phosphorylation.

12.1.2 SECONDARY STRUCTURE

These structures are formed by formation of hydrogen bonds with the back-bone of proteins. Secondary structures are made up of substructures such as alpha-helix and beta-strands.[6] These secondary structures are having specific patterns of hydrogen bonds, which are formed with main chain peptide bond.[7]

12.1.3 TERTIARY STRUCTURE

It is formed by a large number of noncovalent interactions between peptides. So the tertiary structure of a protein is defined by atomic coordinates having information about three-dimensional structures.[8] Proteins can perform different functions as they are having a specific three-dimensional structure. While such structures are diverse and seemingly complex, they are composed of recurring, easily recognizable tertiary structure motifs that serve as molecular building blocks. Tertiary structures are mainly determined by the primary structure of proteins. Identification of three-dimensional structure of protein is known as a 3D structure prediction.

12.1.4 QUATERNARY STRUCTURE

It is formed by noncovalent interactions within multiple polypeptides. Quaternary structure is the arrangement of multiple folding of proteins or coiling of multi subunit of proteins. For nucleic acids, the term is less common, but can refer to the higher level organization of DNA in chromatin,[9] including its interactions with histones, or to the interactions between separate RNA units in the ribosome or spliceosome.[10,11,12]

12.1.5 STRUCTURE PREDICTION

It is a method by which we can predict the structure of any biomolecule.[13] On the basis of these structures, one can predict the function and its role in biological research.[14] There are different methods, which include chemical probing, hydroxyl radical probing, nucleotide analog interference mapping (NAIM), and in-line probing.[13] Structure of a DNA mole can be determined by nuclear magnetic resonance and X-Ray crystallography.[15] This structure analysis reveals that DNA can be of different forma A, B, Z, etc.[16] Most common type of DNA is B-DNA, mainly found in normal conditions of cells.[17] Most of the structures are not well defined but are based on statistical prediction.[18–21]

12.1.6   PROTEIN STRUCTURE PREDICTION

Here, we are discussing methods of computational structure prediction.

12.2   COMPUTATIONAL 3D STRUCTURE PREDICTION METHODS

In this method, previous protein structures, present in PDB, act as a starting point (template) (Fig. 12.1). Sequence–sequence similarity and identity is one of the key factors in comparative structure modeling. 3D structure prediction is very important as it helps in novel drug designing and new enzyme designing. There are two different methods of protein 3D structure prediction, these are:

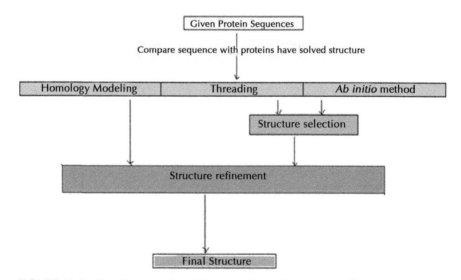

FIGURE 12.1   Flowchart showing different methods of protein modeling.

12.2.1 COMPARATIVE PROTEIN MODELING

1. Homology modeling
2. Protein threading

This method includes modeling of a new protein 3D structure (target) on the basis of previously solved structure (template). This method is based on sequence similarity and identity between template and target sequences. It is an outstanding method as it uses the empirical data available in PDB.

12.2.1.1 HOMOLOGY MODELING

Homology modeling is based on the assumptions that two homologous proteins will share very similar structure, as in the due course of the evolution, structures are more conserved than amino acid sequences. So a good model generation depends upon good alignment between the target and the template. In general, we used to predict a model when sequence identity was more than 30%. Highly homologous sequences will generate a more accurate model (Fig. 12.2). The tools for homology modeling are listed in Table 12.1.

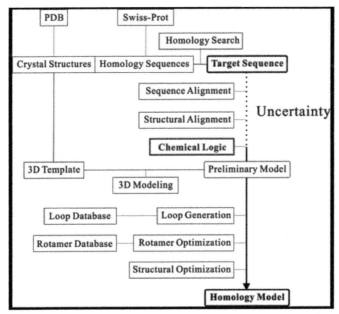

FIGURE 12.2 Flowchart showing steps involved in comparative modeling (Courtesy: www.molfunction.com).

12.2.1.1.1 Steps Involved in Homology Modeling

1. Determination of structurally conserved regions (SCRs) between target (unknown protein) and template protein or reference protein (known protein)
2. Alignment of the amino acid sequence of unknown protein with those of the reference protein within SCRs.
3. Assignment of coordinates in the conserved regions.
4. Prediction of conformations for the rest of the peptide chain, including loops between the SCRs and possibly the N- and C-terminus.
5. Search for the optimum side-chain conformations for residue that differ from those in the reference proteins.
6. Use energy minimization and molecular dynamics to refine the molecular structure so that steric strain introduced during the model-building process can be relieved.

TABLE 12.1 Homology Modeling Tools.

Tool	Prediction Method
3D-JIGSAW Server	Fragment assembly
CPHModel Server	Fragment assembly
EasyModeller (downloadable)	GUI to MODELLER
ESyPred3D Server	Template detection, alignment, 3D modeling
FoldX Server	Energy calculations and protein design
GeneSilico Server	Consensus template search/fragment assembly
LOMETS (server and downloadable)	Local meta threading server
MODELLER (server and downloadable)	Satisfaction of spatial restraints
SWISS-MODEL	Local similarity/fragment assembly
YASARA	Detection of templates, alignment, modeling incl. ligands and oligomers, hybridization of model fragments

12.2.1.2 PROTEIN THREADING

In this method, amino acid sequences (target proteins) were scanned against databases of solved structures and on the basis of matches, particular score will be given. These scores represent the matching between sequence and structure of target and template proteins. This method is also popularly known as fold recognition. This method is used to model the proteins which are having similar fold but they are not homologs. Threading is actually placing and aligning the amino acids of target sequence with that of template sequence. The tools for protein threading are listed in Table 12.2.

TABLE 12.2   Threading Tools.

Tool	Prediction Method
Raptor X software	Remote template detection, single-template and multitemplate threading, totally different from and much better than the old program RAPTOR designed by the same group
3D-PSSM Webserver	3D–1D sequence profiling
HHpred web server	Template detection, alignment, 3D modeling
LOOPP	Multiple methods
MUSTER Webserver	Profile–profile alignment
Phyre and Phyre2 server	Remote template detection, alignment, 3D modeling, multitemplates, ab initio
SUPERFAMILY Web server/standalone	Hidden Markov modeling
SPARKSx/SP Web server series	3D structure modeling by fold recognition according to sequence profiles and structural profiles

12.2.2   AB INITIO/DE NOVO METHOD

Ab initio or de novo protein modeling methods depend upon the physical properties of amino acids rather than previously solved structure. It is based on the fact that protein native structure is having minimum global free energy. Ab initio method or de novo method has two important aspects (Table 12.3):

1. Efficient conformation search
2. Calculation of minimum global free energy.

TABLE 12.3   Prediction Method.

Tool	Prediction Method
QUARK Server	Monte Carlo fragment assembly
I-TASSER Server	Threading fragment structure reassembly
ROBETTA Server	Rosetta homology modeling and ab initio fragment assembly with Ginzu domain prediction
Rosetta@home Downloadable	Distributed computing implementation of Rosetta algorithm
PEP-FOLD Server	De novo approach, based on a HMM

12.2.8 APPLICATION OF PROTEIN MODELING

Protein modeling is used to identify the useful information about the protein of interest. It is used to investigate the structure, dynamics, surface properties and thermodynamics of inorganic, biological, and polymeric systems. Different types of biological activity that have been investigated using molecular modeling include protein folding, enzyme catalysis, protein stability, conformational changes, etc. Few more specific applications are:

1. Identifying active site and binding site
2. Searching, designing, and improving ligand for a given binding site
3. Modeling substrate specificity
4. Prediction of antigenic epitope
5. Protein–protein docking simulation
6. Testing and improving the sequence structure alignment
7. Help in determining the electrostatic potential around the protein,
8. Genome annotation

12.3 CONCLUSION

Although there are only 20 types of amino acids, human body contains thousands of proteins made from these two. Proteins do perform different functions and have particular folding pattern. These patterns are actually a basic functional entity. So proteins with a particular 3D conformation are important to perform any functions. Variety of enzymes, peptides all can be synthesized by computational methods and these methods open the doors of novel research including immunological research.

KEYWORDS

- proteins
- mRNA
- peptide bond
- thermodynamics
- Protein Data Bank

REFERENCES

1. Stoker, H. S. *Organic and Biological Chemistry;* Cengage Learning, 2015; p 371. ISBN 978-1-305-68645-8.
2. Brocchieri, L.; Karlin, S. Protein Length in Eukaryotic and Prokaryotic Proteomes. *Nucleic Acids Res.* **2005,** *33*(10), 3390–3400.
3. Sanger, F.; Tuppy, H. The Amino-acid Sequence in the Phenylalanyl Chain of Insulin. I. The Identification of Lower Peptides from Partial Hydrolysates. *Biochem. J.* **1951,** *49*(4), 463–481.
4. Sanger, F. Chemistry of Insulin. *Science* **1959,** *129*(3359), 1340–1344.
5. How many proteins exist in human body? http://www.innovateus.net/health/how-many-proteins-exist-human-body (accessed 2016).
6. Pauling, L.; Corey, R. B.; Branson, H. R. The Structure of Proteins; Two Hydrogen-Bonded Helical Configurations of the Polypeptide Chain. *Proc. Natl. Acad. Sci. USA* **1951,** *37*(4), 205–211.
7. IUPAC. *Compendium of Chemical Terminology,* 2nd ed. (the "Gold Book", 1997.) Online corrected version: (2006) Tertiary Structure.
8. Sipski, M.; Leonide, W.; Thomas, E. Probing DNA Quaternary Ordering with Circular Dichroism Spectroscopy: Studies of Equine Sperm Chromosomal Fibers. *Biopolymers* **1977,** *16*(3), 573–582.
9. Noller, H. F. Structure of Ribosomal RNA. *Annu. Rev. Biochem.* **1984,** *53,* 119–162.
10. Nissen, P.; Ippolito, J. A; Ban, N.; Moore, P. B.; Steitz, T. A. *RNA Tertiary Interactions in the Large Ribosomal Subunit: The A-minor Motif. Proc. Natl. Acad. Sci.* **2001,** *98*(9), 4899–903.
11. Teunissen, A. W. M. *RNA Structure Probing: Biochemical Structure Analysis of Auto-immune-related RNA Molecules*; 1979; pp 1–27.
12. Franklin, R. E.; Gosling, R. G. The Structure of Sodium Thymonucleate Fibres I. The Influence of Water Content. *Acta Crystallogr.* **1953,** *6*, 673.
13. Franklin, R. E. Gosling, R. G. The Structure of Sodium Thymonucleate Fibers II. The Cylindrically Symmetrical Patterson Function. *Acta Crystallogr.* **1953,** *6*, 678.
14. Franklin, R.; Gosling, R. G. Molecular Configuration in Sodium Thymonucleate. *Nature* **1953,** *171*(4356), 740–741.
15. Wilkins, M. H. F.; Stokes, A. R.; Wilson, H. R. Molecular Structure of Deoxypentose Nucleic Acids. *Nature* **1953,** *171*(4356), 738–740.
16. Leslie, A. G.; Arnott, S.; Chandrasekaran, R.; Ratliff, R. L. Polymorphism of DNA Double Helices. *J. Mol. Biol.* **1980,** *143*(1), 49–72.
17. Baianu, I. C. Structural Order and Partial Disorder in Biological systems. *Bull. Math. Biol.* **1980,** *42*(1), 137–41.
18. Hosemann, R.; Bagchi, R. N. *Direct Analysis of Diffraction by Matter.* North-Holland Publs.: Amsterdam, New York, 1962.
19. Baianu, I.C. X-ray Scattering by Partially Disordered Membrane Systems. *Acta Crystallogr. A* **1978,** *34*, 751–753.
20. Pace, N. R.; Thomas, B. C.; Woese, C. R. *Probing RNA Structure, Function, and History by Comparative Analysis;* Cold Spring Harbor Laboratory Press: Harbor, N.Y.,1999; pp. 113–117.
21. https://www.rcsb.org/pdb/home/home.do

CHAPTER 13

THE CONDENSATION REACTIONS OF 1-CHLORO-2,3,4,6-TETRA-*O*-ACETYL-α-D-GLUCO (GALACTO) PYRANOSE WITH HETEROCYCLIC AMINES

N. SIDAMONIDZE*, R. GAKHOKIDZE, and R. VARDIASHVILI

Department of Chemistry, Ivane Javakhishvili Tbilisi State University, Ilia Chavchavadze Ave., 0128 Tbilisi, Georgia

*Corresponding author. E-mail: sidamonidzeneli@yahoo.com

CONTENTS

ABSTRACT

The condensation reaction of α-chloro-2,3,4,6-tetra-O-acetyl-α-D-gluco (galacto)pyranose with 4,4,8,8-tetramethyl-2,3,6,7-dibenzo-9-oxabicyclo-(3,3,1)-nonan-1-N-(4-methyltiazolylethyl-amino)-5-ol in the presence of the silver carbonate catalyst has been studied for the first time. The experimental result has been obtained corresponding dibenzooxabicyclo-thiazolylamino containing 1,2-*trans*-glucosides. Their structure was determined by physi-cochemical methods of analysis.

13.1 INTRODUCTION

In modern chemical, biological, and pharmacological researches, specific significance is attributed to synthesis of new type of biologically active, highly efficient, and low-toxic substances and their further application. With this in view, the role of natural and synthetic O-glycosides is significant. These compounds are interesting not only with the pure chemical (synthesis, modification) but also with the biological point of view.

Therapeutic effect of glycosides on the organism is mainly conditioned by aglycones (nonsugar part of glycosides). The presence of traces of sugar contributes to the improvement of solubility, decrease of toxicity, conduc-tivity in biological membranes, which contributes to the creation of favor-able conditions for a decrease of active concentration of some hard pharma-cologic preparations and increase of the range of therapeutic effect.[1–3]

In certain previous studies,[4,5] we have proposed a convenient method for synthesizing new types of heterocyclic derivatives of 1,2-*trans*-glycosides.

In the continuation of these works, we synthesized 1-N-(4-methylthia zolylathylamino)-5-O-(2,3,4,6-tetra-O-acetyl-β-D-glucopyranosyl)-4,4,8,8-tetramethyl-2,3,6,7-dibenzo-9-oxabicyclo-(3,3,1)-nona-ne [4] and 1-N-(4-methylthiazolylathylamino)-5-O-(2,3,4,6-tetra-O-acetyl-β-D-galactopyranosyl)-4,4,8,8-tetramethyl-2,3,6,7-dibenzo-9-oxabicyclo-(3,3,1)-nonane [5] by condensation of 1-chloro-2,3,4,6-tetra-O-acetyl-α-D-glucopyranose [1] and α-chloro-2,3,4,6-tetra-O-acetyl-α-D-galactopyranose [2] with 4,4,8,8-tetramethyl-2,3,6,7-dibenzo-9-oxabicyclo-(3,3,1)-nonan-1-N-(β-methyltiazolinethylamino)-5-ol [3] at room temperature in the pres-ence of freshly prepared Ag_2CO_3 catalyst in ether solution, according to Scheme 13.1:

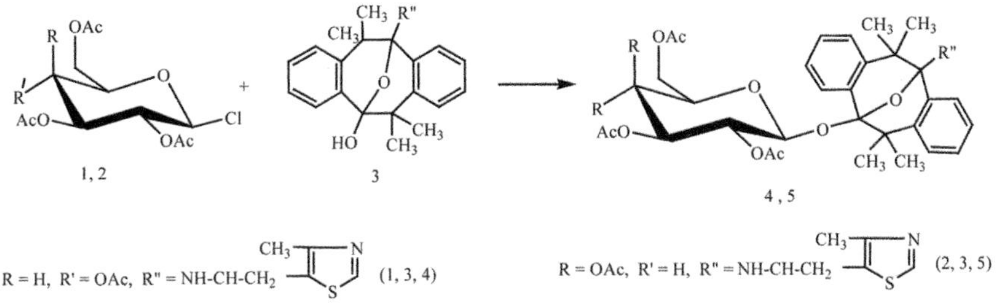

SCHEME 13.1

The course of reactions was monitored by TLC. The reactions took 13–14 h to produce mainly 1,2-*trans*-glycosides [4,5], although small quantities of the 1,2-*cis*-isomers were also observed. The products were yellow crystalline compounds that were soluble in $CHCl_3$ and alcohol (MeOH, EtOH).

The structures of obtained compounds were established by physical–chemical methods of analysis.

Physical–chemical characteristics of synthesized compounds are presented in the Table 13.1.

We performed quantum-chemical computation using CS MOPAC (Chem 3D Ultra-version 8.03) in order to justify theoretically the direction of the condensation of 1-chloro-2,3,4,6-tetra-*O*-acetyl-α-D-gluco(galacto)pyranose [1,2] with the 4-methylthiazolylamino derivative [3]. Each computation by the AM1 method was preceded by optimization of the compounds, that is, minimization of the energy by molecular (MM) and quantum-chemical method.

The model condensation was the reaction of 1-chloro-2,3,4,6-tetra-*O*-acetyl-α-D-glucopyranose [1] with 4,4,8,8-tetramethyl-2,3,6,7-dibenzo-9-oxabicyclo-(3,3,1)-nonan-1-*N*-(β-methyltiazolinethylamino)-5-ol [3]. Two possible reaction pathways were examined that formed the 1,2-*trans*-glycoside (Scheme 13.2, structure I) and 1,2-*cis*-glycoside (Scheme 13.2, structure II).

The calculated heats of formation of the products showed that the probability of generating structure I was greatest, $\Delta H_f = -922.77$ kJ/mol (structure II, $\Delta H_f = -844.88$ kJ/mol). This was confirmed by PMR spectroscopy. PMR spectra of 4 and 5 showed resonances of anomeric proton H-1 bonded to C-1 at δ = 4.45–5.60 and splitting as a result of coupling with H atoms on C-2 into two lines with SSCC $J_{1,2} = 8$ Hz. This value was typical of axial–axial placement of the coupled atoms (1,2-*trans*-glycoside).

TABLE 13.1 Physical–Chemical Characteristics of Synthesized Compounds.

Compound	mp°C	$[\alpha]_D$, (c,CHCl₃)	R_f	Formula	Elemental Analysis Found,% Calculated,%				Yield %
					C	H	N	S	
4	152–153.5	+31.6° (0.52; $t = 18°C$)	0.72*	$C_{40}H_{48}O_{11}N_2S$	62.65	6.15	3.34	3.95	65.4
					62.82	6.28	3.66	4.18	
5	137–139	+39.5° (0.61; $t = 20°C$)	0.52**	$C_{40}H_{48}O_{11}N_2S$	61.95	6.04	3.29	3.53	58.9
					62.82	6.28	3.66	4.18	

*Chloroform–methanol.
**Chloroform–methanol.

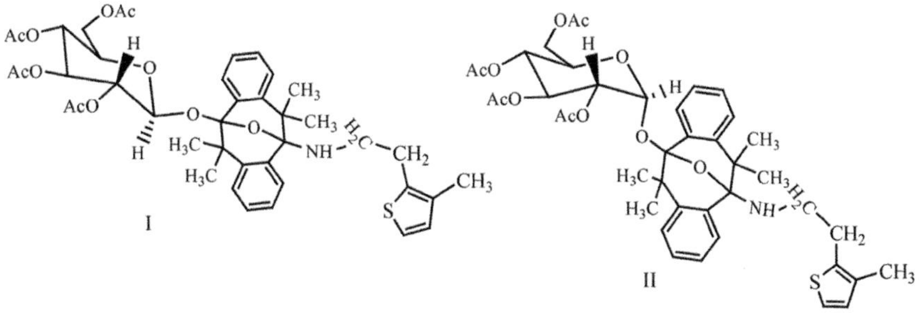

SCHEME 13.2

The condensation reaction of α-chloro-2,3,4,6-tetra-O-acetyl-α-D-gluco(galacto)pyranose with 4,4,8,8-tetramethyl-2,3,6,7-dibenzo-9-oxabicyclo-(3,3,1)-nonan-1-N-(4-methylthiazolylethylamino)-5-ol in the presence of the silver carbonate is nucleophilic substitution reaction that occurs through an SN2 mechanism. The direction of the reaction depends on the relative configuration of C1 and C2 in the initial acylated chloroglucose and chlorogalactose and on the acceptor of the released HCl. Condensation of 1,2-*cis*-acylglycosylhalides with alcohols in the presence of Ag_2CO_3 occurred mainly with C1 configuration inversion, resulting in formation of 1,2-*trans*-glycosides.

13.2 EXPERIMENTAL

The IR spectra were obtained on a UR-20 spectrometer in KBr disks. The ^{1}H NMR spectrum was recorded in $CDCl_3$ on a Bruker WM-250 spectrometer (250 MHz) with TMS internal standard; the ^{13}C NMR spectrum was taken on Bruker AM-300 spectrometer (75 MHz) in $CDCl_3$. The purity of the compounds obtained and the R_f values were determined on Silufol UV-254 using solvent systems CHCl3:CH3OH (19:1 system a; 3:2 system b; 3:1 system c). The optical rotation was measured on an SU-3 general-purpose saccharimeter at $20 \pm 2°C$.

The preparation of 1-chloro-2,3,4,6-tetra-O-acetyl-α-D-gluco(galacto) pyranose [1,2] has been reported.[6]

The preparation of 4,4,8,8-tetramethyl-2,3,6,7-dibenzo-9-oxabicyclo-(3,3,1)-nonan-1-N-(β-methylthiazolinethylamino)-5-ol [3] was carried out by the literature method.[7]

1-N-(4-methylthiazolylathylamino)-5-O-(2,3,4,6-tetra-O-acetyl-β-D-glucopyranosyl)-4,4,8,8-tetramethyl-2,3,6,7-dibenzo-9-oxabicyclo-

(3,3,1)-nonane [4]. A mixture of α-chloro-2,3,4,6-tetra-O-acetyl-α-D-glucopyranose (0.73 g, 0.002 mol) in anhydrous ether (50 mL) and 4,4,8,8-tetramethyl-2,3,6,7-dibenzo-9-oxabicyclo-(3,3,1)-nonan-1-N-(β-methyltiazolinethylamino)-5-ol (1.08 g, 0.0025 mol) was treated with freshly prepared Ag_2CO_3 (0.11 g). The reaction was carried out under N_2 with constant stirring for 13 h (30–35°C). The mixture was filtered and evaporated. The resulting syrup was dissolved in $CHCl_3$, treated with activated carbon, and again evaporated in vacuum. Separation over a column (system c, silica gel L50/100) produced a chromatographically pure product (1.0 g, 65.44%). R_f 0.72 (system a), mp 152–153.5°C, $[\alpha]_D^{18}$ + 31.6° (c 0.52, $CHCl_3$).

IR spectrum (ν, cm^{-1}): 1370–1390 (gem-dimethyl), 1050–1180 (C–O–C), 2910 (NH), 1725 (C=O), 680 (C–S–C), 2977, 1340 (CH_3), 580–620 (C–H$_{ar}$).

PMR spectrum (δ, ppm, J/Hz): 4.45 (1H, d, $J_{1,2}$ = 8.0, H-1); 5.0 (1H, dd, $J_{2,1}$ = 8.0, $J_{2,3}$ = 9.5, H-2); 3.72 (1H, dd, $J_{3,2}$ = 9.5, $J_{3,4}$ = 3.0, H-3); 5.20 (1H, dd, $J_{4,3}$ = 3.0, $J_{4,5}$ = 9.5, H-4); 3.88 (1H, ddd, $J_{5,4}$ = 9.5, $J_{5,6'}$ = 5.0, $J_{5,6''}$ = 2.5, H-5); 4.14 (1H, H-6', dd, $J_{6',6''}$ = 12, $J_{6'5}$ = 2.5, CH_2OCOCH_3); 4.20 (1H, H-6'', dd, $J_{6'6''}$ = 12, $J_{6''5}$ = 5.0, CH_2OCOCH_3); 2.21, 2.10, 1.98, 1.93 (12H, m, 4CO–CH3); 9.98 (1H, s, –CH thiazolyne); 7.15–7.6 (8H, m, aromatic protons); 3.2–2.2 (4H, m, NH–CH_2–CH_2); 3.35 (1H, s, NH); 1.44 and 1.47 (12H, s, gem-dimethl), 4CH_3.

1-N-(4-methylthiazolylathylamino)-5-O-(2,3,4,6-tetra-O-acetyl-β-D-galactopyranosyl)-4,4,8,8-tetramethyl-2,3,6,7-dibenzo-9-oxabicyclo-(3,3,1)-nonane [5] was prepared analogously. A mixture of α-chloro-2,3,4,6-tetra-O-acetyl-α-D-glucopyranose (0.73 g, 0.002 mol) in anhydrous ether (50 mL) and 4,4,8,8-tetramethyl-2,3,6,7-dibenzo-9-oxabicyclo-(3,3,1)-nonan-1-N-(β-methyltiazolinethylamino)-5-ol (1.08 g, 0.0025 mol) was treated with freshly prepared Ag_2CO_3 (0.11 g). The reaction was carried out under N_2 with constant stirring for 14 h (30–35°C). The yield of product 5 (0.9 g, 58.9%). R_f 0.52 (system b), mp 137–139°C, $[\alpha]_D^{20}$ + 39.5° (c) 0.61, $CHCl_3$).

IR spectrum (ν, cm^{-1}): 1370–1380 (gem-dimethyl), 1040–1120 (C–O–C), 2923 (NH), 1715 (C=O), 720 (C–S–C), 2854, 1627 (CH_3), 580–600 (C–H$_{ar}$).

PMR spectrum (δ, ppm, J/Hz): 5.60 (1H, d, $J_{1,2}$ = 8.0, H-1); 5.28 (1H, dd, $J_{2,1}$ = 8.0, $J_{2,3}$ = 9.5, H-2); 4.95 (1H, dd, $J_{3,2}$ = 9.5, $J_{3,4}$ = 3.0, H-3); 5.30 (1H, dd, $J_{4,3}$ = 3.0, $J_{4,5}$ = 9.5, H-4); 3.94 (1H, ddd, $J_{5,4}$ = 9.5, $J_{5,6'}$ = 5.0, $J_{5,6''}$ = 2.5, H-5); 4.20 (1H, H-6', dd, $J_{6',6''}$ = 12, $J_{6'5}$ = 2.5, CH_2OCOCH_3); 4.20 (1H, H-6'', dd, $J_{6'6''}$ = 12, $J_{6''5}$ = 5.0, CH_2OCOCH_3); 2.08, 2.10, 1.98, 1.93 (12H, m, 4CO–CH3); 8.90 (1H, s, –CH thiazolyne); 7.1–7.4 (8H, m, aromatic protons); 3.0–2.4 (4H, m, NH–CH_2–CH_2); 3.30 (1H, s, NH); 1.43 and 1.45 (12H, s, gem-dimethyl), 4CH_3.

KEYWORDS

- 4-methylthiazolylamino derivatives
- silver carbonate
- 1,2-*trans*-glucosides
- biological membranes
- chlorogalactose

REFERENCES

1. Kren, V.; Martínková, L. *Curr. Med. Chem.* **2001,** *11,* 1303–1328.
2. Heinrich, M.; Barnes, J.; Gibbons, S.; Williamson, E. M. *Fundamentals of Pharmacognosy and Phytotherapy,* 1st ed.; Churchill Livingstone: Edinburgh, 2004.
3. Sidamonidze, N.; Vardiashvili, R.; Isakadze, M.; Janiashvili, L.; Lomtatidze, Z. *Chem. Pharm. J.* **2007,** *41*(3), 14–15.
4. Sidamonidze, N. N.; Gakhokidze, R. A.; Ramishvili, M. A.; Samsonia, T. G. *Georg. Chem. J.* **2007,** *7*(1), 44–48.
5. Sidamonidze, N. N.; Vardiashvili, R. O.; Gverdciteli, M. I.; Gakhokidze, R. A. Synthesis of Some Dibenzooxabicycloamino Containing 1,2-trans-glucosides. *Chem. Nat. Comp.* **2009,** *45*(4), 231–238.
6. Freudenberg, K.; Soff, K. *Ind. Engg. Chem. Ber.* **1936,** *69,* 1245–1247.
7. Lagidze, R. M.; Iremadze, N. G.; Kiriakidi, A. S. *Bull. Georg. Acad. Sci.* **2003,** *4,* 3–8.

PART III
Nanoscale Technology

CHAPTER 14

3D SIMULATION OF ELECTROSPUN NANOFIBER JET

SH. MAGHSOODLOU*

Department of Textile Engineering, University of Guilan, Rasht, Iran

Corresponding author. E-mail: sh.maghsoodlou@gmail.com

CONTENTS

ABSTRACT

To optimize electrospinning process, the B-spline collocation methods and a new ordinary differential equation (ODE) solver based on B-spline quasi-interpolation are developed. The problem consists of nonlinear ordinary differential equations. To solve the system of ODE, a 3D simulation obtained by applying quartic B-spline collocation method. To achieve this, the Reneker's model (i.e., bead-spring model) applied and the governing equations were numerically simulated by new ODEs solver without using perturbation equations in x and y directions. Most likely, this technique can represent the results of the random perturbation. The results show that it could be possible to build mathematically a real-time simulation.

14.1 INTRODUCTION

The physical reasoning problem, such as electrospinning phenomena has usually required a representational apparatus that can deal with the vast amount of physical knowledge that is used in reasoning tasks.[1]

Mathematical and theoretical *modeling and simulating* procedure will permit to offer an in-depth prediction of electrospun fiber properties and morphology. Utilizing a model to express the effect of electrospinning parameters, it will assist the researchers in making an easy and systematic way of presenting the influence of variables and by means of that, the process can be controlled.[2]

Similar to most scientific problems, electrospinning is inherently nonlinear so it does not have analytical solution and should be solved using other methods. The B-spline collocation methods and B-spline quasi-interpolation have been applied to solve many problems similar to this. In this chapter, cubic and quartic B-spline collocation method and cubic B-spline quasi-interpolation was used for solving problem.[3,4]

The numerical simulation of the electrospun nanofiber process was observed from momentum equation. The Reneker's model, which has been chosen in this study, focuses on the whipping jet part and assumes the jet as a slender viscous object.

The main novelty in this research assignment will be the application of this mathematical model for electrospinning simulation, which numerically solves the equations by B-spline quasi-interpolation ordinary differential equations (ODEs) solver. Additionally, perturbation function was assumed to the initial equations without using perturbation equations in x and y directions. Also, a minimum applicable time was considered for running simulation program.

14.2 EQUATIONS OF MATHEMATICAL MODEL

For modeling the fiber motion, bead-spring model (Reneker's model) was chosen, which describes the fiber as a chain that consisted of beads connected by springs with a viscoelastic model.[5-7] The momentum equation for the motion of the ith bead can be express as eq 14.1.

$$m\frac{d^2 r_i}{dt^2} = \sum_{\substack{j=1 \\ j\neq i}}^{N} \frac{e^2}{R_{ij}^2}(r_i - r_j) - e\frac{V_0}{h}\hat{k} + \frac{\pi a_{ui}^2(\overline{\sigma}_{ui} + G\ln(l_{ui}))}{l_{ui}}(r_{i+1} - r_i)$$

$$-\frac{\pi a_{bi}^2(\overline{\sigma}_{bi} + G\ln(l_{bi}))}{l_{bi}}(r_i - r_{i-1}) - \frac{\alpha\pi a_a^2 k_i}{\sqrt{(x_i^2 + y_i^2)}}[ix_i + jy_i]$$

(14.1)

The meaning of "sign" in eq 14.1 is as follows eq 14.2.

$$sign(function) = \begin{cases} 1 & x \succ 0 \\ 0 & x = 0 \\ -1 & x \prec 0 \end{cases}$$

(14.2)

Also, in this model, a single perturbation is added by inserting an initial bead of i by eq 14.3.

$$\begin{bmatrix} x_i \\ y_j \end{bmatrix} = 10^{-3}L\begin{bmatrix} \sin(\omega t) \\ \cos(\omega t) \end{bmatrix}$$

(14.3)

In this formula L can be defined as eq 14.4.

$$L = \left(\frac{4e^2}{\pi d_0^2 G}\right)^{\frac{1}{2}}$$

(14.4)

But in this study to reach the random perturbation, these equations cannot be considered for the simulation proposes alone. Nevertheless, in the calculation, the air drag force and gravity force are neglected. Considering the fact that both space and time are perturbations dependent. Therefore, the development of whipping instability can be expected. To achieve a higher accuracy in simulation, a new solver, B-spline quasi-interpolation, by error $O(h^5)$ was developed to replace common solver (Runge–Kutta).

14.3 NEW SOLVER: B-SPLINE QUASI-INTERPOLATION ODES SOLVER

Consider partition $\pi = \{x_0, x_1,\ldots, x_n\}$ of interval $[a,b]$ and suppose that $X_n = (x_j)^{n+d}_{j=-d}$ subject to $x_{-d} = x_{-d+1} =\ldots= x_{-1} = a,\; x_n = x_{n+1} = \ldots = x_{n+d} = b$. According to recurrence relation of B-spline,[8] the jth B-spline of degree d for the knot sequence X_n is denoted by $B_{Xn,j,d}$ or B_j and can be defined as eqs 14.5 and 14.6.

$$B_{X_{n,j,d}}(r) = \omega_{X_{n,j,d}} B_{X_{n,j,d-1}}(r) + \left(1 - \omega_{X_{n,j+1,d}}\right) B_{X_{n,j+1,d-1}}(r) \tag{14.5}$$

$$\omega_{X_{n,j,d}}(r) = \frac{r - x_j}{x_{j+d-1} - x_j}, \quad B_{X_{n,j,d}}(r) = \begin{cases} 1, & \to x_j \le r < x_{j+1} \\ 0, & \to otherwise \end{cases} \tag{14.6}$$

With these notations, the support of $B_{Xn,j,d}$ is $\mathrm{Supp}(B_{Xn,j,d}) = [x_{j-d-1}, x_j]$ and the sets $= \{B_1, B_2,\ldots, B_{n+d}\}$ form a basis over the region $a \le x \le b$. The univariate B-spline quasi-interpolants can be defined as operators of the form in eq 14.7.[9]

$$Q_d f = \sum_{j=1}^{n+d} \mu_j (f) B_j \tag{14.7}$$

For $f \in C^{d+1}(I)$ we have, $\|f - Q_d f\|_\infty = O(h^4)$. Let $f_i = f(x_i), j = 0, 1,\ldots, n$, the coefficient functional for cubic QI ($d = 3$) are given respectively[9,10] in eq 14.8.

$$\mu_1 (f) = f_0$$
$$\mu_2 (f) = \frac{1}{18}(7f_0 + 18f_1 - 9f_2 + 2f_3)$$
$$\mu_j (f) = \frac{1}{6}(-f_{j-3} + 8f_{j-2} - f_{j-1}), \quad 3 \le j \le n+1 \tag{14.8}$$
$$\mu_{n+2} (f) = \frac{1}{18}(2f_{n-3} - 9f_{n-2} + 18f_{n-1} + 7f_n)$$
$$\mu_{n+3} (f) = f_n$$

Now, let f be a given function of x, y where y is a function of x and y' is the first derivative with respect to x. Consider the first order of ordinary differential equation as follows eq 14.9. Where the initial condition y_0 is a given number and the answer of y is unique.

$$y'(x) = f(x, y), \quad y(0) = y_0 \tag{14.9}$$

Using cubic QI as an approximation of $F(x)$: $= f(x, y(x))$ in eq 14.9 and integrating in the interval $[x_i, x_{i+1}]$, $I = 0, 1, \ldots, n-1$, we have eq 14.10.

$$\int_{x_i}^{x_{i+1}} y'(x) \, dx = \int_{x_i}^{x_{i+1}} \sum_{j=1}^{n+3} \mu_j(F) B_j(x) \, dx \tag{14.10}$$

And from B-spline properties we have eq 14.11. Where $y_i = y(x_i)$.

$$y_{i+1} - y_i = \sum_{k=1}^{4} \mu_{i+k}(F) \int_{x_i}^{x_{i+1}} B_{i+k}(x) \, dx \tag{14.11}$$

From eq 14.10 the following ODEs solver will be achieved. Where $F_i = F(x_i)$, $i = 0, 1, \ldots, n-1$. The nonlinear system eq 14.12 is formed of n nonlinear equations in n unknowns $y_1, y_2, \ldots, y_n$, that can be solved using trust-region dogleg method.[11,12]

$$y_1 = y_0 + h\left(\frac{3}{8}F_0 + \frac{19}{24}F_1 - \frac{5}{24}F_2 + \frac{1}{24}F_3\right)$$

$$y_2 = y_1 + h\left(-\frac{7}{144}F_0 + \frac{41}{72}F_1 + \frac{1}{2}F_2 - \frac{1}{72}F_3 - \frac{1}{144}F_4\right)$$

$$y_{i+1} = y_i + h\left(-\frac{1}{144}F_{i-2} - \frac{1}{48}F_{i-1} + \frac{19}{36}F_i + \frac{19}{36}F_{i+1} - \frac{1}{48}F_{i+2} - \frac{1}{144}F_{i+3}\right) \quad i = 2, 3, \ldots, n-3 \tag{14.12}$$

$$y_n = y_{n-1} + h\left(\frac{1}{24}F_{n-3} - \frac{5}{24}F_{n-2} + \frac{19}{24}F_{n-1} + \frac{3}{8}F_n\right)$$

Note that since we have eq 14.13.

$$\left\| \int_{x_i}^{x_{i+1}} \left((F)(x) - Q_3 F(x)\right) dx \right\| \le h \| F - Q_3 F \| \tag{14.13}$$

Then from eq 14.7 the truncation error of present method will be $O(h^5)$.

14.4   SIMULATION ANALYSIS

According to the mathematical model described above, the time evolution of the jet whipping instability was determined by the following procedure: At $t = 0$, the initial whipping jet includes three beads, bead 1, 2, and 3. The distance was set to be a small distance, say, $H/10{,}000$. Other initial conditions, including the stresses $\sigma_{i-1,I}$ and $\sigma_{i,i+1}$ and the initial velocity of bead i, dRi/dt, are set to be zero. For a given time, t, eq 14.14 was solved numerically, using B-spline quasi-interpolation algorithm. All the variables related to bead i, including the stresses $\sigma_{i-1,i}$ and $\sigma_{i,i+1}$, the position r_i, the length of the jet segment $l_{i-1,i}$, were obtained simultaneously. The new values of all the variables at time $t + \Delta t$ were calculated numerically. We denoted the last bead, pulled out of the spinneret by $i = N$. When the distance between this bead and the spinneret became long enough, say $H/5000$, a new bead $i = N + 1$ was inserted at a small distance, $H/10{,}000$, from the previous bead. Now, the positions of all beads can be traced and obtain the path of the jet during the time. As the jet passes through the collector, the calculation stopped. A comparison between the present model and other works is presented in Table 14.1. Due to application of cubic B-spline quasi-interpolation solver for the simulation, the number of beads studied could be more whereas the simulation time decreased significantly.

14.5   DISCUSSION

The jet segment length increases as time develops. It demonstrates that the jet is stretched as it moves downward from the initial position to the collector. The results of the simulation of the electrospinning process and the variation of the jet radius can be seen in Figure 14.1.

Figure 14.1 illustrated the results of the model data output throughout the calculation of the jet path. As described earlier, new beads were inserted into the model when the distance between the last bead and nozzle (located at 20 cm in this case) exceeded the set value. As the beads progress downwards, the perturbation added to the x and y coordinates began to grow as it fully developed into the bending instability, at which point the loops continued to grow outward as the jet moves down. It showed that the longitudinal stress caused by the external electric field acting on the charge carried by the jet stabilized the straight jet for some distance. Then, a lateral perturbation grew in response to the repulsive forces between adjacent elements of the charge carried by the jet. The motion of a segment of the jet grew rapidly into an

TABLE 14.1 A Comparison Between the Present Studies with the Other Works.

Researcher	Equations for investigating model	Method	Time	Year	Reference
Karra	Dimensionless continuity momentum	Lattice Boltzmann method	$N = 100$ $\bar{t} = 0.183$ $\bar{t} = 0.179$	2007	6
Thompson et al.	Governing quasi-one-dimensional continuity momentum and charge conservation	Lagrangian parameter	$t = 0.02$ s	2007	13
Zeng	Dimensionless continuity momentum	Fourth-order Runge–Kutta	$N = 100$ $\bar{t} = 4.471$	2009	5
Lauricella	Three-dimensional continuity momentum	First-order accurate Euler	Final time: 10^{-1} s Time step: 10^{-8} s	2015	14
		Second-order accurate Heun (Runge–Kutta)	Final time: 2 s Time step: 10^{-7} s		
		Fourth-order accurate Runge–Kutta	Final time: 0.5 s Time step: 10^{-8} s		
Present study	Three-dimensional continuity momentum	Cubic B-spline quasi-interpolation	$N = 150$ $T = 0–10^{-9}$ s Time Step: 10^{-14} s	—	—

electrically driven bending instability. In the second section, the radius diagram was obtained from the pathway. It can be seen that by solvent evaporation, the radius decreased.

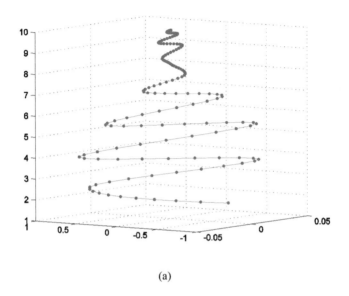

(a)

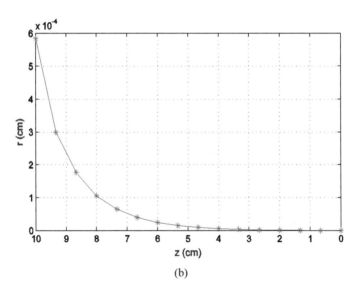

(b)

FIGURE 14.1 (a) The path calculated for $N = 150$ at times ranging from 0 to 10^{-9} s, with step time period 10^{-14} and (b) the radius diagram obtained from pathway.

14.6 CONCLUSIONS

In this article, the Reneker's mathematical model was used to optimize electrospinning process. Meanwhile, the quartic B-spline collocation method is used to find more accurate value of $\sigma = f''(0)$. The cubic and quartic B-spline collocation methods and a new ODEs solver based on B-spline quasi-interpolation are developed and used to solve the problem. All of these methods led to a system of nonlinear equations where the unknowns are obtained by using trust-region dogleg method. This model was employed to describe the dynamic behavior of the electrospun jet in instability part without using leading perturbation equations. The results of the bending instability phenomenon along with simulation of the model can be obtained.

KEYWORDS

- **B-spline approximation method**
- **electrospinning process**
- **Reneker's model**
- **electrospinning simulation improvement**
- **viscoelastic model**

REFERENCES

1. Bhaskar, R.; Nigam, A. Qualitative Physics Using Dimensional Analysis. *Artif. Intell.* **1990,** *45*(1), 73–111.
2. Greenfeld, I.; Arinstein, A.; Fezzaa, K.; Rafailovich, M. H.; Zussman, E. Polymer Dynamics in Semidilute Solution During Electrospinning: A Simple Model and Experimental Observations. *Phys. Rev.* **2011,** *84*(4), 41806–41815.
3. Chawla, T. C.; Leaf, G.; Chen, W. A Collocation Method Using B-splines for One-Dimensional Heat or Mass-Transfer-Controlled Moving Boundary Problems. *Nucl. Eng. Des.* **1975,** *35*(2), 163–180.
4. Saka, B.; Dag, I. A Collocation Method for the Numerical Solution of the RLW Equation Using Cubic B-spline Basis. *Arab. J. Sci. Eng.* **2005,** *30*(1) 39–50.
5. Zeng, Y.; Pei, Z.; Wang, X.; Chen, S. In *Numerical Simulation of Whipping Process in Electrospinning,* WSEAS International Conference Proceedings. Mathematics and Computers in Science and Engineering, Greece, 2009, pp 309–317.
6. Karra, S. *Modeling Electrospinning Process and a Numerical Scheme using Lattice Boltzmann Method to Simulate Viscoelastic Fluid Flows;* A & M University: Texas, 2007.

7. Dasri, T. Mathematical Models of Bead-spring Jets During Electrospinning for Fabrication of Nanofibers. *Walailak J. Sci. Technol.* **2012**, *9*(4), 287–296.

8. Howarth, L. On the Solution of the Laminar Boundary Layer Equations. *Math. Phys. Eng. Sci.* **1938**, *164*(919), 547–579.

9. Aminikhah, H.; Jamalian, A. An Analytical Approximation for Boundary Layer Flow Convection Heat and Mass Transfer Over a Flat Plate. *J. Math. Comput. Sci.* **2012**, *5*(4), 241–257.

10. Duan, Q.; Djidjeli, K.; Price, W. G.; Twizell, E. H. Weighted Rational Cubic Spline Interpolation and Its Application. *J. Comput. Appl. Math.* **2000**, *117*(2), 121–135.

11. Prenter, P. M. *Splines and Variational Methods;* John Wiley & Sons: New York, 2008; pp 95–167.

12. Byrd, R. H.; Schnabel, R. B.; Shultz, G. A. A Trust Region Algorithm for Nonlinearly Constrained Optimization SIAM. *J. Numer. Anal.* **1987**, *24*(5), 1152–1170.

13. Thompson, C. J.; Chase, G. G.; Yarin, A. L.; Reneker, D. H. Effects of Parameters on Nanofiber Diameter Determined from Electrospinning Model. *Polymer* **2007**, *48*(23), 6913–6922.

14. Lauricella, M.; Pontrelli, G.; Coluzza, I.; Pisignano, D.; Succi, S. JETSPIN: A Specific-Purpose Open-Source Software for Simulations of Nanofiber Electrospinning. *Comput. Phys. Commun.* **2015**, *197*, 227–238.

CHAPTER 15

SOME ASPECTS OF CUBIC B-SPLINE APPROXIMATION METHOD FOR 3D SIMULATION OF ELECTROSPUN NANOFIBER JET

SH. MAGHSOODLOU*

Department of Textile Engineering, University of Guilan, Rasht, Iran

Corresponding author. E-mail: sh.maghsoodlou@gmail.com

CONTENTS

ABSTRACT

Development and validation of electrospinning process require long periods of time and high costs to determine the optimal system. In this study, an operational approach to experimental optimization "Taguchi method," where each variable is tested at every level of the other variables, was used. Then, the B-spline collocation methods and a new ODEs solver based on B-spline quasi-interpolation are developed to study the problem of electrospinning process. The problem is a system of nonlinear ordinary differential equations. A 3D simulation was obtained by applying quartic B-spline collocation method and utilized to solve the system of ODE. For this purpose, the Reneker's model (bead–spring model) was selected. The governing equations were numerically simulated by new ODEs solver without using perturbation equations in x and y directions. Finally, a comparison between experimental and simulation data was done by using paired t-test which showed an acceptable accuracy of the simulation.

15.1 INTRODUCTION

Electrospinning can be used as a straightforward method to produce polymer fibers with nanoscale diameters. Although the setup for electrospinning is ordinary and proper, the spinning mechanism itself is quite complicated and the factors that govern fiber formation are not well understood. Therefore, the methods for controlling the shape, structure, and uniformity of electrospun fibers have not yet been fully investigated.[1,2] Different research groups have systematically studied the effects of different electrospinning parameters (solution feeding rate, applied voltage, working distance, and needle size) and polymer solution properties (concentration, viscosity and conductivity, solvent type) on fiber diameter and morphology by doing experiments.[3–7]

The properties of the polymer solution and processing conditions such as polymer concentration, applied voltage, solution feeding rate, and nozzle–collector distance have the most significant variables' influence in the electrospinning process and the resultant nanofiber morphology, when compared with other variables. The force balance between the surface tension, gravitational force, electron repulsion between the charged jets, and electric force to the opposite electrode will determine the jet behavior ejecting from a drop. While the polymer solution concentration has leading effect on obtained electrospun nanofibers, the intercommunication between concentration and flow rate needs to be worked out for a smooth and continuous nanofiber generation.[8–10] Applied electric field is one of the most important parameters in the electrospinning process due to its direct influence on the dynamics of the fluid flow. In the electrospinning process, a high voltage is introduced into a polymer solution such that charges are induced within the fluid. However, the changes in the applied voltage will be reflected on the shape of the suspending droplet at the nozzle of the spinneret, its surface charge, dripping rate, velocity of the flowing fluid, and hence on the structural morphology of electrospun fibers. The applied voltage has been shown to have a less significant effect on the morphology and diameter of electrospun nanofibers than the solution concentration. Depending upon the feeding rate and viscosity of the polymeric solution, a higher applied voltage may be required to keep the stability of jet. As the supplied voltage has an influence on the stretching and acceleration of the jet, it will have an influence on the morphology of the fibers obtained.[11,12]

Because a proper electrospun nanofiber requires a delicate balance of forces for stable nanofiber formation regime, and due to the complex interrelationships among variables affecting the electrospinning process, the design of experiment (DoE) is a common method used for the optimization of variables affecting the nanofiber manufacturing process.[13] DoE is a collection of statistical and mathematical techniques useful for the statistical modeling and analysis of problems in which a response of interest is affected by several variables, with the objective being to optimize this response. Among the DoE methods, a full-factorial design (FFD) is often considered unusable due to its requirement of a large number of experiments.[8] So by carefully choosing the parameters, a fractional factorial design can be useful for preventing the drawbacks of FFD. Taguchi design is one of the most applicable methods in this way due to different levels can be chosen for parameters. Taguchi designs use orthogonal arrays, which estimate the effects of factors on the response mean and variation. An orthogonal array means the design is balanced so that factor levels are weighted equally. Because of this, each factor can be assessed independently of all the other factors, so the effect of one factor does not affect the estimation of a different factor. This can reduce the time and cost associated with the experiment when fractionated designs are used.[14,15]

On the other hand, the physical reasoning problem, such as electrospinning phenomena, has usually required a representational apparatus that can deal with the vast amount of physical knowledge that is used in reasoning tasks.[16] For prediction of electrospun fiber properties and morphology, modeling and simulating are used. Utilizing a model to express the effect of electrospinning parameters will assist researchers in making an easy and systematic way of presenting the influence of variables and by means of that the process can be controlled.[17] Electrospinning similar to most scientific problems is inherently of nonlinearity so it does not have analytical solution and should be solved using other methods. The B-spline collocation methods and B-spline quasi-interpolation have been applied to solve many problems similar to this. In this chapter, cubic and quartic B-spline collocation method and cubic B-spline quasi-interpolation were used for solving the problem.[18,19] The numerical simulation of the electrospun nanofibers process results from momentum equation. The Reneker model, which has been chosen in this study, focuses on the whipping jet part and assumes the jet as a slender viscous object. The main novelty in this research assignment will be to use this mathematical model for electrospinning simulation which numerically solves the equations by B-spline quasi-interpolation ODEs solver. Additionally, perturbation function was assumed to the initial equations without using perturbation equations in x and y directions. Also, a minimum applicable time was considered for running simulation program.

A quick review is summarized in Table 15.1. As it can be seen in the table in the present study, in comparing with other researches, because of using cubic B-spline quasi-interpolation solver for the simulation, the number of considered beads could be increased and the simulation time obviously decreased as much as possible.

15.2   EXPERIMENTAL DETAILS

15.2.1   MATERIALS

Polyacrylonitrile (PAN) powder (Mw = 10^5 g/mol) consisting of 93.7 wt% acrylonitrile (AN) and 6.3 wt% methylacrylate (MA) was supplied with Polyacryl Co. (Isfahan, Iran) and N,N-dimethylformamide (DMF) was obtained from Merck, respectively, as polymer and solvent.

15.2.2   DESIGN OF EXPERIMENTS

DoE method has a particular quality to study the effect of simultaneously changing a wide range of parameters. The DoE method is an important tool for the planning optimization of experimental research. DoE displays an important role in estimating the effect of several variables, and whether these specific variables need to be evaluated separately, simultaneously, or as a combination of the two.[24] In this context, Taguchi experimental design

TABLE 15.1 Comparisons Between the Present Studies with the Other Works.

Researcher	Equations for investigating model	Method	Time		Year	Reference
Karra	Dimensionless continuity momentum	Lattice Boltzmann method	$N = 100$		2007	20
			$\bar{t} = 0.183$			
			$\bar{t} = 0.179$			
Thompson et al.	Governing quasi-one-dimensional continuity momentum and charge conservation	Lagrangian parameter	$T = 0.02$ s		2007	21
Zeng	Dimensionless continuity momentum	Fourth-order Runge–Kutta	$N = 100$		2009	22
			$\bar{t} = 4.471$			
Lauricella	Three-dimensional continuity momentum	First-order accurate Euler	Final time: 10^{-1} s	Time step: 10^{-8} s	2015	23
		Second-order accurate Heun (Runge–Kutta)	Final time: 2 s	Time step: 10^{-7} s		
		Fourth-order accurate Runge–Kutta	Final time: 0.5 s	Time step: 10^{-8} s		
Present study	Three-dimensional continuity momentum	Cubic B-spline quasi-interpolation	$N = 120$		—	—
			$T = 0$–10^{-9} s			
			Time step: 10^{-14} s			

approach is a very simple technique by which one can optimize device parameters by very less number of experiments.[25] This technique is based on "orthogonal array" experiments, which causes fewer "variance" for the experiment with "optimum setting" of controlling parameter. A very useful statistical term "signal-to-noise ratios (S/N)," which is log function of desired output, serves to find out the most influential parameters to obtain the desired output. The optimization procedure includes determining the best controlling factor so as the output gives the targeted value. There are three types of signals-to-noise ratios (S/N ratio) for optimization including: smaller the better, larger the better, and nominal the best. Each of these methods could be used for optimization depending on the optimization necessity. The analysis of variance (ANOVA) on the collected data obtained from Taguchi's DoE method could be useful to optimize the performance characteristic of each factor.[26]

In this study, the DoEs by using Taguchi's method was conducted. Since four parameters consisting of solution concentration, solution feeding rate, applied voltage, and distance between the tip and the collector (spinning distance) are the most significant parameters in electrospinning process; so they were considered as four factors of design. Six levels were considered for solution concentration factor (8, 10, 12, 14, 16, and 18 (w/v %)). For each of the three other factors, three levels were considered which included 10, 14, and 18 (kV) as applied voltage, 0.5, 0.75, and 1 (mL/h) as solution feeding rate, and 10, 13, and 15 (cm) as spinning distance. Finally, by employing MINITAB 14 software, the L_{18} arrays of Taguchi method were used to design electrospinning experiments of preparing PAN nanofibers.

15.2.3 PREPARING PAN SOLUTIONS

PAN/DMF solutions at different concentrations according to Table 15.1 were prepared by dissolving PAN grains in DMF and stirring for 24 h at room temperature in an electromagnetic stirrer, stirring at 1000 rpm.

15.2.4 ELECTROSPINNING PROCESS

In each run detailed in Table 15.2, the prepared solution was placed in a 2-mL syringe. The metallic needle (23G) with an inner diameter of 0.6 mm and a length of 10 mm was used for ejecting the polymer solutions toward the collector. The applied voltage was generated by connecting the needle and the collector plate to the positive and negative terminals of the high voltage supply, respectively. Finally, by considering the required spinning distance, the electrospinning processes were conducted and electrospun nanofibers and composite nanofibers were collected.

15.2.5 CHARACTERIZATION

The morphology and fiber diameter of the electrospun nanofibers were observed by Camscan MV2300 SEM equipment. To take an SEM image of the nanofiber pattern, the samples were coated in gold by E5200 AUTO SPUTTER COATER equipment to make them conductive. ImageJ software version 1.49s was used for measuring the fiber diameter from the SEM micrographs. The mean fiber diameters were calculated from 100 measurements of each sample.

15.3 EQUATIONS OF MATHEMATICAL MODEL

For modeling the fiber motion (Fig. 15.1), bead–spring model (Reneker's model) was chosen, which describes the fiber as a chain that consisted of beads connected by springs with a viscoelastic model.[20,22,27]

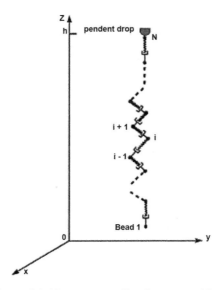

FIGURE 15.1 Fluid jet modeled by a system of beads connected by viscoelastic elements.

The momentum equation for the motion of the ith bead can be expressed as:

$$m\frac{d^2r_i}{dt^2} = \sum_{\substack{j=1\\j\neq i}}^{N}\frac{e^2}{R_{ij}^2}(r_i - r_j) - e\frac{V_0}{h}\hat{k} + \frac{\pi a_{ui}^2(\sigma_{ui} + G\ln(l_{ui}))}{l_{ui}}(r_{i+1} - r_i) - \frac{\pi a_{bi}^2(\sigma_{bi} + G\ln(l_{bi}))}{l_{bi}}$$

$$(r_i - r_{i-1}) - \frac{\alpha\pi a_{av}^2 k_i}{\sqrt{(x_i^2 + y_i^2)}}sign[ix_i + jy_i]$$

(15.1)

The meaning of "sign" in eq 15.1 is as follows:

$$sign\,(function\,) = \begin{cases} 1 & x \succ 0 \\ 0 & x = 0 \\ -1 & x \prec 0 \end{cases}$$

(15.2)

In this model, a single perturbation is added by inserting an initial bead of i by:

$$\begin{bmatrix} x_i \\ y_i \end{bmatrix} = 10^{-3}L\begin{bmatrix} \sin(\omega t) \\ \cos(\omega t) \end{bmatrix}$$

(15.3)

In this formula, L can be defined as:

$$L = \left(\frac{4e^2}{\pi d_0^2 G}\right)^{\frac{1}{2}}$$

(15.4)

The result can be seen in Figure 15.2.

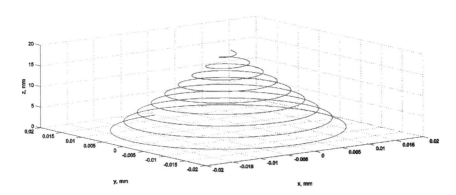

FIGURE 15.2 Shape of the jet with circular perturbations (isometric view).

But in this study for getting to the random perturbation, these equations did not use the simulation process. Additionally, in the calculation, the air drag force and gravity force are neglected as both space and time are dependent perturbations. Therefore, the development of whipping instability occurs. For getting higher accuracy in simulation instead of the common solver (Runge–Kutta), a new solver, B-spline quasi-interpolation, by error $O(h^5)$ was developed.

15.4   NEW SOLVER: B-SPLINE QUASI-INTERPOLATION ODES SOLVER

Consider partition $\pi = \{x_0, x_1, ..., x_n\}$ of interval $[a, b]$ and suppose that $X_n = (x_j)_{j=-d}^{n+d}$ subject to $x_{-d} = x_{-d+1} = ... = x_{-1} = a,$ $x_n = x_{n+1} = ... = x_{n+d} = b.$ According to recurrence relation of B-spline,[28] the jth B-spline of degree d for the knot sequence X_n is denoted by $B_{X_{n,j,d}}$ or B_j and can be defined as follows:

$$B_{X_{n,j,d}}(r) = \omega_{X_{n,j,d}} B_{X_{n,j,d-1}}(r) + (1 - \omega_{X_{n,j+1,d}}) B_{X_{n,j+1,d-1}}(r) \tag{15.5}$$

where

$$\omega_{X_{n,j,d}}(r) = \frac{r - x_j}{x_{j+d-1} - x_j}, \quad B_{X_{n,j,0}}(r) = \begin{cases} 1, & x_j \leq r < x_{j+1} \\ 0, & \rightarrow otherwise \end{cases} \tag{15.6}$$

With these notations, the support of $B_{X_{n,j,d}}$ is supp $(B_{X_{n,j,d}}) = [x_{j-d-1}, x_j]$ and the set $s = \{B_1, B_2, ... B_n+_d\}$ forms a basis over the region $a \leq x \leq b$. The univariate B-spline quasi-interpolants can be defined as operators of the form:[29]

$$Q_d f = \sum_{j=1}^{n+d} \mu_j(f) B_j \tag{15.7}$$

For $f \in C^{d+1}(I)$, we have $\|f - Q_d f\|_\infty = O(h^{d+1})$.

Let $f_j = f(x_j)$, $j = 0,1,...,n$, the coefficient functions for cubic QI $(d = 3)$ are as follows:[29,30]

$$\begin{cases} \mu_1(f) = f_0 \\ \mu_2(f) = \dfrac{1}{18}\left(7f_0 + 18f_1 - 9f_2 + 2f_3\right) \\ \mu_j(f) = \dfrac{1}{6}\left(-f_{j-3} + 8f_{j-2} - f_{j-1}\right), \quad 3 \le j \le n+1 \\ \mu_{n+2}(f) = \dfrac{1}{18}\left(2f_{n-3} - 9f_{n-2} + 18f_{n-1} + 7f_n\right) \\ \mu_{n+3}(f) = f_n \end{cases} \qquad (15.8)$$

Now, let f be a given function of x, y, where y is a function of x and y' is the first derivative with respect to x. Consider the first order of ordinary differential equation as follows:

$$y'(x) = f(x, y), \quad y(0) = y_0$$

where the initial condition y_0 is a given number and the answer of y is unique. Using cubic QI as an approximation of $F(x) := f(x, y(x))$ in eq 15.9 and integrating in the interval $[x_i, x_{i+1}]$, $i = 0,1,\ldots,n-1$, we have

$$\int_{x_i}^{x_{i+1}} y'(x)dx = \int_{x_i}^{x_{i+1}} \sum_{j=1}^{n+3} \mu_j(F)B_j(x)dx \qquad (15.10)$$

And from B-spline properties we have

$$y_{i+1} - y_i = \sum_{k=1}^{4} \mu_{i+k}(F)\int_{x_i}^{x_{i+1}} B_{i+k}(x)dx \qquad (15.11)$$

where $y_i = y(x_i)$. From eq 15.10, the following ODEs solver will be achieved

$$\begin{cases} y_1 = y_0 + h\left(\dfrac{3}{8}F_0 + \dfrac{19}{24}F_1 - \dfrac{5}{24}F_2 + \dfrac{1}{24}F_3\right) \\ y_2 = y_1 + h\left(-\dfrac{7}{144}F_0 + \dfrac{41}{72}F_1 + \dfrac{1}{2}F_2 - \dfrac{1}{72}F_3 - \dfrac{1}{144}F_4\right) \\ y_{i+1} = y_i + h\left(-\dfrac{1}{144}F_{i-2} - \dfrac{1}{48}F_{i-1} + \dfrac{19}{36}F_i + \dfrac{19}{36}F_{i+1} - \dfrac{1}{48}F_{i+2} - \dfrac{1}{144}F_{i+3}\right) \quad i = 2,3,\ldots,n-3 \\ y_{n-1} = y_{n-2} + h\left(-\dfrac{1}{144}F_{n-4} - \dfrac{1}{72}F_{n-3} + \dfrac{1}{2}F_{n-2} + \dfrac{41}{72}F_{n-1} - \dfrac{7}{144}F_n\right) \\ y_n = y_{n-1} + h\left(\dfrac{1}{24}F_{n-3} - \dfrac{5}{24}F_{n-2} + \dfrac{19}{24}F_{n-1} + \dfrac{3}{8}F_n\right) \end{cases} \qquad (15.12)$$

where $F_i = F(x_i)$, $i = 0,1,\ldots,n-1$. The nonlinear system (15.12) is formed of n nonlinear equations in n unknowns $y_1, y_2, \ldots, y_n$ that can be solved using trust-region-dogleg method.[31,32] Note that since we have

$$\left\| \int_{x_i}^{x_{i+1}} \left(F(x) - Q_3 F(x) \right) dx \right\| \leq h \left\| F - Q_3 F \right\| \tag{15.13}$$

then from eq 15.7, the truncation error of present method will be .

15.5 RESULTS AND DISCUSSION

15.5.1 *EXPERIMENTAL INVESTIGATIONS*

The samples of obtaining SEM images of electrospun nanofibers (Figs. 15.2 and 15.3) show that the surface of produced nanofibers is smooth and nanofibers are fairly well aligned.

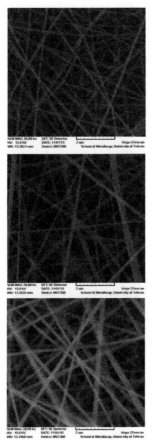

FIGURE 15.3 SEM images of obtained samples of PAN (8% w/v) nanofibers (magnification 20 kX).

The results of mean diameter measurements in each trail are shown in Table 15.2. For investigating the results, Taguchi offers using the analysis of variance, popularly known as ANOVA. In addition to the experimental design, we also used MINITAB software for the purpose of analysis. The obtained results showed that at a confidence level of 99%, only concentration parameter has $F_{0.01}$ more than computational F which means that there are significant changes in this parameter. Therefore, concentration parameter has the most effect on the final diameter of nanofibers. As expected, by increasing the concentration of polymer solution, the mean diameter of nanofibers increases, while the considered levels of three other parameters had no significant effect on diameters. In the following text, the same result is shown by simulation plots studies.

TABLE 15.2 Applied Amount of Factors According to L_{18} Arrays of Taguchi Method for Electrospinning Process with Mean Diameter for Each Case.

Trial number	Solution concentration (% w/v)	Applied voltage (kV)	Feeding rate (mL/h)	Spinning distance (cm)	Mean diameter (nm)
1	8	10	0.5	10	106.88
2	8	14	0.75	13	106.66
3	8	18	1	15	120.07
4	10	10	0.5	13	267.18
5	10	14	0.75	15	255.43
6	10	18	1	10	294.48
7	12	10	0.75	10	440.55
8	12	14	1	13	468.35
9	12	18	0.5	15	431.33
10	14	10	1	15	590.52
11	14	14	0.5	10	572.45
12	14	18	0.75	13	638.75
13	16	10	0.75	15	825.16
14	16	14	1	10	769.66
15	16	18	0.5	13	718.26
16	18	10	1	13	956.58
17	18	14	0.5	15	1152.29
18	18	18	0.75	10	986.89

15.5.2 SIMULATION ANALYSIS

According to the mathematical model described above, the time evolution of the jet whipping instability was determined by the following procedure: At $t = 0$, the initial whipping jet includes three beads, bead 1, 2, and 3. The distance was set to be a small distance, say, $H/10,000$. Other initial conditions, including the stresses $\sigma_{i-1,i}$ and $\sigma_{i,i+1}$ and the initial velocity of bead i, dRi/dt, are set to be zero. For a given time, t, eq 15.1 was solved numerically using B-spline quasi-interpolation algorithm. All the variables related to bead i, including the stresses $\sigma_{i-1,i}$ and $\sigma_{i,i+1}$, the position r_i, the length of the jet segment $l_{i-1,i}$, were obtained simultaneously. The new values of all the variables at time $t + \Delta t$ were calculated numerically. We denoted the last bead, pulled out of the spinneret by $i = N$. When the distance between this bead and the spinneret became long enough, say $H/5000$, a new bead $i = N + 1$ was inserted at a small distance, $H/10,000$, from the previous bead. With this work, the positions of all beads could be followed and the path of the jet during the time could be obtained. As the jet arrived at the collector, the calculation stopped.

The jet segment length increases as time develops. It demonstrates that the jet is stretched as it moves downward from the initial position to the collector. The electrospinning jet flowed continuously from the surface of the drop when the applied electrical force overcame the surface tension. The jet surface is covered by the frozen charges. The concept of "frozen" charges is applied, where the motions of the free charges at jet surface are defined by convection only, and they do not move along or across the jet by themselves. The jet moved straight away from the orifice for some distance and then became unstable and was bent into coiled loops as a result of the interaction of repulsion forces between charges trapped in the jet and stabilizing forces of surface tension, viscoelastic forces, and external electric field. The nanofiber is stretched while its diameter is reduced, and the solvent is evaporated. The determination of the behavior of the jet path in the vicinity of the onset of the primary electrical bending instability is important for the orderly collection of the nanofibers produced by electrospinning.[8,33–36]

Electrospun jet instabilities have been described in literature as "whipping" instabilities. Whipping process should result in oscillatory motion of viscoelastic jet particles which the experimental observations of unstable electrospun jets seem to confirm it. Presented numerical model was used to assess the nature of jet instabilities by evaluating jet particle trajectories. Discrete approach allows tracking the path of individual particles of the jet. For this purpose, each single parameter has been calculated for running the model. The data can be seen in Table 15.3.

TABLE 15.3   Calculation Parameters for Electrospinning of PAN Used in This Model.

No	C	H	V0	Q	N	r0	μ	γ	m	G	e	a0	t	h
–	%	cm	V	mL/h	–	g/cm³	g·m⁻¹·s⁻¹	dyn/cm	g	kg/m.s²	kg·m³/s²	cm	s	s
1	8	10	10,000	0/5	10	1.18	5.60×10^{-7}	59.60	1.6	6.30×10^{3}	1.12×10^{9}	0.035	$0\text{-}10^{-7}$	10^{-12}
2	8	13	14,000	0/75	10	1.18	5.60×10^{-7}	69.12	1.6	6.30×10^{3}	1.12×10^{9}	0.035	$0\text{-}10^{-7}$	10^{-12}
3	8	15	18,000	1	10	1.18	5.60×10^{-7}	85.83	1.6	6.30×10^{3}	1.12×10^{9}	0.035	$0\text{-}10^{-7}$	10^{-12}
4	10	13	10,000	0/5	10	1.18	1.42×10^{-6}	35.27	2	15.97×10^{3}	7.89×10^{8}	0.035	$0\text{-}10^{-7}$	10^{-12}
5	10	15	14,000	0/75	10	1.18	1.42×10^{-6}	51.92	2	15.97×10^{3}	7.89×10^{8}	0.035	$0\text{-}10^{-7}$	10^{-12}
6	10	10	18,000	1	10	1.18	1.42×10^{-6}	193.11	2	15.97×10^{3}	7.89×10^{8}	0.035	$0\text{-}10^{-7}$	10^{-12}
7	12	10	10,000	0/75	10	1.18	3.03×10^{-6}	59.60	2.4	34.08×10^{3}	5.92×10^{8}	0.035	$0\text{-}10^{-7}$	10^{-12}
8	12	13	14,000	1	10	1.18	3.03×10^{-6}	69.12	2.4	34.08×10^{3}	5.92×10^{8}	0.035	$0\text{-}10^{-7}$	10^{-12}
9	12	15	18,000	0/5	10	1.18	3.03×10^{-6}	85.83	2.4	34.08×10^{3}	5.92×10^{8}	0.035	$0\text{-}10^{-7}$	10^{-12}
10	14	15	10,000	1	10	1.18	5.75×10^{-6}	26.49	2.8	64.67×10^{3}	4.64×10^{8}	0.035	$0\text{-}10^{-7}$	10^{-12}
11	14	10	14,000	0/5	10	1.18	5.75×10^{-6}	116.82	2.8	64.67×10^{3}	4.64×10^{8}	0.035	$0\text{-}10^{-7}$	10^{-12}
12	14	13	18,000	0/75	10	1.18	5.75×10^{-6}	114.27	2.8	64.67×10^{3}	4.64×10^{8}	0.035	$0\text{-}10^{-7}$	10^{-12}
13	16	15	10,000	0/75	10	1.18	1.002×10^{-5}	26.49	3.2	116.33×10^{3}	3.82×10^{8}	0.035	$0\text{-}10^{-7}$	10^{-12}
14	16	10	14,000	1	10	1.18	1.002×10^{-5}	116.82	3.2	116.33×10^{3}	3.82×10^{8}	0.035	$0\text{-}10^{-7}$	10^{-12}
15	16	13	18,000	0/5	10	1.18	1.002×10^{-5}	114.27	3.2	116.33×10^{3}	3.82×10^{8}	0.035	$0\text{-}10^{-7}$	10^{-12}
16	18	13	10,000	1	10	1.18	1.636×10^{-5}	35.27	3.6	189.94×10^{3}	3.17×10^{8}	0.035	$0\text{-}10^{-7}$	10^{-12}
17	18	15	14,000	0/5	10	1.18	1.636×10^{-5}	51.92	3.6	189.94×10^{3}	3.17×10^{8}	0.035	$0\text{-}10^{-7}$	10^{-12}
18	18	10	18,000	0/75	10	1.18	1.636×10^{-5}	193.11	3.6	189.94×10^{3}	3.17×10^{8}	0.035	$0\text{-}10^{-7}$	10^{-12}

Figure 15.4 illustrates the results of the model data output throughout the calculation of the jet path. As it can be seen, the development of jet path is followed by formation of elliptic and circular loops as the jet expands. As described earlier, new beads were inserted into the calculation when the distance between the last bead and nozzle exceeded the set value. The fluid first begins to drip out of the nozzle. The dripping frequency increases during time and the stable jet transits to unstable form. The jet is deemed stable when no surface perturbations in the conical base region exist.[37] As the beads progress downwards, the perturbation added to the x and y coordinates began to grow as it fully developed into the bending instability, at which point the loops continued to grow outward as the jet moves down. It showed that the longitudinal stress caused by the external electric field acting on the charge carried by the jet stabilized the straight jet for some distance. Then a lateral perturbation grew in response to the repulsive forces between adjacent elements of the charge carried by the jet. The motion of a segment of the jet grew rapidly into an electrically driven bending instability.

Figure 15.4 also demonstrates that random perturbations may cause the change of direction of the jet issuing from the nozzle during the loop development. It can be claimed that this approach for implementation of perturbations is more natural than periodic because there is no physical mechanism of circular instabilities in electrospinning process, and the perturbations are stochastic.

The fluid jet in the straight segment of the path, and the more solid nanofibers in the coils of the primary electrical bending instability are collected on surface. The diameter and characteristic path of the jet depended on various parameters. A variety of structures of loops, both conglutinated and not, associated with the instabilities were created.[33] In Figure 15.5, the area of each collection can be observed.

The interaction between the electron repulsion between the charged jets and the electric force opposite to the electrode may determine the area covered by the electrospun fibers and its grayscale light intensity.[10] Electrospun nanofibers are most commonly collected as randomly oriented or parallel-aligned mats. Randomly oriented fiber mats result when a simple static collecting surface is used, and parallel-aligned mats have been collected by several methods.[38]

With periodical perturbations, the jet deposits on the substrate as thin circular stripe at a fixed distance from axis of the envelope cone which disagrees with the experimental observations, where deposited fibers form solid round or elliptic area on the collecting surface with decrease of fiber distribution density from the center to the edge of the area. On the other

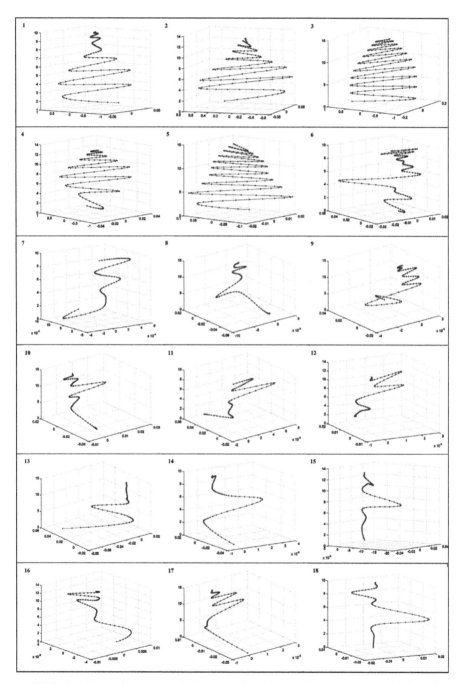

FIGURE 15.4 Development of the jet path of the electrospinning process modeling with random perturbations for each case (isometric view).

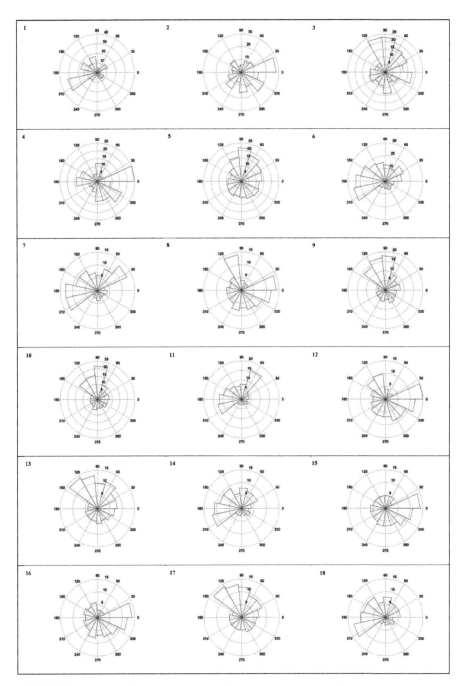

FIGURE 15.5   Angle histogram for mass distribution of deposited fibers.

hand, the randomly perturbed jet simulation results in nonlinear distribution of deposited fibers with bulk of fibers collected near the axis of envelope cone (Fig. 15.5). Due to the randomness, the fiber distribution is roughly equal in all directions on collecting surface, and the randomly perturbed jet forms circle deposition area.

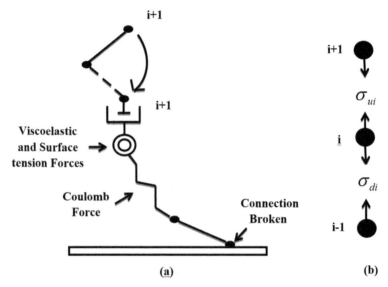

FIGURE 15.6   (a) Broken connection between bead $i - 1$ and bead i, axis units are in cm and bead $i + 1$ passes bead i on its way downward and (b) expanded view of stress for beads.

As it is presented in Figure 15.6, the schematic of forces in this model terminates to stretch the jet. Now, for better understanding of the force acting on beads, the stress between bead i and bead $i - 1$ (below stress) and the stress between bead i and bead $i + 1$ (upper stress) is illustrated in Figure 15.7.

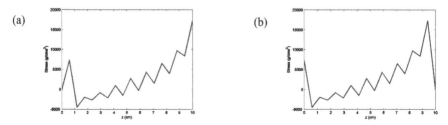

FIGURE 15.7   (a) Below stress in z direction from syringe to collector and (b) upper stress in z direction from syringe to collector.

The jets are stabilized by the buildup of viscoelastic stresses due to the stretching of the polymer during spinning.[39] As it can be seen in Figure 15.7, the below and upper stress acts against each other so that the jet stretches and when the bead reaches the collector, the stress which has been increased in the path, decreases sharply.

Now, the effect on the jet diameter is shown in the following plots.

The results of Figure 15.8 are summarized as follows:

1. As the jet sustains huge stretching and elongation due to the electric forces, the strenuously bending loops on the jet surface area increase significantly and accelerate solvent evaporation. The solvent concentration rapidly decreases in the bending loops, only a few centimeters below the beginning of the bending instability. Once the polymer concentration reaches about 90%, the jet continues to elongate, but at a much lower rate, as seen in the radius of the jet loops. The slower elongation rate is due to the increase in viscosity and elastic modulus of the solution at higher polymer concentrations. The form of the plots is similar to the other studies.[21,37,39,40]

2. By investigation on diameter plots, it can be obviously concluded that by increasing the concentration, the diameters grow. But the other parameters have no significant effect on the diameter in each concentration. These results were also seen in experiments and the other research studies.[3,4,8,9,11,21]

3. In lower concentrations such as 8 (w/v) and 10 (w/v), by increasing flow rate, surface tension, viscosity, modulus, as it expected the diameter incremented.[12]

4. When the concentration is 12 (w/v), it can be concluded that for initial diameter, the result is against the final diameter. In initial diameter, voltage has more influence than feeding rate because lower voltage with mean feeding rate lead to the maximum diameter. But for final diameter, the feeding rate plays a critical role. As it is mentioned before, applied electric field has a direct influence on the dynamics of the fluid flow and the initial decrease in nanofiber diameter is attributed to a higher degree of jet stretching in correlation to increased charge repulsion within the jet and a strong external electric field as a consequence of an increase in the applied voltage.[11,12] The point is that the applied voltage may affect the fiber diameter in two different ways during the same electrospinning process. On the one hand, a higher applied voltage pushes more polymer solution from the needle and causes the formation of a larger diameter initial

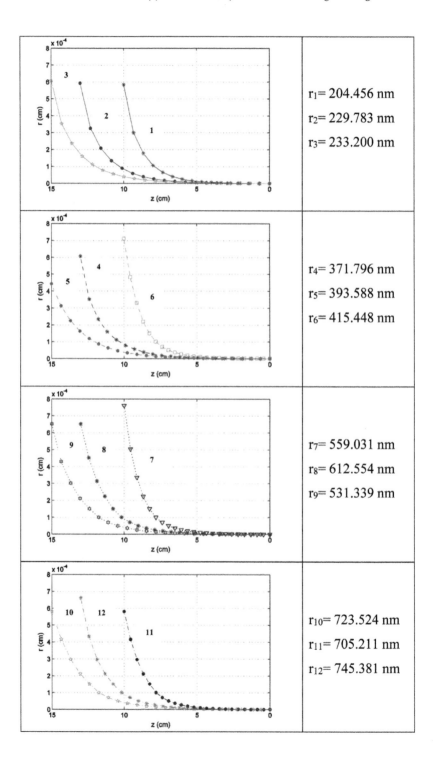

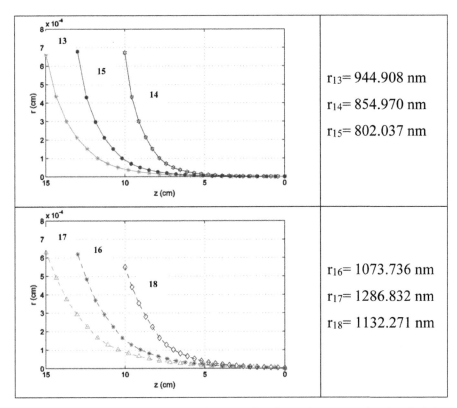

FIGURE 15.8 Comparison between diameter plots for each concentration by obtaining final diameters.

jet, which in turn facilitates the formation of larger diameter fibers. On the other hand, a higher applied voltage renders a higher density of surface charges on the initial jet and hence promotes the jet to split into smaller jets, resulting in small diameter fibers. Both effects can occur at the same time. If the first effect dominates, large diameter fibers are obtained, but if the second effect dominates, small diameter fibers are formed.[3] The second one can be obviously seen in the ninth condition (the biggest voltage and the lowest feeding rate) in which the most voltage results to the less final diameter. The most flow rate terminates to the utmost diameter. The flow rate of the polymer solution through a capillary influences the nanofiber diameter, porosity, and geometry of the electrospun nanofibers.[12] When the feeding rate increases, more solution will be ejected from the needle tip during the same time.[9,41]

5. Concentrations 14 (w/v) and 16 (w/v) show the effect of modulus, viscosity, and charge on diameter. In this concentration, the highest viscosity with the lower charge beside the mid amount of feeding rate, voltage, and surface tension lead to the maximum initial and final diameters. And the lowest flow rate results in the lowest diameter. In concentration 16 (w/v), the least flow rate and charge alongside the maximum voltage, viscosity, and modulus terminate to the highest initial diameter and the lowest final diameter. And the fewest voltage, viscosity, and modulus besides the maximum of distance spinning and charge lead to the minimum initial diameter and the maximum final diameter. As discussed previously, increasing the applied voltage gives rise to higher electrostatic repulsion forces between the needle tip and the collector, which in turn provides higher drawing stress in the jet. Theoretical analysis reveals that higher voltage favors the whipping instability and suppresses the axisymmetric instabilities, which was thought to be the mechanism for this phenomenon.[8,9,41]

6. In the maximum concentration, there is no specific proceeding.

15.5.3 SIMULATION VERIFICATION

A paired sample t-test is used to determine whether there is a significant difference between the average values of the same measurement made under two different conditions. Both measurements are made on each unit in a sample, and the test is based on the paired differences between these two values. The usual null hypothesis is that the difference in the mean values is zero. The null hypothesis for the paired sample t-test is

$$H0:\ d = \mu 1 - \mu 2 = 0 \qquad\qquad (15.14)$$

where d is the mean value of the difference.

This null hypothesis is tested against one of the following alternative hypotheses, depending on the question posed:

$$\begin{cases} H1:\ d = 0 \\ H1:\ d > 0 \\ H1:\ d < 0 \end{cases} \qquad\qquad (15.15)$$

The paired sample t-test is a more powerful alternative to a two-sample procedure, such as the two-sample t-test, but can only be used when we have matched samples.[42,43]

The objective was to see if diameters obtained from model simulation have any difference by experimental data at a 95% level of confidence. For this purpose, the paired t-test was utilized. The results are shown in Table 15.4.

Applying the paired t-test formula gives: $t = 1.23$ and sig (two-tailed) = 0.235, for this evidence, it can be concluded that there is no difference between the two samples at the 95% level.

TABLE 15.4   Paired Samples Test for Diameters Data.

Cases	Mean of diameter	Mean difference	T-value	P-value
Model data	1156.95	617.976	1.23	0.235
Experimental data	538.97			

15.6 CONCLUSION

For performing the electrospinning process analysis, using simple models and comparing with experimental results would be useful. In this study, a Taguchi design was used for experimental optimization and then the Reneker's mathematical model was reviewed in developing the electrospinning process. The quartic B-spline collocation method was used to find more accurate value of $\sigma = f''(0)$. The cubic and quartic B-spline collocation methods and a new ODEs solver based on B-spline quasi-interpolation are developed and used to solve the problem. All of these methods leaded to a system of nonlinear equations that the unknowns are obtained by using trust-region-dogleg method. This model was used to describe the dynamic behavior of the electrospun jet in instability part without using leading perturbation equations. The results of the bending instability phenomenon with simulated model were presented and the nanofiber diameters were calculated. The importance of the perturbation was obvious in the results of simulations which terminated to different collection area for nanofibers. Applying paired t-test showed that the accuracy of simulation is good enough because there was no difference between obtained diameters from experiments and simulation results in 95% confidence.

KEYWORDS

- B-spline approximation method
- Taguchi method
- Reneker's model
- electrospinning simulation improvement
- optimization

REFERENCES

1. Kong, C. S., et al. Electrospinning Mechanism for Producing Nanoscale Polymer Fibers. *J. Macromol. Sci. Part B: Phys.* **2010,** *49*(1), 122–131.
2. Kong, C. S., et al. Electrospinning Instabilities in the Drop Formation and Multi-jet Ejection Part I: Various Concentrations of PVA (Polyvinyl Alcohol) Polymer Solution. *J. Macromol. Sci. Part B* **2011,** *50*(3), 517–527.
3. Tong, H. W.; Wang, M. Electrospinning of Poly(hydroxybutyrate-co-hydroxyvalerate) Fibrous Scaffolds for Tissue Engineering Applications: Effects of Electrospinning Parameters and Solution Properties. *J. Macromol. Sci. Part B* **2011,** *50*(8), 1535–1558.
4. Forouharshad, M., et al. Manufacture and Characterization of Poly(butylene Tere-phthalate) Nanofibers by Electrospinning. *J. Macromol. Sci. Part B* **2010,** *49*(4), 833–842.
5. Branciforti, M. C., et al. Characterization of Nano-structured Poly(D, L-lactic Acid) Nonwoven Mats Obtained from Different Solutions by Electrospinning. *J. Macromol. Sci. Part B* **2009,** *48*(6),1222–1240.
6. Moghadam, H. B., et al. Computational-based Approach for Predicting Porosity of Elec-trospun Nanofiber Mats Using Response Surface Methodology and Artificial Neural Network Methods. *J. Macromol. Sci. Part B* **2015,** *54*(11), 1404–1425.
7. Nasouri, K., et al. Modeling and Optimization of Electrospun PAN Nanofiber Diameter Using Response Surface Methodology and Artificial Neural Networks. *J. Appl. Polym. Sci.* **2012,** *126*(1), 127–135.
8. Nasouri, K.; Shoushtari, A. M.; Mojtahedi, M. R. M. Evaluation of Effective Electros-pinning Parameters Controlling Polyvinylpyrrolidone Nanofibers Surface Morphology via Response Surface Methodology. *Fibers Polym.* **2015,** *16*(9), 1941–1954.
9. Aljehani, A. K., et al. Effect of Electrospinning Parameters on Nanofiber Diameter Made of Poly(vinyl Alcohol) as Determined by Atomic Force Microscopy. In *Biomed-ical Engineering (MECBME)*; IEEE: Middle East, 2014.
10. Kong, C. S., et al. Interference Between the Charged Jets in Electrospinning of Poly-vinyl Alcohol. *J. Macromol. Sci. Part B* **2009,** *48*(1),77–91.
11. Jacobs, V.; Anandjiwala, R. D.; Maaza, M. The Influence of Electrospinning Parameters on the Structural Morphology and Diameter of Electrospun Nanofibers. *J. Appl. Polym. Sci.* **2010,** *115*(5), 3130–3136.

12. Pillay, V., et al. A Review of the Effect of Processing Variables on the Fabrication of Electrospun Nanofibers for Drug Delivery Applications. *J. Nanomater.* **2013,** 22.

13. Ziabari, M.; Mottaghitalab, V.; Haghi, A. K. A New Approach for Optimization of Electrospun Nanofiber Formation Process. *Korean J. Chem. Eng.* **2010,** *27*(1), 340–354.

14. Packianather, M. S.; Drake, P. R.; Rowlands, H. Optimizing the Parameters of Multilayered Feedforward Neural Networks Through Taguchi Design of Experiments. *Quality Reliab. Eng. Int.* **2000,** *16*(6), 461–473.

15. Ballantyne, K. N.; Van, O. R. A.; Mitchell, R. J. Reduce Optimisation Time and Effort: Taguchi Experimental Design Methods. *Forensic Sci. Int. Genet.* (Supplement Series) **2008,** *1*(1), 7–8.

16. Bhaskar, R.; Nigam, A. Qualitative Physics Using Dimensional Analysis. *Artif. Intell.* **1990,** *45*(1), 73–111.

17. Greenfeld, I., et al. Polymer Dynamics in Semidilute Solution during Electrospinning: A Simple Model and Experimental Observations. *Phys. Rev. E* **2011,** *84*(4), 041806.

18. Chawla, T. C.; Leaf, G.; Chen, W. A Collocation Method Using B-splines for One-dimensional Heat or Mass-transfer-controlled Moving Boundary Problems. *Nucl. Eng. Des.* **1975,** *35*(2), 163–180.

19. Saka, B.; Dag, I. A Collocation Method for the Numerical Solution of the RLW Equation Using Cubic B-spline Basis. *Arab. J. Sci. Eng.* **2005,** *30*(1), 39–50.

20. Karra, S. *Modeling Electrospinning Process and a Numerical Scheme Using Lattice Boltzmann Method to Simulate Viscoelastic Fluid Flows*; Texas A & M University: USA, 2007.

21. Thompson, C. J., et al. Effects of Parameters on Nanofiber Diameter Determined from Electrospinning Model. *Polymer* **2007,** *48*(23), 6913–6922.

22. Zeng, Y., et al. In *Numerical Simulation of Whipping Process in Electrospinning*, WSEAS International Conference. Proceedings. Mathematics and Computers in Science and Engineering, 2009; World Scientific and Engineering Academy and Society.

23. Lauricella, M., et al. *JETSPIN:* A Specific-purpose Open-source Software for Simulations of Nanofiber Electrospinning. *Comput. Phys. Commun.* **2015,** *197*, 227–238.

24. Khanlou, H. M., et al. Prediction and Optimization of Electrospinning Parameters for Polymethyl Methacrylate Nanofiber Fabrication Using Response Surface Methodology and Artificial Neural Networks. *Neural Comput. Appl.* **2014,** *25*(3–4), 767–777.

25. Yazdanpanah, M., et al. Optimization of Electrospinning Process of Poly(vinyl Alcohol) via Response Surface Methodology (RSM) Based on the Central Composite Design. *Curr. Chem. Lett.* **2014,** *3*(3),175–182.

26. Amini, N., et al. Morphological Optimization of Electrospun Polyacrylamide/MWCNTs Nanocomposite Nanofibers Using Taguchi's Experimental Design. *Int. J. Adv. Manufact. Technol.* **2013,** *69*(1–4), 139–146.

27. Dasri, T. Mathematical Models of Bead–Spring Jets During Electrospinning for Fabrication of Nanofibers. *Walailak J. Sci. Technol.* **2012,** *9*(4), 287–296.

28. Howarth, L. In *On the Solution of the Laminar Boundary Layer Equations*, Proceedings of the Royal Society of London A: Mathematical, Physical and Engineering Sciences, 1938. The Royal Society.

29. Aminikhah, H.; Jamalian, A. An Analytical Approximation for Boundary Layer Flow Convection Heat and Mass Transfer Over a Flat Plate. *J. Math. Comput. Sci.* **2012,** *5*(4), 241–257.

30. Duan, Q., et al. Weighted Rational Cubic Spline Interpolation and its Application. *J. Comput. Appl. Math.* **2000,** *117*(2), 121–135.

31. Prenter, P. M. *Splines and Variational Methods*; Courier Corporation: USA, 2008; p 325.
32. Byrd, R. H.; Schnabel, R. B.; Shultz, G. A. A Trust Region Algorithm for Nonlinearly Constrained Optimization. SIAM *J. Numer. Anal.* **1987,** *24*(5),1152–1170.
33. Reneker, D. H.; Han, T. In *Electrical Bending and Mechanical Buckling Instabilities in Electrospinning Jets*, MRS Proceedings, 2006, Cambridge University Press.
34. Reneker, D. H., et al. Bending Instability of Electrically Charged Liquid Jets of Polymer Solutions in Electrospinning. *J. Appl. Phys.* **2000,** *87*, 4531–4547.
35. Bhattacharyya, P., et al. *An Examination of the Whipping Instability of Viscoelastic Jets in Electrospinning*, APS Meeting Abstracts, 2007.
36. Reneker, D. H., et al. Electrospinning of Nanofibers from Polymer Solutions and Melts. *Adv. Appl. Mech.* **2007,** *41*, 343–346.
37. Shin, Y. M., et al. Experimental Characterization of Electrospinning: The Electrically Forced Jet and Instabilities. *Polymer* **2001,** *42*(25), 09955–09967.
38. Beachley, V.; Wen, X. Effect of Electrospinning Parameters on the Nanofiber Diameter and Length. *Mater. Sci. Eng. C* **2009,** *29*(3),663–668.
39. Carroll, C. P., et al. Nanofibers from Electrically Driven Viscoelastic Jets: Modeling and Experiments. *Korea Aust. Rheol. J.* **2008,** *20*, 153–164.
40. Yarin, A. L.; Koombhongse, S.; Reneker, D. H. Taylor Cone and Jetting from Liquid Droplets in Electrospinning of Nanofibers. *J. Appl. Phys.* **2001,** *90*(9), 4836–4846.
41. Zuo, W., et al. Experimental Study on Relationship Between Jet Instability and Formation of Beaded Fibers During Electrospinning. *Polym. Eng. Sci.* **2005,** *45*(5), 704–709.
42. Harvey, I. D.; Leybourne, S. J.; Taylor, A. M. R. A Simple, Robust and Powerful Test of the Trend Hypothesis. *J. Econ.* **2007,** *141*, 1302–1330.
43. Demšar, J. Statistical Comparisons of Classifiers over Multiple Data Sets. *J. Mach. Learn. Res.* **2006,** *7*, 1–30.

CHAPTER 16

SYNTHESIS OF 1,4-CIS POLYBUTADIENE BY THE HETEROGENIZED DITHIOSYSTEM ON BASE OF NANOSIZE MONTMORILLONITE

FIZULI A. NASIROV, NAZIL F. JANIBAYOV*, SEVDA R. RAFIYEVA, and GULARA N. HASANOVA

Institute of Petrochemical Processes of Azerbaijan, National Academy of Sciences, Khojaly Ave., 30, AZ 1025 Baku, Azerbaijan

*Corresponding author. E-mail: j.nazil@yahoo.com

CONTENTS

ABSTRACT

The problems of phosphorosulfurization of montmorillonite, preparation of metal complex dithiophosphates and results of their investigation as the heterogenized catalytic systems in combination with aluminum organic compounds in the polymerization process of butadiene are considered. It has been established that Ni-,Co-, and Nd-containing dithiosystem are the high active and high selective catalysts in this process.

16.1 INTRODUCTION

The organic metal dithiophosphates, in particular nickel, cobalt, lanthanides in combination with aluminum organic compounds recently are widely investigated as the homogeneous catalytic systems in the processes of oligomerization, cooligomerization, and polymerization of olefin and diene hydrocarbons.[1–4]

Metal dithioderivatives also show a high efficiency as the chemical additive such as polyfunctional stabilizers of polymers, additives for motor oil.[5–8]

The processes of oligomerization, polymerization of olefins and dienes processed with application of the homogeneous catalytic systems of Ziegler–Natta type. The homogeneous catalysis has a number of advantages concluding in high selectivity, large reaction proceeding rate, etc. But simultaneously it has a number of essential defects: single use of the catalysts, application of a large quantity of solvent, necessity of washing of purposeful products by water from catalysts owing to what a large number of quantities of waste waters containing ions of heavy metals, deteriorating ecological situation of production in formed. In addition, it is extremely difficult to create a continuous technology or production, to carry out a process in the gas phase, etc.

The most perspective direction is the creation of heterogenized catalytic systems processing by advantages of both homogeneous and heterogeneous catalysis.

Only in last year, there appeared reports about the use of modified zeolites in the process of oligomerization and polymerization of unsaturated hydrocarbons.[9–12]

We have previously synthesized dithiophosphorylated (DTP) zeolites (HY, HLaY, CaA, and mordenite etc.), which in combination with the aluminum organic compounds showed a high activity and selectivity as the catalyst systems in the polymerization process of butadiene.[10–12]

With the aim of increasing assortment of bearers for preparation of heterogenized catalytic systems, in this work, it has been chosen the widely spread montmorillonite (MMT) with its preliminary treatment.

As it has been established earlier, phosphorosulfurization of zeolites proceeds with participation of hydroxyl groups available in zeolite with P_2S_5, as a result of which O, O-di-substituted dithiophosphorus acids are prepared, and further on this basis, metal-complex compounds are prepared. The assumed route of reactions in simplified version can be presented as:

where, M—Ni, Co, and Nd.

In the case of 3-valent metal, a structure has more complex character.

16.2 EXPERIMENTAL

The polymer grade (PG) MMT was obtained from Nanocor (USA). PG MMTs are high purity aluminosilicate minerals referred to as phyllosilicates. They are intended for use as additives to hydrophilic polymers such as polyvinilalcohols, polysaccharides, and polyacrylic acids. When fully dispersed in these host polymers, they create a new category of composite materials called nanocomposites.

MMTs have a sheet-type or platey structure. Although their dimensions in the length and width directions can be measured in hundreds of nanometers, the minerals thickness is only 1 nm.

As a result individual sheets have aspect ratios (L/w) varying from 200 to 1000, with a majority of platelets in the 200–400 range after purification.

The theoretical formula and structure for MMT are:

$$M^+_y(Al_{2-y}Mg_y)(Si_4)O_{10}(OH)_2 \cdot nH_2O$$

The preliminarily dried MMT was placed into glass reactor equipped with a mixer, reflux condenser, gas outlet line, and a neutralizer H_2S.

MMT emulsion in m-xylene to which in mixing drop wise was added in portions crushed P_2S_5. The intensive isolation of H_2S was immediately observed. After giving all quantity of P_2S_5, the temperature was raised to 135–140°C and the process was carried out to complete isolation of H_2S, which was catched by aqueous $CdCl_2$. On quantity of formed precipitate CdS a quantity of isolated H_2S was calculated.

For complete removal of H_2S the process was carried out under small vacuum (580–600 mm Hg), and on completion of reaction the reaction mixture was blown by nitrogen for 2 h, then 2–3 times washed by warm ethanol and dried under vacuum at 80°C. As a result it has been prepared MMT–DTP with content of 3.17% phosphor.

For preparation of metal containing compounds, MMT–DTP as suspensions in ethanol was placed in the abovementioned reactor and at T = 80–90°C the alcohol solution of chloride of the corresponding metal was given to it and for 4–5 h. was intensively mixed. After cooling the reaction product was filtered, was washed with hot ethyl alcohol and dried under vacuum at 69–70°C.

Has been determining the phosphorus and metal content in the metal-complexes:

	P (%)	Metal (%)
MMT–DTP–Ni	1.87	2.01
MMT–DTP–Co	1.85	2.29
MMT–DTP–Nd	2.00	3.72

Further, as well as initial MMT–DTP and MMT–DTP-metals on its basis have been used by the methods of infrared (IR) and differential thermal analysis (DTA).

16.2.1 INFRARED SPECTROSCOPY

IR-spectrum (Fourier)—spectrometer Spectrum BX. The investigations in the spectra of the initial MMT observed a peak in the field of 1055 cm^{-1}, and inherent vibrations of bond Si–O–Si is observed. The peak in the field of 770 cm^{-1} showed imperfections in MMT skeleton. The peak in the field of 3400 and 1620 cm^{-1} is referred to valence and deformation vibrations of OH-groups, correspondingly.

After dithiophosphorylation (MMT–DTP), the absorption band of Si–O–Si groups and increases by 1600 cm⁻¹ and new peak which indicates to amorphization of the sample is appeared. An intensification of the absorption band in the field of 1190 cm⁻¹ indicates to formation of P–O bond. The band in the field of 1380 cm⁻¹ is also referred to valence vibrations of this group. Also observed $v_\mu - s$—789.34 cm⁻¹ and $v_\mu - s$—545.21 cm⁻¹.

In the spectrum of MMT–DTP–Ni, the absorption peaks presenting in the initial MMT are kept and are additionally fixed the absorption bands in the field of 1600 and 1000 cm⁻¹, the last one can be referred to vibrations of group Ni–S, P–S in 796.5 cm⁻¹ and P $=$ S—521.9 cm⁻¹.

In the spectra of Co-containing sample in the field of 1500, 3400, and 1600 cm⁻¹ the absorption bands are detected. In the field of 1470 and 1390 cm⁻¹ the new absorption bands are appeared. These vibration bands and also the band in the field of 1160 cm⁻¹ can be accepted as formation of Co complex, $v_\mu - s$—795.5 cm⁻¹, $v_\mu - s$—522.4 cm⁻¹.

In the spectra of Nd-containing sample: $v_s - xa$—896 cm⁻¹, $v_\mu - s$—522.71, and $v_\mu - s$— 797.4 cm⁻¹.

A conservation of the absorption bands in the field of 3600–3400 cm⁻¹ in the spectra of the MMT–DTP and MMT–DTP–metal indicates to the fact not all OH groups entered the reaction with P_2S_5.

16.2.2 DIFFERENTIAL THERMAL ANALYSIS

Q-1500D derivatigraph (Hungary) firm MOM. As follows from thermographic curves of the initial MMT, M-DSP, and M-DTP–metal on its basis as after the dithiophosphorylation as well as after preparation of metal-complexes in temperature properties of the initial MMT the considerable changes occur. In the initial MMT, there is an endothermic peak at 68°C, after dithiophosphorylation (MMT–DTP), this peak is observed already at 107°C, it is simultaneously appeared an endothermic peak at 235°C. In these samples, before endothermic peaks, the exothermic peaks in the field of 75°C are observed.

The peaks of such view are also observed in Ni- and Co-containing samples.

The endothermic peaks observed after dithiophosphorylation (107°C) and in metal-complexes (Ni—115°C, Co—120°C, and Nd—110°C) for 30–40°C are differed from initial MMT.

The same changes are observed in the glass transition temperature (TG)-curves. It can be established according to these curves that the initial MMT is more stable than the MMT–DTP and MMT–DTP–metal on its basis. So, if the initial MMT loses 10% initial weight at 175°C, then for MMT–DTP it occurs at 115°C and for metal-complexes Ni—100°C, Co—96°C, and Nd—122°C.

This confirms that in structure of MMT both after dithiophosphorylation and after preparation of metal-complexes the considerable changes occur.

16.2.3   INVESTIGATION OF THE Ni-, Co-, AND Nd-CONTAINING MMT–DTP AS THE CATALYTIC SYSTEM IN THE POLYMERIZATION OF BUTADIENE

The synthesized heterogenized metal-complex dithiosystems have been used as a component of catalysts in the polymerization process of butadiene in the gas phase or liquid phases in suspension of catalyst. Diethyl aluminum chloride (DEAC) has been used as a cocatalyst. The results of the investigation are presented in Table 16.1.

The polymerization of butadiene in suspension of the heterogenized component of catalyst is carried out similar to homogeneous process with addition of the corresponding quantity of cocatalyst.

For carrying out polymerization process in the gas phase the catalyst is made as follows: necessary quantities of heterogeneous component of catalyst and 10% working solution of DEAC in a current of gaseous butadiene at need ratio Al:Me is introduced into a reactor preliminarily tuned to work with metalorganic catalysts. The polymerization of butadiene is carried out in the gas phase for definite time period. After polymerization the prepared polymer is not needed in washing from catalyst, as unlike in contrast existing catalytic systems, the residues of the catalytic dithiosystems in the process of storage and applications of polybutadiene play a role of antioxidant against its light and thermooxidation.

As follows from results of Table 16.1 the heterogenized catalytic system on the basis of MMT–DTP-Co + DEAC (metal content on bearer 0.025%, wt) at polymerization of butadiene in the gas phase leads to the formation high-molecular polybutadiene with content of 1,4-cis-links 97% and characteristic viscosity $[\eta] = 3.2$ dL/g with yield 97% and productivity 2700 kg PBD/g Co · h.

In the presence of the same catalyst in suspension of heterogeneous component in toluene it also formed a high-molecular 1,4-cis polybutadiene with yield 90%, productivity PBD 1280 kg PBD/g Co · h. The prepared

TABLE 16.1 Results of Investigation of the Catalytic Systems Based on the MMT–DTP–Me in the Polymerization Processes in the Gas and Liquid (in Suspension of Catalyst) Phases. Cocatalyst—DEAC, Al:Me = 100:1, T = 25°C, Solvent for Liquid-phase Process—Toluene.

No.	Catalyst	Polymerization conditions	Metal content on bearer, [Me],%(wt)	Time, τ, (min)	Butadiene conversion (%)	Catalyst productivity (kg PBD/gMe · h)	Characteristic viscosity, 96% and [η] dL/g	Microstructure (%)		
								1,4-cis	1,4-trans	1,2-links
1.	MMT–DTP–Co	Gas phase	1.85	120	97	2700	3.2	97	2	1
2.	MMT–DTP–Co	Suspension	1.87	90	90.0	1280	2.0	95	3	2
3.	MMT–DTP–Ni	Gas phase	1.85	90	96.0	1540	0.85	92	5	3
4.	MMT–DTP–Nd	Gas phase	1.85	90	99	2800	2.7	2	2	1
5.	CTP–Co*	Homogeneous suspension	$4.0 \cdot 10^{-4}$	30	92.0	380	1.62	93	5	2
6.	DTP–Ni*	Homogeneous suspension	$4.0 \cdot 10^{-4}$	10	94.0	250	0.20	80	17	3
7.	Nd–cat**	Gas phase	$3.2 \cdot 10^{-2}$	120	91.7	500	5.2	97	2	1

Note: *Homogeneous catalyst of IPCP of Azerbaijan National Academy of Sciences.

**Neodymium-containing catalyst heterogenized on silica gel of Berlin Technical University (Germany).

polymer with characteristic viscosity $[\eta] = 2.0$ dL/g contains 95% 1,4-cis links.

The activity of the heterogenized component of catalyst MMT–DTP–Ni in combination with cocatalyst DEAC was investigated in the polymerization process of butadiene in gas phase. The catalyst containing 0.025% (wt) metal on bearer allows to prepare relatively high-molecular polybutadiene with characteristic viscosity 0.85 dL/g with yield 96% and productivity 1540 kg PBD/g Ni · h. It is necessary to not especially that in the presence of similar homogeneous nickel-containing catalysts, can be only get the low-molecular polybutadiene with characteristic viscosity within the ranges of 0.05–0.5 dL/g content of 1,4-cis-links within the ranges of 80%. A possibility of synthesis of polybutadienes containing 1,4-cis links higher as 90%, with use of the heterogenized nickel-containing catalysts opens wide possibilities for their industrial application.

MMT–Nd + DEAC catalytic system shows more activity and selectivity—yield of PBD—99%, characteristic viscosity—2.7, and productivity—2800 kg PBD/g Ni · h.

For comparison, in Table 16.1 the data on polymerization of butadiene with use of the homogeneous nickel- and cobalt-containing catalytic dithiosystem DTP-Ni + DEAC and CTP-Co + DEAC developed at the IPCP Azerbaijan National Academy of Sciences[3] and neodymium-containing catalyst now developing in Berlin Technical University (Germany), heterogenized on silicagel are presented.[13] As follows from presented data, the heterogenized catalysts MMT–DTP–Me + DEAC under comparable conditions on productivity (1540–2800 kg PBD/g Me · h.) exceed considerably both their homogeneous (380 kg PBD/g Me · h.), and heterogeneous (500 kg PBD/g Me · h.) analogues.

In Tables 16.1 and 16.2 some data on investigation of influence of various factors (metal content on bearer, ratio Al:Me, temperature, and reaction time) on yield of polymer, productivity of catalysts, and characteristics of polybutadiene are presented. As follow from Table 16.1 in the gas-phase polymerization an increase of cobalt content on bearer from 0.025 to 0.050% leads to the increase in both activity (from 92% to 96%), and productivity (from 2350 to 3020 kg PBD/g Co · h.) of heterogeneous catalyst. In this case some decrease of value of characteristic viscosity (from 2.8 to 2.3 dL/g) and 1,4-cis-links (from 96% to 94%) is observed. An analogous dependence is observed at polymerization in suspension of MMT–Co.

In the presence of the heterogenized catalyst MMT–Ni + DEAC an increase of nickel content on bearer from 0.025 to 0.050% (wt) leads to the increased yield of polymer (from 96% for 90 min to 97% for 60 min), but, in this case, a productivity of catalyst is increased from 1540 to 2550 kg PBB/g Ni · h. Simultaneously, with this a characteristic viscosity (from 0.85 to 0.5 dL g) and content of 1,4-cis-links (from 92% to 88%) are decreased.

At metal content on bearer equal to 0.050% (wt) an increase of ratio Al:Me from 25:1 to 100:1 leads to the increase in yield of polymer (from 92% to 99% in a case of cobalt-containing catalyst and from 90% to 96% in use of nickel catalyst) and productivity of catalyst (from 2350 to 3850 kg PBD/g Co · h—for cobalt and from 1820 to 2650 kg PBD/g Ni · h—for nickel) (Table 16.2).

TABLE 16.2 Influence of Various Factors on Polymerization Process of Butadiene and on Characteristics of the Prepared Polymers.

No.	Catalyst	Al:Me	Temperature, T (°C)	Time, τ (min)	Butadiene conversion (%)	Catalyst productivity (kg PBD/gMe)	Characteristic viscosity,96% and [η] dL/g	Microstructure (%)		
								1,4-cis	1,4-trans	1,2-links
1.	MMT–DTP–Co*	25	25	120	92.0	2350	2.8	96	2	2
2.	MMT–DTP–Co*	50	25	90	96.0	3200	2.5	94	4	2
3.	MMT–DTP–Co*	100	25	90	99.0	3850	2.3	93	5	2
4.	MMT–DTP–Co*	100	50	90	96.0	3550	2.5	94	4	2
5.	MMT–DTP–Ni**	25	25	120	90.0	1820	1.1	93	5	2
6.	MMT–DTP–Ni**	50	25	90	95.0	2175	0.7	92	7	1
7.	MMT–DTP–Ni**	100	25	90	96.0	2650	0.6	85	10	5
8.	MMT–DTP–Ni**	100	50	90	94.0	2360	0.5	90	7	3

Note: *The polymerization has been carried out in gas phase.
**The polymerization has been carried out in toluene solution of catalyst suspension.

In this case there is some decrease of characteristic viscosity (from 2.8 to 2.3 dL/g—for cobalt and from 1.1 to 0.60 dL/g—for nickel) and content of 1,4-cis-links (from 96% to 93%—for cobalt and from 93% to 85%—for nickel).

An increase in temperature and reaction time from 25°C to 50°C and from 60 to 120 min, correspondingly, do not show an essential influence both on activity and on selectivity of catalysts.

16.3   CONCLUSIONS

1. On the basis of nanosize MMT by its phosphorosuphurization with P_2S_5 at 135–140°C the ditiophosphorylated mordenite has been synthesized on the basis of which Ni, Co, and Nd-containing complexes have been prepared.
2. The physical–chemical properties of these complexes have been studied by analytical (content P% and Me%), IR and DTP methods and it has been confirmed that the metals have been chemically connected with bearer matrix through dithiophosphor group.
3. Ni-, Co-, and Nd-containing compound in combination with aluminum organic compounds (DEAC) showed more high activity (conversion of butadiene 96–99%) and selectivity (content of 1,4-cis-links 92–96%) in comparison with analogues Ni- and Co-containing homogeneous systems. It has been established that on Ni-containing heterogenized systems one can get the high-molecular polybuta-diene (characteristic viscosity 0.85 dL/g), which it is impossible to prepare on the analogous homogeneous systems.

KEYWORDS

- dithiophosphorylated montmorillonite
- heterogenized catalytic dithiosystems
- polymerization of butadiene
- Ziegler–Natta type
- infrared spectroscopy

REFERENCES

1. Kubasov, A. A. Zeolites in Catalysis. *Soros Educ. J.* **2000,** *6*(6), 44–51.
2. Janibayov, N. F.; Nasirov, F. A.; Markova, Y. I.; Mamedov, M. X.; Mamedov, R. A. *World Forum on Polymer Applications and Theory POLYCAR—2003;* Texas A&M University: USA, 2003; pp 104–107.
3. Nasirov, F. A.; Azizov, A. H; Janibayov, N. F. Development of Investigations at IPCP in the Field of Development of Bifunctional Catalysts–Stabilizers for Process of Prepara-tion and Stabilization of Polymers. *Process. Petrochem. Oil-refin.* **2008,** *3–4* (34–35), 84–97.

4. Nasirov, F. A.; Azizov, A. H.; Janibayov, N. F. Organic Dithioderivative of Metals in the Processes of Petrochemical. *Azerb. Oil Econ.* **2008,** *8*, 53–59.

5. Application 1203805, European Patent Office, 2002.

6. Application 1361263, European Patent Office, 2003.

7. Mamedov, M. K.; Markova, Y. I.; Rafiyeva, S. R.; Tagiyeva, A. M.; Janibayov, N. F. O,O-dicycloalkyl phenyl Dithiophosphates of Metals as Additives for Motor Oil. *Chem. Probl.* **2011,** *1*, 105–110.

8. Markova, Y. I.; Mamedov, M. K.; Rafiyeva, S. R.; Janibayov, N. F. Synthesis of Nickel-metal Dithiophosphates. *Appl. Chem.* **2012,** *85*(2), 75–81.

9. Azerbaijan, Patent I-2011059, European Patent Office, 2011.

10. Janibayov, N. F.; Nasirov, F. A.; Rafiyeva, S. R.; Markova, Y. I.; Hasanova, G. N. In *Ditio-phosphorylated Zeolites—New Class of Heterogenized Catalytic Systems for Polymerization of Butadiene*, Materials of Russian Congress on Catalysis "ROSKATALIZ", 2011, Moscow, Volume II, p 218.

11. Janibayov, N. F.; Nasirov, F. A.; Rafiyeva, S. R.; Hasanova, G. N.; Kuliyev, A. D.; Asadova, G. G. Ditiophosphorylated Metal-containing Zeolites. *Process. Petrochem. Oil-refin.* **2009,** *10*, 3–4(39–40), 279–283.

12. Janibayov, N. F.; Nasirov, F. A.; Rafiyeva, S. R.; Markova, Y. I.; Hasanova, G. N.; Mamedov, M. K. Heterogenizated Catalyst Systems on the Basis of Dithiophosphorylated Silicagel. *Izvestiya NAS Ga. Ser. Chim.* **2010,** *4*, 421–424.

13. Bartke, M.; Wartmann, A.; Reichert, K.-H. Gas-phase Polymerization of Butadiene. Data Acquisition Using Minireactor Technology and Particle Modeling. *J. Appl. Polym. Sci.* **2003,** *87*(2), 270–279.

PART IV
Selected Topics

CHAPTER 17

RESEARCH OF HYDRODYNAMIC CHARACTERISTICS AND SAMPLING OF AN OPTIMUM DESIGN OF THE BUBBLE-VORTEX APPARATUS

R. R. USMANOVA[1] and G. E. ZAIKOV[2,*]

[1]*Department of Chemical Science, Ufa State Technical University of Aviation, Ufa 450000, Bashkortostan, Russia*

[2]*Department of Polymer Science, Emanuel Institute of Biochemical Physics, Russian Academy of Sciences, Moscow 119991, Russia*

[*]Corresponding author. E-mail: gezaikov@yahoo.com

CONTENTS

ABSTRACT

In this chapter, we show how hydrodynamics of vortex flows allows to make selection of an optimum configuration of working space, and the design of blades, followed by design features of separate knots bubbling-vortex apparatus.

17.1 INTRODUCTION

Apparatuses of whirlwind type are widely used in chemical engineering. All these apparatuses are merged by the general principle: their work is based on centrifugal force use. The scientific and technical literatures do not observe traffic of corpuscles in a zone of walls of the apparatus. Conditions of loss of stability of an eddy flow in the presence of a viscous radial stream are not analyzed. Now the theory of the twirled currents in connection with aspiration to explain the nature of whirlwind effect which remains till now unrevealed intensively develops. Therefore, researches in this direction represent scientific interest.

Application of apparatuses with the twirled traffic of phases in gas cleaning systems is actual. It is necessary to note that engineering protection of a circumambient is based on well-developed chemical engineering.

Necessity and importance of the solution of a problem of raise of efficiency of the gas cleaning, apparatuses based on functional features with the twirled traffic of phases, define an urgency of the given research.

17.2 DEVELOPMENT OF THE CONSTRUCTION BUBBLE-VORTEX APPARATUS

To optimize the bubble-vortex apparatus, experimental studies were conducted. The experiments were performed by a single method[8] of comparative tests on dust collectors bubble-vortex apparatus with a cylindrical chamber 0.6 m and a diameter of 0.2 and 0.4 m bubble-vortex machine with adjustable blades in accordance with Figure 17.1. It comprises a cylindrical chamber 1 inlet pipe 2. The cylindrical chamber 1 is three swirl gas flow, which is a four blades, curved sinusoidal curve. Adjusting the blades 2 is done by turning the eccentrics, sealed with a cylindrical chamber 1 through the spring washers and lock nuts.[1–6]

One swirler is located before the midpoint of the gas flow nozzle 4, and other is located after peripheral nozzle 5, which serves scrubbing liquid. Removal of dispersed particles produced by sludge overflow pipe 6 chip catcher 7.

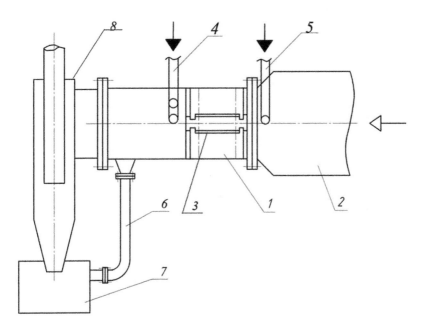

1—cylindrical chamber, 2—inlet pipe, 3—swirl, 4—central nozzle, 5—peripheral nozzles, 6—overflow pipe cuttings, 7—chip catcher, and 8—cyclone.

FIGURE 17.1   Bubble-vortex apparatus with adjustable blades.[5]

Bubble-vortex machine with adjustable blades works as follows. The dusty gas is fed into a cylindrical chamber 1, where the swirl 3 with blades mounted in the radial grooves of the rod, deflects the flow and gives it a rotation. Under the action occurring at the same centrifugal forces, dispersed particles are moved to the sides of the unit. For adjusting the blades at the entrance and exit of each blade 3 have two tabs. With their help, the blade is in contact with a pair of eccentrics. Eccentrics are turning vanes on the inlet and outlet sections of the cylindrical chamber 1 in different directions. With this, blade 3 is installed in the position corresponding to the most efficient gas purification.

17.3 THEORETICAL BASES OF MODELING

Analytical and numerical methods of calculation can be applied to the description of hydrodynamic characteristics of traffic of streams in working space of the apparatus.[7]

The description of traffic of the twirled stream is based on an analytical method on one of following approaches[9,10]:

1. The stream is represented in the form of superposition of a flat current on potential twirl. The design procedure is based on use of empirical factors; it provides only water resistance definition.
2. Bernoulli's theorem application is the presence of an extreme as one of components of speed of the twirled stream at conservation along radius of the working chamber of an angular momentum. An approach deficiency, a coarse schematization of a current, is the absence of the account of features of traffic of streams in an axial zone of the working chamber.
3. Use of Bernoulli's theorem for the description of traffic of a liquid in the chamber of spiral type. This method demands preliminary definition of the characteristics depending on geometry of the working chamber, but he does not allow to define all components of full speed.

The specified deficiencies lead to considerable restriction of a scope of these methods for the description and calculation of vortex flows.

To calculate the trajectories of the particles, we need to know their equations of motion. Such a problem for some particular case is solved by the author.[9]

We introduce a system of coordinates OXYZ. Its axis is directed along the OZ axis of symmetry scrubber (Fig. 17.2). Law of motion of dust particles in the fixed coordinate system OXYZ can be written as follows:

$$m \frac{\overline{dv'_p}}{dt} = \overrightarrow{F_{st}},$$
(17.1)

where m—mass of the particle.

dv_p—velocity of the particle.

F_{st}—aerodynamic force.

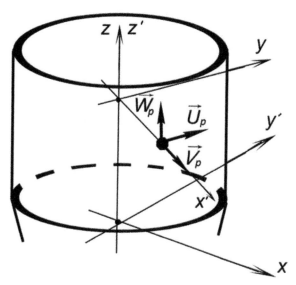

FIGURE 17.2   The velocity vector of the particle.

For the calculations, it is necessary to present the vector eq 17.1 motion in scalar form. Position of the particle will be given by its cylindrical coordinates $(r; \varphi; z)$. Velocity of a particle is defined by three components: U_p—tangential, V_p—radial, and W_p—axial velocity.

We take a coordinate system O′X′Y′Z′; let O′X′ passes axis through the particle itself, and the axis O′Z′ lies on the axis OZ. Adopted reference system moves forward along the axis OZ W_p speed and rotates around an angular velocity.

$$\omega(t) = \frac{U_p}{r_p}. \tag{17.2}$$

The equation of motion of a particle of mass $m = \frac{1}{6}\pi\rho_p d^3{}_p$; coordinate system O′X′Y′Z′ becomes:

$$m\frac{d\vec{v_p}}{dt} = \vec{F_{st}} - m\vec{a_0} + m\left[\vec{r_p} \cdot \vec{\omega}\right] + m\left[\vec{r_p} \cdot \vec{\omega}\right] + m\left[\vec{\omega} \cdot \left[\vec{r_p} \cdot \vec{\omega}\right]\right] + 2m\left[\vec{v_p} \cdot \vec{\omega}\right]$$

or

$$\frac{d\vec{v_p}}{dt} = \frac{1}{m}\vec{F_{st}} - \vec{a_0} + \left[\vec{r_p} \cdot \vec{\omega}\right] + \left[\vec{\omega} \cdot \left[\vec{r_p} \cdot \vec{\omega}\right]\right] + 2\left[\vec{v_p} \cdot \vec{\omega}\right], \tag{17.3}$$

where a_0—translational acceleration vector of the reference frame.

dv_p—velocity of the particle.

r_p—the radius vector of the particle.

$\left[\vec{r}_p' \cdot \vec{\omega}\right]$—acceleration due to unevenness of rotation.

$\left[\vec{\omega}\left[\vec{r}_p' \cdot \vec{\omega}\right]\right]$—the centrifugal acceleration.

$2\left[\vec{v}_p \cdot \vec{\omega}\right]$—Coriolis acceleration.

The first term on the right-hand side of eq 17.3 is the force acting on the particle of gas flow, and is given by Stokes

$$F_{st} = 3\pi\mu_g d_p \left[\vec{v}_g - \vec{v}_p\right], \tag{17.4}$$

where μ_g—dynamic viscosity of the gas.

The second term (3) is defined as

$$\frac{dW_p}{dt}\vec{e}_z = \frac{dW_p}{dt}\vec{e}_{z'}.$$

Convert the remaining terms:

$$\left[\vec{r}_p' \cdot \vec{\omega}\right] = \left[\vec{r}_p' \cdot \frac{d\vec{\omega}}{dt}\right] = \left[\vec{r}_p' \cdot \frac{d}{dt}\left(\frac{U_p}{r_p}\vec{e}_{z'}\right)\right] = -r_p\left(\frac{1}{r_p}\frac{dU_p}{dt} - \frac{U_p}{r^2_p}V_x\right)\vec{e}_y = \left(-\frac{dU_p}{dt} + \frac{U_p V_p}{r_p}\right)\vec{e}_y,$$

$$\left[\vec{\omega}\left[r_p^{-1} \cdot \vec{\omega}\right]\right] = \frac{U_p^2}{r_x}\left[\vec{e}_z \cdot \left[\vec{e}_x \cdot \vec{e}_z\right]\right] = -\frac{U_p^2}{r_p}\left[\vec{e}_z \cdot \vec{e}_v\right] = \frac{U_p^2}{r_p}\vec{e}_x,$$

$$2\left[\vec{v}_p \cdot \vec{\omega}\right] = 2v_{x'}\left[\vec{e}_{x'} \cdot \vec{\omega}\right] = 2v_{x'}\frac{U_x}{r_p}\left[\vec{e}_{x'} \cdot \vec{e}_{z'}\right] = \left(-2\frac{U_p V_p}{r_p}\right)\vec{e}_{y'}.$$

where $\hat{e}_p, \hat{e}_v, \hat{e}_z$—vectors of the reference frame and used the fact that

$$\vec{r}_p = \vec{e}_x \cdot r_+ \cdot v_x = V_p.$$

Substituting these expressions in the equation of motion (17.3),

$$m\frac{d\vec{v}_p'}{dt} = \vec{F}_{st} - m\vec{a}_0' + m\left[\vec{r}_p' \cdot \vec{\omega}\right] + m\left[\vec{r}_p' \cdot \vec{\omega}\right] + m\left[\vec{\omega} \cdot \left[\vec{r}_p' \cdot \vec{\omega}\right]\right] + 2m\left[\vec{v}_p \cdot \vec{\omega}\right]$$

or

$$\frac{d\vec{v_p}}{dt} = \frac{1}{m}\vec{F_{st}} - \vec{a_0'} + \left[\vec{r_p'}\cdot\vec{\omega}\right] + \left[\vec{\omega}\cdot\left[\vec{r_p'}\cdot\vec{\omega}\right]\right] + 2\left[\vec{v_p}\cdot\vec{\omega}\right]$$

We write this equation in the projections on the axes of the coordinate system O'X'Y'Z':

$$\begin{cases} \dfrac{dV_{x'}}{dt} = \dfrac{1}{m}F_{stx'} + \dfrac{U_p^2}{r_p} \\[2mm] 0 = \dfrac{1}{m}F_{sty} - \dfrac{dU_p}{dt} - \dfrac{U_p V_p}{r_p} \\[2mm] 0 = \dfrac{1}{m}F_{stz} - \dfrac{dW_p}{dt} \end{cases}$$

$$\begin{cases} \dfrac{dV_p}{dt} = \dfrac{1}{m}F_{stx} + \dfrac{U_p^2}{r_p} \\[2mm] \dfrac{dU_p}{dt} = \dfrac{1}{m}F_{sty} - \dfrac{U_p V_p}{r_p} \\[2mm] \dfrac{dW_p}{dt} = \dfrac{1}{m}F_{stz} \end{cases} \qquad (17.5)$$

We have the equation of motion of a particle in a rotating gas flow projected on the axis of the cylindrical coordinate system.

Substituting eqs 17.2 and 17.4 in eq 17.5, we obtain the system of equations of motion of the particle:

$$\begin{cases} \dfrac{dV_p}{dt} = \dfrac{18}{\tilde{n}_p d^2 p}\left(V_g - V_p\right) + \dfrac{U_p^2}{r_p} \\[2mm] \dfrac{dU_p}{dt} = \dfrac{18}{\tilde{n}_\cdot d_\div^2}\left(U_g - U_p\right)\dfrac{U_p V_p}{r_p} \\[2mm] \dfrac{dW_p}{dt} = \dfrac{18}{\tilde{n}_p d_p^2}\left(W_g - W_p\right) \end{cases} \qquad (17.6)$$

17.4 COMPUTER MODELING

Computer modeling was spent also for studying of hydrodynamics of a gas stream in a swept volume bubble-vortex apparatus a various configuration

[1–4] at installation of blades of various types (Fig. 17.3) with a different slope of shovels α and their quantity $n = 4$.

At installation, in the device of four direct shovels, the increase in speed of a gas stream at shovels is observed. Speed decreases at once at passage on a swept volume. In the swept volume center, there are stagnation zones. At installation, in the device of four sinusoidal shovels, the uniform distribution of speeds on all swept volume of the whirlwind apparatus with insignificant formation of stagnation zones in the center is observed. At increase in a slope of shovels (60°), speed in a swept volume considerably decreases; at shovels, local eddyings are formed. In the center, formation of zones of reduction of speed is observed. At perforation of shovels, speed is slightly stabilized by volume.

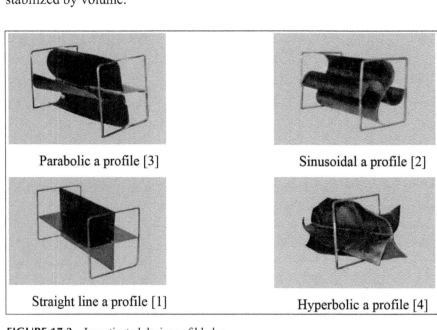

Parabolic a profile [3]   Sinusoidal a profile [2]

Straight line a profile [1]   Hyperbolic a profile [4]

FIGURE 17.3   Investigated designs of blades.

Parabolic profile (Fig. 17.4a):

- A uniform distribution of a gas stream in the bottom part of the appa-ratus without its twisting
- Stagnation zones in the overhead part of working space of the apparatus
- Low volume of traffic of a gas stream in separation to a zone and in a zone of tap of small corpuscles

Sinusoidal profile (Fig. 17.4b):

- An intensive twisting of a gas stream in the bottom part of the apparatus
- An intensive whirling motion of gas on apparatus altitude
- High volume of traffic of a gas stream in separation to a zone and a zone of tap of small corpuscles

Straight-line profile and hyperbolic profile (Fig. 17.4c and d):

- A uniform distribution of a gas stream on the apparatus center
- Low intensity of a twisting of a gas stream

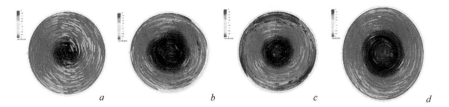

$\underline{a}$ – α = 30 °, n = 4 Parabolic profile; b – α = 0 °, n = 4 Straight lines a profile;

$\underline{c}$ – α = 30 °, n = 4 Sinusoidal profile; d – α = 60 °, n = 4 Hyperbolic profile

a – α = 30°, n = 4 parabolic profile; b – α = 0°, n = 4 straight-line profile; c – α = 30°, n = 4 sinusoidal profile; d – α = 60°, n = 4 hyperbolic profile.

FIGURE 17.4 The image of a field of speeds of a gas stream at following configurations blades (top view).

- Presence of stagnation zones near apparatus walls
- Low volume of traffic of a gas stream in separation to a zone and a zone of tap of small corpuscles

17.5 ANALYSIS OF RESULTS AND THE RECOMMENDATION FOR SAMPLING OF AN OPTIMUM DESIGN OF THE APPARATUS

The analysis of results of researches has allowed to define regularity of agency of a design bubble-vortex apparatus on distribution of speed of traffic of a gas stream.

The configuration of blades makes following impact on gas cleaning process:

- At increase in a slope of shovels, the peripheral velocity of a gas stream on cross section of working space decreases.
- At increase in a slope of shovels, chaotic eddying of a gas stream at the separating device is formed.
- At increase in number of shovels, there is a nonuniform velocity distribution—speed of a gas stream at an entry is much more than at an exit from the apparatus.
- At increase in a slope of shovels, the big areas of stagnation zones of a dispersoid are formed.

From the presented designs of shovels, sinusoidal blades are optimum. They provide necessary distribution of speed of a whirlwind gas stream in working space bubble-vortex apparatus.

17.6 INDUSTRIAL APPLICATION OF VORTICAL DEVICES

It is settlement—theoretical and design—design works were carried out with reference to operating conditions of manufacture hypochlorite calcium of Joint-Stock Company "Caustic."

On Joint-Stock Company "Caustic" the device is offered to be used for clearing smoke gases of furnaces of roasting 1 (Fig. 17.5).

According to the technological scheme, departing from the furnace of roasting 1 gases at temperature 550°C act in the bubbling-vortical device 2. Here on an irrigation, 1–3% solution of limy milk (pH = 11.5–12.5) move. Separated slime acts in a drum—slake 3; clarification and cooling of limy milk happens in the filter-sediment bowl 4: from which it feeds on irrigation. The cleared gas stream smoke exhauster 7 is thrown out in an atmosphere.

Thus, introduction of bubbling—the vortical device will allow to solve a problem of clearing of smoke gases with return of all caught dust in the form of slime in branches of slaking lime, that provides captives manufactures.

Bubbling—the vortical device is supposed to mount in vent dust removal systems with the purpose of economy of the areas of industrial premises, thus hydraulic resistance of system does not exceed 500 Pa, power inputs on clearing of gas in three times below, than in known devices.

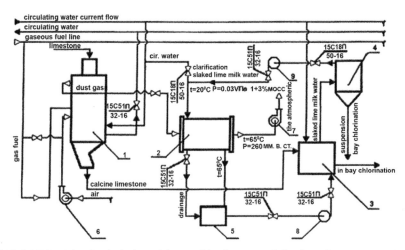

FIGURE 17.5   The circuit design of the combined system of clearing of gas.

Application bubbling—the vortical device allows to achieve an intensification of process of gas purification with reduction of a gassed condition
of air pool.

At the organization of clearing of gas under the offered circuit design, it
is necessary to carry out tap of a low-purity locking liquid from a cavity of
the car 1 in system of regeneration 5 or on emission. It is necessary to direct
equal quantity of the clarified or pure actuating medium to a zone of the
blowing machine 8.

It is experimentally installed that the extent of trapping of polluting firm
impurity is high enough. At certain regimes, efficiency attains 100% in
limiting accuracies of measurements.

17.7 CONCLUSIONS

Thus, the gained results of research of hydrodynamics of vortex flows allow
to make selection of an optimum configuration of working space, a design
of blades, and design features of separate knots Bubbling-vortex apparatus.
Necessary efficiency of air purification depending on demands to composition of gas emissions is thus provided.

The basic advantages whirlwind Bubbling-vortex apparatus:

•   Possibility of essential decrease of overall dimensions (altitude) of
    working space

- Increase in a time of a finding of corpuscles in working space of the whirlwind apparatus
- Possibility of traffic control of corpuscles in working space bubbling-vortex apparatus
- Creation possibility in volume of working space bubbling-vortex apparatus intensive turbulence
- Adaptability to manufacture and simplicity of manufacturing
- Possibility of sweeping change of constructive technological parameters if necessary

KEYWORDS

- **bubbling-vortex apparatus**
- **blade twist**
- **extent swirler**
- **gas washer**
- **vortical devices**

REFERENCES

1. Usmanova, R. R. Bubbling—The Vortical Device. The Patent of the Russian Federation the Invention No. 2182843, May 27, 2002 (The Bulletin No. 15).
2. Usmanova, R. R. Bubbling—The Vortical Device with an Axial Sprinkler. The Patent of the Russian Federation the Invention No. 2316383, February 10, 2008 (The Bulletin No. 4).
3. Usmanova, R. R. Bubbling–Swirling Apparatus with Parabolic Swirler. The Patent of the Russian Federation the Invention No. 2382680, February 27, 2010 (The Bulletin No. 6).
4. Usmanova, R. R. Bubbling–Swirling Apparatus with Conical Swirler. The Patent of the Russian Federation the Invention No. 2403951, November 20, 2010 (The Bulletin No. 320).
5. Usmanova, R. R. Bubbling—The Vortical Device with Adjustable Blades. The Patent of the Russian Federation the Invention No. 2234358, August 20, 2004 (The Bulletin No. 230).
6. Kochevsky, A. N.; Nenja, V. G. Modern the Approach to Modeling and Calculation of Currents of a Liquid in Bladed Hydromachines. *Bull. Sum. Gu.* **2003**, *13*(59), 195–210 (in Russian).

7. Aksenov, A. A.; Dyadkin, A. A.; Gudzovsky, A. V. Numerical Simulation of Car Tire Aquaplaning. In *Computational Fluid Dynamics*; John Wiley & Sons: USA, 1996; pp 815–820.

8. Vatin, N. I.; Strelets, K. I. *Air Purification by Means of Apparatuses of Type the Cyclone Separator;* Bull. Sum. Gu (Russia): St. Petersburg, 2003.

9. Zajcik, L. I.; Pershukov, V. A. Problem of Modeling Gas–liquid Turbulent Flows with Changes of Phase. In *Mechanics of a Liquid and Gas*; Bull. Sum. Gu (Russia), 1996; Vol. 5, pp 3–19 (in Russian).

10. Rodionov, A. I.; Klushin, V. N.; Sister, V. G. *Processes of Ecological Safety*; N. Bochkarevs Publishing House: Kaluga, 2000.

CHAPTER 18

EMISSION OF CLUSTERS DURING N$^+$, O$^+$, N$_2^+$, O$_2^+$, Ar$^+$ IONS BOMBARDMENT OF C, NaCl, Mg, Al, Si, S, Ti, GAAS SURFACES

GEORGE MESKHI*

Faculty of Engineering, Agrarian and Natural Sciences, Samtskhe-Javakheti State University, 106 Rustaveli Str., 0800 Akhaltsikhe, Georgia

*Corresponding author. E-mail: george.meskhi@yahoo.com

CONTENTS

ABSTRACT

A group of atoms bound together by interatomic forces is called an atomic cluster. Clusters are formed of matter intermediation between atoms (or molecules) and the solids. The novelty of atomic cluster physics arises mostly from the fact that cluster properties explain the transition from single atoms or molecules to the solid state. Modern experimental techniques have made it possible to study this transition. Cluster emission phenomena are of scientific and technological interest because of the potential of clusters to form materials with new chemical and physical properties. Consequently, it is important to get high intensity of different cluster flux for their investigations and interactions with matters. The most widely used method of clusters production is the adiabatic expansion of a gas or a vapor through a small nozzle into vacuum. The method of laser evaporation is used very often as well. Clusters from solids can be easily obtained by sputtering surfaces using depositing energy through ions, electrons, or photons onto a surface. In the present work, emission of clusters during ion bombardment of surfaces is investigated. Now the volume of researches on studying the mechanism of formation of clusters is significant and continues to grow, but for a while, an overview of the processes resulting in emission clusters at ion sputtering of surfaces is absent. Especially, it is complicated to determinate the mechanism of emission of positive and negative clusters and their excited states, and despite the multitude of works that are represented in this direction, till now there is no certain view toward process of formation of a charge and exited states of cluster. Investigation of dissociation processes of atomic clusters in different states during production and emission is also very important. With this purpose in the present work, sputtering of various surfaces in a wide range of energy bombardment by various primary ions under different bombarding conditions is presented in experimental investigations of cluster ions emission due to the experimental difficulties encountered in the analysis of sputtered neutral particles.

In the present work, measurements of the negative and positive cluster ions yields from C, NaCl, Mg, Al, Si, S, Ti, and GaAs surfaces under 0.5–50 keV N^+, O^+, N_2^+, O_2^+, Ar^+ bombardment at different experimental conditions are reported. Cluster ions yields are investigated depending on number of atoms in cluster at different conditions of experimental researches. The emission of small ($n < 10$), "light" clusters occurs by a collective motion during the development of the collision cascade after impact. Energy transferred to the surface is strongly directional and can lead to the simultaneous emission of a group of neighboring surface atoms, which in some cases will remain bounded and form a cluster after emission.

18.1 INTRODUCTION

The bombardment of solid with atoms or ions, having energies in the keV range, leads to the release of a variety of secondary particles from the surface. It is well known that not only does the sputtered flux consist of monatomic target particles, electrons, and radiation in the wide range of energy but depending on the experimental conditions, a certain fraction of the emitted flux will consist of neutral and ionized clusters. Also, it is well known that the flux of particles released from a solid surface under energetic ion bombardment (sputtering) contains agglomerates of several or many atoms as well as single atomic species. Originally, information on the yield and energy distribution of clusters came mainly from cluster ions due to the experimental difficulties encountered in the analysis of sputtered neutral atomic particles.

Numerous studies have been devoted to the investigation of cluster emission in sputtering, a review of which is found in work of Gerhard (1995), Benninghoven (1993), Betz and Husinsky (2004), Hofer (1991), Urbassek (2006), and Staudt (2000). Much of this work was conducted on metallic samples as model systems for purely collisional sputtering conditions, and most of the published data has been collected for charged clusters (secondary cluster ions).The interpretation of experimental results can be extremely useful to obtain information about the physical and chemical state of the investigated surface, provided the mechanisms leading to the formation and/or ejection of clusters during the sputtering process are sufficiently well understood. Atomic clusters are atomic compounds with a well-defined number n (the number of atoms—size of clusters) of constituents; they allow study of the evolution of a physical and chemical property from the atom towards the solid surface. However, as the number of atoms in the system increases, atomic clusters acquire more and more specific properties making them unique physical objects different from both single molecules and from the solid state.[1-7]

In the macroscopic world, the physical properties of a material are independent of a sample's size. However, when sample dimensions are made sufficiently small, the properties of a cluster of atoms must ultimately depart from those in the bulk, and evolve as a function of size. Physical and chemical properties (geometric and electronic structure, superconductivity and magnetism, chemical reactivity and catalysts, absorption, etc.) can change drastically (by a few orders of magnitude) from one cluster size to the next. A large number of fundamental questions need to be answered: How do these structures nucleate and grow? How do clusters transform from one structure to another as successive atoms are added during growth? At what size does the bulk structure prevail? Given that atomic structure plays a critical role in determining all cluster properties, it is remarkable that these questions remain unanswered—the inherent difficulties in both the theoretical and experimental determination of cluster structure have been major obstacles.[8] As a result, much structural information has been obtained indirectly, from experimental observations of other cluster properties, which has led to incomplete and sometimes ambiguous interpretation. In addition, studies of the cluster yields due to surface excitation will lead and have led to a better understanding of the excitation mechanism and sputtering process.

By increasing the cluster size, one can observe the emergence of the physical features in the system, such as Plasmon excitations, electron conduction band formation, superconductivity and superfluidity, phase transitions, fission, and many more. Most of these many-body phenomena exist in solid state but are absent for single atoms.

Atomic clusters, as new physical objects, possess some properties, which are distinctive characteristics of these systems. The cluster geometry turns out to be an important feature of clusters, influencing their stability and vice versa. The determination of the most stable cluster forms is not a trivial task and the solution of this problem is different for various types of cluster. This problem is closely connected to the problem of cluster magic numbers.[9] The formation of a sequence of cluster magic numbers should be closely connected to the mechanisms of cluster formation and growth. It is natural to expect that one can explain the magic numbers sequence and find the most stable cluster isomers by modeling mechanisms of cluster assembly and growth, that is, the fusion process of atomic clusters.

The science of clusters is a highly interdisciplinary field. Atomic clusters concern astrophysicists, atomic and molecular physicists, chemists, molecular biologists, solid-state physicists, nuclear physicists, plasma physicists, and technologists all of whom see them as a branch of their subjects, but cluster physics is a new subject in its own right. A cluster of a given size may have many different isomers, some of which may be close in energy. Clusters offer exciting prospects for designing new materials, owing to the strong dependence of their electronic properties on their size and structure. Clusters are considered as promising catalysts because they exhibit a higher reactivity than bulk materials. Even noble metals such as gold and platinum become highly reactive as clusters.

18.2 METHODS OF RESEARCH

Using the method of mass spectrometry of secondary positive and negative ions (SIMS), cluster ions yields are measured during primary ion sputtering of surfaces. The measurements of positive and negative cluster ions yields $Y_n^{+,-}$ at bombardment by N^+, O^+, N_2^+, O_2^+, Ar^+ ions of Al, GaAs surfaces in energy range of the primary ions E_0 = 5–50 keV, and also research behavior of ions yields depending on pressure of oxygen in the chamber of interaction, are executed on experimental arrangement at the University of Salford (England) consisting of secondary ions mass spectrometry method in addition with energy analyzer.[10] Yields $Y_n^{+,-}$ at bombardment of C, NaCl, Mg, Al, Si, S, and Ti surfaces by Ar^+ ions, in an energy range E_0=0.5–10 keV and depending on an angle of bombardment—α of primary ions Ar^+, counted from normal to a researched surface are measured on installation developed at Ivane Javakhishvili Tbilisi State University for research of solid surfaces by a method SIMS on base to the UHV arrangement.[11,12]

The N^+, O^+, N_2^+, O_2^+, Ar^+ ions are formed in a plasma type ions source with the high-frequency discharge, in energy range E_0 = 5–50 keV, are accelerated, allocated on mass by mass monochromator, collimated by slits and bombarded along normal researched surfaces. Emitting thus the SIMS get in the hemispherical energy analyzer with radius 24 cm and further act in quadruple mass spectrometer such as Finnigan 750 for the analysis of mass in a range 1–750 a.m.u. Using monochromator in visual spectra light, emission from excitation states of the emitted atomic particles during bombardment is investigated as well. Typical intensity of a primary ions beam is changing from some tens $\mu A/cm^2$ to a few mA/cm^2. The all measurements of cluster ions emission from solids in both cases of using methods[2,3] were performed at UHV conditions ($P = 10^{-9}–10^{-10}$ torr). For investigation of cluster ions yields depending on a condition of researched surfaces in the interaction chamber, the oxygen up to pressure $Po_2 = 10^{-6}$ torr was filled. The time of stabilization of pressure O_2 and time of establishment a condition of balance of cluster ions yield at letting-to-oxygen gas in the interaction chamber were experimentally determined.[10] In the work,[11,12] the primary ions beam of Ar^+ with energy E_0 = 0.5–10 keV and density of a current a few mA/cm^2 collides with the researched sample surface along with the normal of the surface. The minimal diameter of etch pit by ion beam makes ~0.5 mm. Emitted during bombardment, the SIMS are analyzed by monopole mass spectrometer directed under 45° angle to the surfaces of the sample. As a source of the primary ions, duoplasmatron-type ion source is applied, which represents a discharge plasma source with the cold cathode and double contraction of plasma. The source of the primary ions is equipped with electronic optics, which provides focusing and scanning of the ion beam on the surface on the area 5×5 mm^2 for providing the balancing formation of the edges of a crater at the depth profile analysis. The secondary ions are analyzed by monopole mass spectrometer MX7304A, which is complemented with electronic optics for the stretch of the secondary ions and is altered for a method of the SIMS. The range of registered mass is 1–380 a.m.u. The technique also allows to investigate yields of secondary ions depending on an angle of incident of primary ions in range $\alpha = 0$–$45°$. The errors of relative measurements of cluster ions yields in both methods are made (~10–20%) and at low intensities of cluster ions, yields are determined within the factor 2.

18.3   RESULTS AND DISCUSSIONS

The most easily accessible experimental information on the cluster formation process is the cluster partial yield or the cluster size distribution. In our investigations, the measurement of the charged cluster distribution method has been used. The method suffers from the fact that the probability of cluster ionization increases with cluster size.[5] In the case of metal clusters, a rather large difference in the cluster distribution has been reported depending on the method used.[3,9] In the case of nonmetallic clusters, the ionization probability is relatively constant with cluster size.

In Figure 18.1, the cluster ion yields $Y_n^{-,+}$ formed under Ar$^+$ ions bombardment of C, NaCl, Mg, Al, Si, S, Ti, GaAs surfaces at E$_0$ = 8 keV by dependence on number of atoms in cluster—n are presented.[12,13] It is visible that yields of cluster ions Na$_n^+$ ($n = 1, 2$), Mg$_n^+$ ($n =1,...,4$), Al$_n^+$ ($n = 1, ...,$ 4), Si$_n^+$ ($n= 1, ..., 5$), S$_n^+$ ($n = 1, 2$), and Ga$_n^+$ ($n = 1, 2, 3$) monotonously decreases with increase of n. The yield of clusters decreases strongly with cluster size, typically the yield of clusters with size $n+1$ is lower by one order of magnitude than for size n, but there is good evidence that large clusters are relatively more abundant and that the relative yield of clusters shows a power-law dependence of the cluster yield on the size of cluster, the number n of atoms in the cluster.[3,16,17] From the results presented below and previous work,[7] it is observed that a power law cluster yield distribution according to

$$Y(n) \propto n^{-\delta} \tag{18.1}$$

is a general phenomenon. The observed value of $-\delta$ varies between -2 and -15 depending upon sputtering conditions such as the incident kinetic energy, projectile charge, and angle of incidence of the projectile and on the material properties of the target surface. For example, in Ar^+–Cu case, the calculated value of $\delta = 4.5$ from the work of Betz 2004 is in reasonable agreement with experimental results. In our case, for different solids, δ has the value of about 4–8, and it has to be, and depends on an electronic configuration of the emitted clusters. Besides, the considerably large number of light (small) clusters is caused from fragmentation of big size clusters as well.

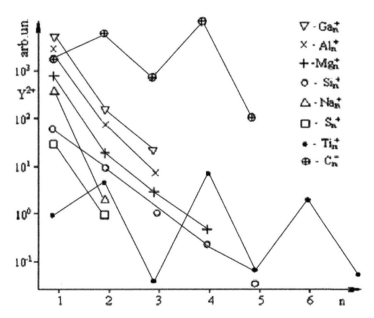

FIGURE 18.1   Cluster ion yields during Ar^+–C, NaCl, Mg, Al, Si, S, Ti, GaAs at $E_0 = 8$ keV.

Only two of the published models predict a power law cluster yield distribution. The first is a shock wave initiated process of Bitensky and Parilis.[23] The second model that predicts a power law cluster yield distribution is the model by Urbassek[8] which is applicable to sputtering by both singly and highly charged ions. In this "equilibrium" model, a highly energized region of the surface undergoes a "liquid–gas" phase transition upon expanding into vacuum. If the phase transition happens near the critical point (where inter-particle binding is just balanced by the kinetic energy), fluctuations are high enough to produce high yields of large clusters. In this model, the clusters are assumed to be in equilibrium with each other and monoatomic species. The cluster yield depends on the energy deposited into the near-surface volume.

Reaching the critical point requires the kinetic energy of the target atoms be high, so that chemical bonding loses its importance and the system becomes fluid. The equilibrium model predicts transitions from an exponential decay to power law decay as the phase transition occurs closer to the critical point. The dependence of the cluster yield $Y(n)$ on the cluster size, n, is

$$Y(n) = Y_0 n^{-\delta} \exp[-\Delta Gn - 4\pi n^{2/3} r^2 \sigma / kT], \tag{18.2}$$

where ΔG is the difference of the Gibbs free energies of the liquid and gas phase, k is Boltzmann's constant, T is the temperature of the energized region, Y_0 is the sputter yield, r is the cluster radius, σ is the surface tension, and $-\delta$ is the critical exponent.[8] At equilibrium ΔG is zero and at the critical point the surface tension vanishes. It follows that predicted cluster size distributions at the critical point are very similar for the equilibrium and shock wave models. The equilibrium model can explain changes in the power law exponent over a narrow range.[24]

Differently, the yield versus size dependency exhibits an odd–even oscillation for Ti_n^+ ($n = 1, ..., 7$) and C_n^- ($n = 1,..., 5$) (Fig. 18.1).[16,17] Such behavior was observed by many authors for various surfaces (for $[Ag_n^+]$, $[Ag_n^-]$,[9] and $[Cu_n^-]$[18,19]), and can be explained by recombination model of clusters formation. For Ti, C, Ag, and Cu surfaces, positive and negative cluster ions emission are dominating. At insignificant changes of bombardment angle—α approximately on about $10°$ observable for the Ti structure it is retained (see Fig. 18.2), however, the ratio of clusters yield with even and odd n varies.

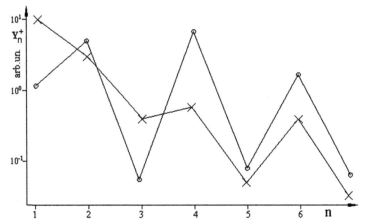

FIGURE 18.2   Behavior of yields Y_n^+ during Ar^+—Ti at $\alpha = 0°$ (0) and $\alpha = 10°$ (x); $E_0 = 8$ keV.

The determination of the most stable cluster forms is not a trivial task, and the solution of this problem is different for various types of cluster. The cluster geometry turns out to be an important feature of clusters, influencing their stability. This problem is closely connected to the problem of cluster magic numbers. The differences in the interatomic potentials and pairing forces lead to the significant differences in structure of clusters, their mass spectra, and their magic numbers. The origin of the cluster magic numbers is different and is based on the principles of quantum mechanics. This feature of light clusters make them qualitatively similar to atomic nuclei for which quantum shell effects play the crucial role in determining their properties.

In the present work, decrease in a ratio of yield of one and two atomic ions $Y_{12}^+ = Y_1^+/Y_2^+$ with increase in atomic number of researched targets Na, Mg, Al, Si, (200, 60, 40, and 10) was observed correspondingly. Such behavior of the ratio can be explained by various electronic configurations of different emitted particles.[20]

In Figure 18.3, the yields of secondary Al^+, Al_2^+, Al_3^+, and Al_4^+ particles at bombarding of the Al by N^+, N_2^+, O^+, O_2^+, Ar^+ ions with energies $E_0 = 5$–50 keV were presented. The yields Y_n^+ ($n = 1, 2, 3, 4$) practically does not change in the investigated range of energy of bombardment, like atoms of Al at sputtering.[20,21] From Figure 18.3, it is visible that the yields Y_n^+ ($n = 1, 2, 3,$) is less in case of O^+, N^+–Al, than in O_2^+, N_2^+–Al case; it can be caused by

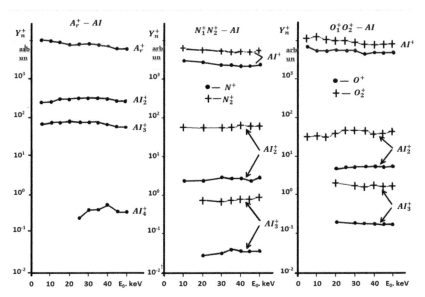

FIGURE 18.3 The yields of Al_n^+ ions during Ar^+, O_2^+, N_2^+, O^+, N^+–Al at $E_0 = 40$ keV.

a big penetration O^+, N^+ in Al, and that leads to reduction of clusters yields. In Table 18.1, the averaged, maximum values of Y_n at the range of saturation are given and corresponding ratios of Y_2^+/Y_1^+, Y_3^+/Y_1^+, and Y_3^+/Y_2^+ are also presented. In our experimental investigations, approximately the same, such as Al_n^+ in case of Ar^+, O_2^+, N_2^+, O^+, N^+–Al clusters, yields dependence on primary ions energy E_0 was observed for Na, Mg, Si, S, Ga, and As clusters during Ar^+–NaCl, Mg, Al, Si, S, GaAs interactions in the energy range $E_0 = 0.5$–50 keV.

TABLE 18.1 Experimental Data and Results.

Collisions	Y_1^+	Y_2^+	Y_3^+	Y_2^+/Y_1^+	Y_3^+/Y_1^+	Y_3^+/Y_2^+
Ar^+–Al	8×10^3	3×10^2	60	4×10^{-2}	8×10^{-3}	0.2
N^+–Al	2.5×10^3	$2.8 \times$	3×10^{-2}	1×10^{-3}	1×10^{-5}	1×10^{-2}
N_2^+–Al	5.5×10^3	60	0.8	1×10^{-2}	1.5×10^{-4}	1×10^{-2}
O^+–Al	5.7×10^3	6.0	0.18	1×10^{-3}	3×10^{-5}	3×10^{-2}
O_2^+–Al	1×10^4	40	2.0	4×10^{-3}	2×10^{-4}	5×10^{-2}

In Figure 18.4, the yields of secondary Al^+, Al_2^+, and Al_3^+ particles during bombarding of the Al by Ar^+, O_2^+, N_2^+, O^+, N^+, ions at the energy $E_0 = 40$ keV were presented. From Figure 18.4, significant reduction of clusters yields in cases of N^+–Al and O^+–Al, especially in case of N^+–Al with increase of size of clusters is visible. Besides, bombarding surfaces by ions of oxygen

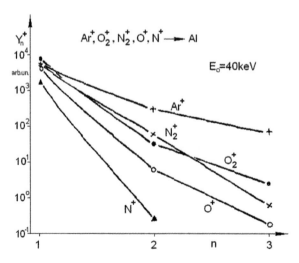

FIGURE 18.4 The yields of Al_n^+ cluster ions during Ar^+, O_2^+, N_2^+, O^+, N^+–Al at $E_0 = 40$ keV.

conducts to oxidation of aluminum, and that increases probability of emission of secondary particles. Because of a multielectronic configuration of ions Ar^+, they interacted with atoms of aluminum of a near-surface layer and therefore in this case, the emission of clusters is considerable. As it was noted above, such behavior can be caused by a big penetration O^+, N^+ in Al in difference of Ar^+, and that leads to reduction of clusters yields.[20]

In Figure 18.5, the yields of a different ions (single negative and positive, multicharged, cluster, and molecular) and radiations during Ar^+–Al at different ($Po_2 = 5 \times 10^{-10}$–1×10^{-6} torr) pressures of oxygen in the chamber of interaction at $E_o = 40$ keV were also presented.[15,22] The increase at 2–3 orders of magnitude of AlO^+ and AlO^{2+} cluster ions yields with the increase pressure of oxygen in the camera of interaction indicates process of oxidation of a surface of aluminum. From Figure 18.5, it is visible that the yields of cluster ions of Al^{2+} and Al^{3+}, just as well as multicharged Al^{2+} and Al^{3+} ions by two to three times decreases with increase of oxygen pressure Po_2.

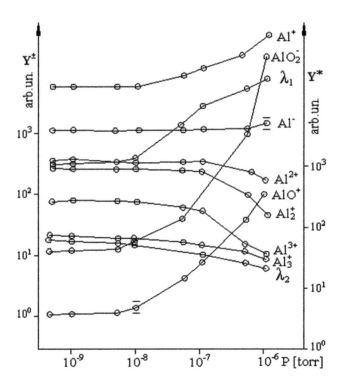

FIGURE 18.5 The yields of ions ($Y_n^\pm$) and radiations (Y^*) during Ar^+–Al at different pressure of oxygen Po_2 in the chamber of interaction; $E_o = 40$ keV.

As for yields of the photons which are formed at decay of the excited states of atoms (Al)*–(λ_1) and ions (Al$^+$)*–(λ_2) of Al, they behave differently: the yields of atoms in the excited states increases considerably as the number of the excited ions poorly decreases with increase of Po_2. The exception represents behavior of yields of negative ions of aluminum (Al) which practically does not change with change in Po_2. The behaviors of the emitted particles and radiation that are presented in Figure 18.5 can be connected with changes of electronic structure of a surface of aluminum at oxidation.[22]

The yield of ions Al$^+$, Al$_2^+$, and Al$_3^+$ is measured at angles of bombardment $\alpha = 0°$ and $\alpha = 45°$ by ions Ar$^+$ in an energy range $E_0 = 5$–9 keV and the ratio of ions yields $Y_{21} = Y_2^+/Y_1^+$ and $Y_{31} = Y_3^+/Y_1^+$ is presented in Figure 18.6.[13,14] It is visible that values of ratio Y_{21} and Y_{31} at angles $\alpha = 45°$ are above than at $\alpha = 0°$. From the data, yields of various ions follows the weak dependence of yield on α in a range of angles (0°–45°). Many experimental investigations of primary ions bombardment of solids specify that in the above range of angles of incident, yields of emitted ions and radiation practically do not change. Increase of yield of cluster ions Al$_n^+$ ($n = 2, 3$) at increase of an angle α can be connected to reduction of average depth of exit, and that with increase of probability of their emission.[20,22] Ion yields of Al$^+$, and also the molecules of AlH$^+$, AlO$^+$, AlOH$^+$ are increased approximately twice during transfer of projectile from N$^+$ to N$_2^+$ and from O$^+$ to O$_2^+$, in difference of the Al$_2^+$ and Al$_3^+$. The yields of Al$_2$O$^+$ increase more than by an

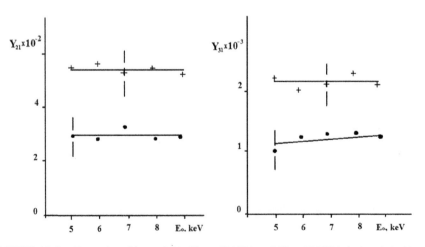

FIGURE 18.6 The ratio of ions yields $Y_{21} = Y_2^+/Y_1^+$ and $Y_{31} = Y_3^+/Y_1^+$ during Ar$^+$–Al at various angle of bombardment: $\alpha = 45°$ (+) and $\alpha = 0°$ (0).

order value during the replacement of the bombarding atomic ions N^+, O^+ by molecular ions N_2^+, O_2^+. Large ion yield (especially Al_3^+) in the case $Ar^+–Al$ is also observed. With the energy of ion bombardment of N_2^+, O_2^+ (tens of keV) surface of Al, one should assume that with the first step of collision, the dissociation of ions N_2^+, O_2^+ occurs and the doubled intensities (atom + ion) are obtained with the redistribution between them primary E_0 energies.[16,17,22]

After emission of clusters, some difficult processes take place (fragmentation, ionization, de-excitation) and consequently, it is complicated to determine the mechanisms of clusters forming.[8] It is reasonable to assume that the mechanisms of the emission of clusters, especially the output of the "light" ($n \leq 10$) or large ($n \geq 10$), are distinguished. The experimentally observed light clusters also consist of the number of clusters after fragmentation of large clusters.

Calculation by Betz 2004, Hofer 1991, and Urbassek 1989 shows that the sputtering in the form of atoms occurs for about hundreds fs after bombardment, the emission of light clusters during the first ps, and detection for tens of μs after emission. A detailed analysis reveals that most of these "nascent" excited clusters exhibit an extremely high degree of strong internal (vibrational) excitation (the medium energy of the excitation of ≈ 1 eV/atom) and will therefore rapidly decompose on their path away from the surface. Due to the short (picosecond) time scale of the corresponding fragmentation processes, the experimentally detected clusters basically represent the "final" (meta)stable end products. According to Betz 2004, Al_2^+ are formed from the adjacent atoms of surface layer and Al_3^+ are emitted during the first ps after bombarding, therefore, the simultaneous (ion atom) collision of molecular ion plays important role. Depending on the method of formation, different excited states of cluster and the fragmentation are produced along the different channels. It occurs with different probabilities.[7] Also, it is considered that the transfer of the part of the internal excited energy of cluster into the electronic subsystem, and the exchange interaction of the electron structure of surface and electronic subsystem of the emitted complex determine "final" state of cluster.

According to molecular dynamics (MD) computer simulations,[3] the atomized flow at sputtering contains 92% atomic particles from the first layer, and 5.1% from the second, 1.3% from the third, and 1.4% from deeper layers of surface. Clusters, especially "lights," are formed from the atoms of near-surface, first layer. About 60–70% of all clusters with $n > 3$ are exclusively from first layer atoms.[3] This supports and confirms previous evidence that sputtered dimers and larger clusters originate predominantly from next-neighbor sites at the surface, indicating that a true double-collision mechanism is responsible for the majority of dimers formed. With shaping of surface appear the so-called ragged connections between the atoms, which are not energetically profitable and approaching the equilibrium of system, where the additional constraints are formed and the united pairs are created—dimers they converge, being moved away from the adjacent atoms.

The main mechanism of cluster emission from metals discussed in the literature is the so-called double-collision mechanism (recombination model, atomic combination, statistical model), which is based on the fact that, for metals and semiconductors, the dissociation energy of a dimer is typically smaller than the surface binding energy (sublimation energy) and, if enough center-of-mass energy is transferred to a dimer, its internal energy is usually so high that it will not survive the emission from the solid intact. No preformed molecules exist. In the double collision mechanism, dimers or larger clusters are formed from independently ejected atoms, if they are emitted close enough in time and space and their relative kinetic energies are less than the binding energy of the dimer (cluster).[1,6]

18.4   CONCLUSIONS

The observed decrease of the output of the cluster ions of Al_2^+ and Al_3^+, from aluminum upon transfer from the bombarding molecular ions (N_2^+, O_2^+) to the atomic (N^+, O^+) can be connected with the formation of the excited states of Al_2^+ and Al_3^+. During de-excitation, the fragmentation of clusters and thus the decrease in their number occur. The formation of the excited states, which lead to the fragmentation of the clusters Al_2^+ and Al_3^+ in the case of the bombardment with molecular ions is, apparently, improbable. Furthermore, from the calculations by Betz 2004, according to the model of dual collision, the simultaneous collision of molecular ion (ion + atom), apparently, conducts to effective removal from the surface layer of the clusters (Al_2^+, Al_3^+). On the basis of these considerations, the role of molecular ions with respect to the atomic in the emission of Al_2^+ and Al_3^+, predominates. Interactions between the electron structure of surface and the electrons of the atoms, which form part of emitted poly atoms, are being determined with shaping of cluster.[20]

It is found by Betz (2004) that two distinct processes can be distinguished which lead to cluster emission under energetic ion bombardment. The first process causes the emission of small ($n < 10$) "light" clusters, which are emitted by a collective motion during the development of the collision cascade within the first picosecond after impact. Thus, emission times of such clusters agree with the emission times of atoms in sputtering and its time-dependent variation must play a dominant role in cluster sputtering. Such a process can be envisioned if, for example, a few layers below the surface, an energetic recoil causes the development of a subcascade. Typically, clusters emitted by this mechanism consist of atoms, which are neighboring in the target and are almost exclusively surface atoms, similar to all sputtered atoms.

Emission of large clusters (cluster sizes of 10 or more atoms), ($n > 10$), as observed experimentally, is a puzzling phenomenon. The cluster yield data, collected recently, provide clear evidence that the emission of larger clusters in sputtering cannot be described by simple statistical models. The emission of such large clusters happens much later (5–10 ps after ion impact). Emission can occur for spike events, where all the energy of the impinging ion is deposited locally in a small volume near to the surface, and the sputtering yield is three to five times the average yield. Difference between the yields of Al_2^+ and Al_3^+ clusters is sensitive to whether atomic (N^+, O^+) or molecular (N_2^+, O_2^+) projectiles are used.[15,21] Two possible explanations for this difference are suggested: (1) the formation of dissociate excitation states of emitted clusters and (2) the effective production of Al_2^+ and Al_3^+ from neighbor atoms in a simultaneous impact of molecular ions (ion + atom).

KEYWORDS

- atomic cluster
- surface
- sputtering
- ions
- emission

REFERENCES

1. Gerhard, Z. *Phys. B* **1995**, *22*, 31.
2. Benninghoven. *Surf. Sci.*, **1973**, *35*, 427.
3. Betz, G.; Husinsky W. *Phil. Trans. R. Soc. Lond. A* **2004**, *362*, 177.
4. Hofer, W. O. *Sputtering by Particle Bombardment III*; Behrisch R., Wittmaack K., Eds.; Springer: USA, 1991.

5. Urbassek, H. M. *Sputtering by Particle Bombardment;* Behrisch, R., Eckstein, W., Eds.; Springer: USA, 2006.

6. Staudt, C.; Heinrich R.; Wucher A. *Nucl. Instrum. Met. Phy. Res. B* **2000**, *164,*, 677.

7. Urbassek, H. M. *Rad. Eff. Defects Solids* **1989**, *109*, 293.

8. Urbassek, H. M. *Nucl. Instrum. Method Phys. Res. B* **1988**, *31*, 541.

9. Urbassek, H. M.; Hofer W. O. *Det kongelige Danske Vid. Selsk. Mat. Fys. Medd.* **1993**, *43*, 97.

10. Armour, D. G.; Jimenez-Rodriguez, J. J.; Barber, C. H.; Snowdon, K.; Hedbavny P. *Vacuum* **1984**, *34*, 2177.

11. Armour, D. G.; Kikiani, B. I.; Meskhi, G. G.; *Bulletin of the Georgian Academy of Sciences* **1994**, *150*(3), 429.

12. Armour, D. G.; Meskhi, G. G.; Kikiani, B. I.; Chrelashvili, G. K. Materials of XI International Conference on *Ion Surface Interaction, ISI-1993*, Moscow, Russia, 1993, Vol. 2, 36.

13. Armour, D. G.; Makorenko, B. M.; Meskhi, G. G. Materials of XVII International Conference on *Ion Surface Interaction, ISI-1991*, Moscow, Russia, 1991, Vol. 2, 37.

14. Dubinskii, I. N.; Kikiani, B. I.; Meskhi, G. G.; Chrelashvili, G. K. Materials of VI International Seminar on *Secondary Ions and Ion-Photon Emission*, Kharkov, 1991, 72.

15. Armour, D. G.; Meskhi, G. G. Materials of IX International *Meeting on Radiation Processing*, Istanbul, 1994, 14.

16. Armour, D. G.; Gorgiladze, B. G.; Meskhi, G. G.; Sichinava, A. V. Materials of XVII International Conference on *Ion Surface Interaction, ISI-2005*, Moscow, Russia, 2005, 203.

17. Kikiani, B.; Lomsadze, R.; Meskhi, G.; Meskhi, L. Materials of VIII International Conference on *the Physics of HCI*, Japan, 1996, 87.

18. Wucher, A.; Wahl M. *Nucl. Instrum. Method Phys. Res. B* **1996**, *83*, 73.

19. 19. Jemilev, N. H. Materials of the XII International Conference *Ion-Surface Interaction*, Moscow, 1995, Vol. 1, 203.

20. Armour, D. G.; Meskhi G. G. Materials of the XVIII International Conference *Ion-Surface Interaction*, Moscow, 2007, Vol. 2, 217.

21. Meskhi, G. G. Materials of the 2nd International Caucasian Symposium on *Polymers and Advanced Materials*, Tbilisi, Georgia, 2010, 11.

22. Meskhi, G. G. Materials of the 4th International Caucasian Symposium on *Polymers and Advanced Materials*, Batumi, Georgia, 2015, 91.

23. Bitensky, I. S.; Parilis, E. S. *Nucl. Inst. Method B* **1987**, *21*, 26.

24. Hamzaa, A. V.; Schenkel, T.; Barnes, A. V. *Eur. Phys. J. D* **1999**, *6*, 83.

CHAPTER 19

ADVANCED OXIDATION PROCESSES, MEMBRANE SEPARATION PROCESSES, AND ENVIRONMENTAL ENGINEERING TECHNIQUES: A VISION FOR THE FUTURE AND A CRITICAL OVERVIEW

SUKANCHAN PALIT*

Department of Chemical Engineering, University of Petroleum and Energy Studies, Bid holi via Premnagar, Dehradun 248007, Uttarakhand, India

Corresponding author. E-mail: sukanchan68@gmail.com; sukanchan92@gmail.com

CONTENTS

ABSTRACT

Environmental engineering science in today's scientific generation is moving at a drastic pace. Vision of science, technological validation, and the visionary scientific advancements will lead a long way in the true realization of environmental sustainability. Sustainability of the environment is a key issue today. Human civilization is slowly and steadily moving toward a greater vision of green science and technology. Ecological imbalance, environmental catastrophes, and industrial water pollution have urged the scientific domain to gear up for new innovative technologies. At this critical juncture of human history and time, the global wastewater challenge is formidable. Today, industrial wastewater treatment stands in the midst of deep introspection and unmitigated crisis. In such a crucial juncture of human history and time, the main urge of human civilization is toward protection of environment. The author skillfully delves deep into the world of novel separation processes such as membrane science and nontraditional environmental engineering techniques such as advanced oxidation processes. Hazardous substances are recalcitrant to primary and secondary treatments; so the need arises for tertiary treatment such as advanced oxidation processes. The author also delineates the immense potential of ozonation and other advanced oxidation processes (AOPs) techniques. Vision of science, the unmitigated challenges, and the scientific urge to excel are the torchbearers toward a greater visionary world of chemical process engineering and environmental engineering science. The aim of this study is to review the use of ozone oxidation, titanium dioxide/UV light process, hydrogen peroxide/UV light process, and Fenton's reaction procedures in wastewater treatment. The pros and cons of the processes are highlighted in detail and precision. The author also deeply observes and informs the readers the success of novel separation processes such as membrane separation processes in separating recalcitrant chemicals in industrial wastewater. The vision and the challenge of this treatise unravel a new chapter in the research and development arena of novel separation processes and nontraditional environmental engineering techniques.

19.1 INTRODUCTION

Human civilization and relevant scientific endeavor in environmental engineering are surpassing visionary boundaries. The challenge of science and technology in today's scientific arena is inimitable. Environment, in today's world, is at a state of immense distress and is facing immense catastrophes. Global water shortage is a burning and vexing issue. The scientific urge toward global water challenge needs to be reemphasized and reenvisioned at each step of scientific endeavor and relevant research pursuit. Research and development initiatives in novel separation processes and nontraditional environmental engineering techniques are far-reaching as well as latent. Today, scientific challenges, vision, and sagacity are opening up new vistas in research in the field of technologies and innovations in environmental engineering science.

A wide variety of recalcitrant organic compounds is found in industrial and municipal wastewater. Some of these compounds (both synthetic organic chemicals and naturally occurring substances) pose severe problems and unmitigated challenges in biological treatment systems due to their resistance to biodegradation or/and toxic effects on microbial processes. As a result, the use of alternative treatment technologies, aiming to mineralize or transform refractory molecules into others that could be further degraded, is a matter of deep comprehension. Among them, advanced oxidation processes (AOPs) have already been used for the treatment of wastewater containing recalcitrant compounds such as pesticides, surfactants, coloring matters, pharmaceuticals, and endocrine-disrupting chemicals. Moreover, they have been successfully and effectively used as pretreatment methods in order to reduce the concentrations of toxic organic compounds that prevent biological wastewater treatment. The success and the purpose of this well-informed treatise go beyond visionary boundaries of scientific and technological frontiers. The author pointedly focuses on the future of separation processes and the successful realization of environmental sustainability.[30-33]

19.2 VISION OF THE TREATISE

The vision and the aim of this treatise are vast and versatile. The challenge of human civilization needs reenvisioning with every step of research pursuit. The science of global water challenge is witnessing a new and dramatic change. This treatise pointedly focuses on the efficacy of AOPs and membrane separation processes with more stress on global water issues and industrial and drinking water problem. The path of human scientific endeavor is replete with scientific imagination. The main mission of this study focuses into the arduous world of AOPs, membrane science, and other nontraditional environmental engineering techniques. Science has a definitive vision of its own. Man's visionary prowess, civilization's progress, and the wide and versatile scientific advancement are in today's world all leading toward a newer goal of environmental engineering science. Deep research in the field of fouling and cleaning and the wide concept of concentration boundary layer are the other facets of this treatise. The author also lucidly delineates the wide world of groundwater heavy metal remediation and decontamination and the imminent dangers of arsenic groundwater contamination.

19.3   NEED AND THE RATIONALE OF THE STUDY

The need and the rationale of this study are wide, vast, and far-reaching. Global water shortage has changed the scientific scenario. Environmental engineering science and chemical process engineering are undergoing disastrous challenges with the growing concern of global climate changes and ecological imbalance. Water technology needs reenvisioning and restructuring with the progress of human civilization; here comes the importance of this study of membrane science and AOPs. Separation phenomenon is witnessing new challenges, and the relevant validation of science is changing the face of chemical process engineering. Provision of basic human needs such as water needs to be on the highest priority in future scientific pursuit. The other area of this endeavor is groundwater quality and heavy metal groundwater remediation and decontamination.

19.4   SCOPE OF THE STUDY

The scope of this study surpasses visionary frontiers. The challenge and purpose of scientific advancements in environmental engineering science need reenvisioning. Global water crisis and its alleviation are the focal points of this study. Green chemistry and green technology are the stress areas of this well-informed study. Application of nanotechnology, the wide world of ozonation, and the future trends of application areas of membrane science are the hallmarks of this study.

Nowadays, due to the increasing presence of recalcitrant molecules, refractory to the microorganisms in the wastewater streams, the conventional biological methods have become futile. Hence, there is a need of newer technologies and newer innovations in the form of oxidation technologies to be combined with the biological methods. The scope of this study encompasses various newer innovative nontraditional technologies such as advanced oxidation techniques. The future trends in the areas of research in chemical oxidation are cavitation, acoustic cavitation, sonochemical reactors, and other advanced oxidation techniques. The concept of sonochemical reactors is new and needs scientific revamping.

19.5 ENVIRONMENTAL SUSTAINABILITY AND THE WIDE VISION FOR THE FUTURE

Environmental sustainability is the coin word of the present scientific generation. Economic development with disregard toward energy and environmental sustainability is futile. The vision of science and the cause of green technology are opening up new vistas of scientific endeavor in the areas of sustainable development. Economic sustainability, social sustainability, and environmental sustainability are veritably interlinked. The cause of global climate change and global water frontiers are opening up new scientific thoughts and scientific introspection.

19.6 GLOBAL WATER CRISIS, ENVIRONMENTAL ENGINEERING CATASTROPHES, AND THE ROAD FORWARD

Global water crisis in today's scientific world is in the path of immense introspection and deep scientific vision. Groundwater heavy metal contamination and remediation is another facet of today's research and development initiative. Environmental calamities and ecological imbalance are the pallbearers toward the realization of environmental sustainability of tomorrow. Human scientific endeavor is in the path of immense reenvisioning and restructuring. Global water shortage and the crisis which follows are changing the visionary mindset of newer scientific generation and the avenues of newer scientific innovation.

19.7 SCIENTIFIC DOCTRINE AND DEFINITION OF ADVANCED OXIDATION PROCESSES

Scientific doctrine and fortitude of AOPs are vast, versatile, and far reaching. The vision of science, the challenge of human scientific progress, and the growing environmental concerns have urged environmental scientists to devise newer technologies and newer innovations. Technological validation, the scientific urge to excel, and the futuristic vision of AOPs are paving the way toward a new eon of science and engineering.

A wide range of organic compounds is detected in industrial and municipal wastewater. Some of these compounds (both synthetic organic chemicals and naturally occurring substances) pose severe problems in biological treatment systems due to their resistance to biodegradation or/and toxic effects on microbial processes. As a result, the use of alternative treatment technologies, aiming to mineralize or transform refractory molecules into others that could be further degraded, is a matter of great concern. Among them, AOPs have already been used for the treatment of wastewater containing recalcitrant organic compounds such as pesticides, surfactants, coloring matters, pharmaceuticals, and endocrine-disrupting chemicals.

The main mechanism of AOPs is the generation of highly reactive free radicals. Hydroxyl radicals (HO·) are effective in destroying organic chemicals because they are reactive electrophiles (electron preferring) that react rapidly and nonselectively with nearly all electron-rich organic compounds.

19.8 CHALLENGE, VISION, AND PURPOSE OF ADVANCED OXIDATION PROCESSES

Science and engineering are moving from one paradigm to another. Protection of environment stands as a major support to the advancement of science. Tertiary treatment of wastewater such as advanced oxidation techniques are the necessity of the hour with the advancement of scientific endeavor. The failure of primary and secondary treatments in degradation of recalcitrant chemicals has urged the scientific domain to yearn and reenvision the domain of advanced oxidation techniques. Nowadays, due to the increasing presence of molecules, refractory to the microorganisms in the wastewater streams, the conventional biological methods cannot be used for complete treatment of the effluent; hence, introduction of newer technologies to degrade these chemicals has become veritably imperative.

19.9 RECENT SCIENTIFIC ENDEAVOR IN THE FIELD OF ADVANCED OXIDATION PROCESSES

Human scientific endeavor in the field of AOPs has no bounds and is veritably surpassing visionary frontiers. The scientific challenge, sagacity, and candor will lead a long way in the true emancipation of environmental engineering techniques and true and effective realization of sustainable development. This treatise pointedly focuses on the effectivity of advanced oxidation techniques with greater stress on ozone oxidation. The science of AOPs is moving from one visionary phase over another. The scientific vision and understanding are moving toward a newer and larger realm of application of advanced oxidation techniques. The author also focuses on the wide and versatile world of novel separation processes such as membrane science.

Gogate et al.[2] reviewed imperative technologies for wastewater treatment along with oxidation technologies at ambient conditions. The challenge of this treatise is veritably widespread. Today, due to the increasing presence of molecules, refractory to the microorganisms in the wastewater streams, the conventional biological methods cannot be used for complete treatment of the effluent; hence, introduction of newer technologies and newer innovations to degrade these refractory molecules into smaller molecules which can be further oxidized, has become very much imperative. In this treatise, the authors widely discuss cavitation, acoustic cavitation, reactors used for the generation of acoustic cavitation, optimum operating parameters for sonochemical reactors, hydrodynamic cavitation, reactors used for the generation of hydrodynamic cavitation, optimum operating conditions, and the work done in recent years. The authors also dealt with photocatalysis, reactors used for photocatalytic oxidation, and again optimum parameters. Technological vision and scientific forays into the science of cavitation are opening up new chapters in the field of advanced oxidation techniques.

Stasinakis[3] dealt lucidly on the use of selected AOPs for wastewater treatment in a mini review. Science of AOPs is today moving toward a newer realm. The aim of this study was to review the use of titanium dioxide/UV light process, hydrogen peroxide/UV light process, and Fenton's reactions in industrial wastewater treatment. The main mechanism of AOPs function is the generation of highly reactive free radicals. Hydroxyl radicals (HO·) are effective in destroying organic chemicals because they are reactive electrophiles (electron preferring) that react rapidly and nonselectively with nearly all electron-rich organic compounds. The author extensively deals with the use of AOPs in industrial wastewater treatment.

Oller et al.[4] reviewed combination of AOPs and biological treatments for wastewater decontamination. Technology in today's world is moving at a rapid pace. The scientific challenges in the field of AOPs are unimaginable. Today, there is a continuously growing worldwide concern for development of water reuse technologies, mainly focused on agriculture and industry. AOPs are considered a highly competitive water treatment technology for the removal of these organic pollutants not treatable by conventional techniques due to their chemical stability and low biodegradability. This chapter reviews recent research combining AOPs (as a pretreatment or post-treatment stage) and bioremediation technologies for the decontamination of a wide range of synthetic industrial wastewater. According to the authors, the main routes for destroying toxic compounds in natural water are biodegradation and photodegradation. Another facet of their research is the industrial wastewater treatment by the combination of AOPs/biotreatment technology.

Chakinala et al.[5] discussed with deep and cogent insight treatment of industrial wastewater effluents using hydrodynamic cavitation and the advanced Fenton's process. The spectacular effects of cavitation in terms of highly reactive free radicals, conditions of high temperature/pressure, and generation of intense turbulence and liquid circulation currents can be used for industrial wastewater treatment.

Comninellus et al.[6] discussed with deep comprehension AOPs for water treatment and its advances and trends for research and development initiatives. AOPs are at its helm in the prime objective and motivation of industrial wastewater treatment. Over the past few decades, multidisciplinary research has been carried out to address a broad spectrum of chemical process engineering such as deep understanding of chemical process fundamentals, elucidation of chemical reaction kinetics, development of new materials, chemical process modeling, and instinctive process integration.

Gogate[7] delineated in detail, in a well-observed review, cavitational reactors for process intensification of chemical processing applications. The challenge of cavitation is immensely growing from the last decade. Cavitational reactors are an innovative and challenging form of multiphase reactors. A design of a pilot-scale sonochemical reactor has been presented, which has to be scaled up to an industrial reactor.

Sires et al.[8] dealt lucidly on electrochemical AOPs (EAOPs) in a deeply innovative review. EAOPs have been developed in recent years for the prevention and remediation of environmental pollution. The focus, in this treatise, was on water streams. These methods are based on the electrochemical generation of a very powerful oxidizing agent, such as hydroxyl radicals (HO·) in solution, which will be able to destroy organics to complete mineralization.

Shemer et al.[9] discussed degradation of pharmaceuticals, Metronidazole via UV, Fenton, and photo-Fenton processes. Degradation rates and removal efficiencies of Metronidazole using UV, UV/H_2O_2, H_2O_2/Fe^{2+}, and $UV/H_2O_2/Fe^{2+}$ were studied in deionized water.

Saharan et al.[10] with cogent insight dealt with hydrodynamic cavitation as an advanced oxidation technique for the degradation of Acid Red 88 dye. The effect of various operating parameters such as inlet pressure, inlet concentration of dye, pH of solution, and addition of H_2O_2 and a catalyst ($Fe-TiO_2$) on the extent of decolorization and mineralization have been studied with the aim of enhancing degradation.

Mendez-Arriaga et al.,[11] in a deeply informed treatise, discussed and researched mineralization enhancement of a recalcitrant pharmaceutical pollutant in water by advanced oxidation hybrid processes. Degradation of the biorecalcitrant pharmaceutical micropollutant ibuprofen (IBP) was carried out by means of several advanced oxidation hybrid configurations.

Gogate et al.[12] discussed in detail cavitation reactors and its efficiency assessment using a model reaction. Acoustic and hydrodynamics cavitation can be used for a variety of applications ranging from biological applications such as cell disruption to chemical reactions such as oxidation of organic pollutants in aqueous effluents, including biorefractory toxic chemicals. Science of AOPs is reaching new heights with the evolution of newer and innovative techniques such as hydrodynamic cavitation. Different equipment used for cavitational effects was compared based on a model reaction (decomposition of potassium iodide resulting in iodine liberation).

Klavarioti et al.[13] discussed with deep comprehension in a review on removal of residual pharmaceuticals from aqueous systems by AOPs. Over the last decades, pharmaceuticals are considered as an emerging environmental problem due to their persistence to the aquatic ecosystems even at low concentrations. AOPs are technologies based on the intermediacy of hydroxyl and other radicals to oxidize recalcitrant, toxic, and nonbiodegradable compounds. In such a crucial situation, this treatise reviews and assesses the effectiveness of various AOPs for pharmaceutical removals in water and wastewater systems.

Azbar et al.[14] compared various AOPs and chemical treatment methods for COD and color removal from a polyester and acetate fiber dyeing effluent. In this treatise, a comparison of various AOPs (ozone, ozone/hydrogen peroxide, hydrogen peroxide/ultraviolet, ozone/hydrogen peroxide/ultraviolet, ferrous/hydrogen peroxide) and other chemical treatment methods for the chemical oxygen demand (COD) and color removal from a polyester and acetate fiber dyeing effluent is undertaken.

19.10   VISION OF OZONE OXIDATION AND OTHER ADVANCED OXIDATION TECHNIQUES

The vision of ozone oxidation and other advanced oxidation techniques are vast, versatile, and far-reaching. AOPs techniques are changing the face of human scientific research pursuit. Scientific regeneration in the field of AOPs needs reenvisioning and also needs to be effectively addressed. Global water initiatives today stand in the midst of deep crisis and, in a similar manner, in the juxtaposition of immense introspection. The challenge of science and technology and the vision forward are evolving into new dimensions of global water solutions. The success of global water solutions and the imminent need for alleviating global water challenges has urged the scientific domain to move toward newer initiatives and innovations.[30–33]

19.11   NONTRADITIONAL ENVIRONMENTAL ENGINEERING TECHNIQUES AND THE SUCCESS OF SCIENTIFIC ENDEAVOR

Nontraditional environmental engineering techniques, in today's world, are witnessing newer and challenging present and the future. The success of scientific endeavor needs to be reenvisioned and revamped with each step of human life. Today, environment is at a state of immense catastrophe. Global water crisis, global climate change, and the recurrent environmental disasters have urged and propelled science to move toward newer solutions and innovations.[30–33]

19.12   SCIENTIFIC DOCTRINE OF MEMBRANE SEPARATION PROCESSES: ITS DEFINITION AND CLASSIFICATIONS

Membrane separation phenomenon stands in the midst of deep scientific comprehension today. Novel separation techniques are changing the face of scientific research pursuit. The classification of membrane separation processes is presented in Table 19.1.

Human scientific endeavor is on the path of newer regeneration and a newer visionary eon. Global water challenges are witnessing drastic changes. Chemical process engineering and environmental engineering science are, in a similar vein, witnessing wide restructuring and reenvisioning globally. The global concern for environment has changed the scenario of human scientific research pursuit.

Global water crisis is a serious and urgent security issue.[31] The undeniable seriousness of global water situation was first brought to the attention of the international community at the 1992 United Nations Conference on Environment and Development in Rio de Janeiro, which came to be known as the Rio summit.[31] Twenty years after the Rio summit, the global situation

TABLE 19.1 Characteristics of Membrane Processes.[1]

Process	Driving force	Retentate	Permeate
Osmosis[1]	Chemical potential	Solutes/water	Water
Dialysis	Concentration difference	Large molecules/water	Small molecules/water
Microfiltration	Pressure	Suspended particles/water	Dissolved solutes/water
Ultrafiltration	Pressure	Large molecules/water	Small molecules/water
Nanofiltration	Pressure	Small molecules/divalent salts/dissociated acids/water	Monovalent ions/undissociated acids/water
Reverse osmosis	Pressure	All solutes/water	Water
Electrodialysis	Voltage/current	Nonionic solutes/water	Ionized solutes/water
Pervaporation[1]	Pressure	Nonvolatile molecules/water	Volatile small molecules/water

with respect to water has improved in some domains, but still has long ways to go in other domains because of the needs and the global concerns of rapidly growing population. According to a Global Report, water scarcity is becoming a major issue on our planet.[31] The most important and vexing questions which attack the human society and human science are rapidly growing population with associated changes in lifestyle and consumption patterns; competition between sectors, such as industry, agriculture, and energy for precious land and water issues; inadequate access to water supply and sanitation services; the failure to envision indigenous water rights, and include marginalized populations in water decision-making processes; matters relating to environment protection; and growing tension over trans-boundary water issues.[31] In this treatise, the author deeply comprehends the success of the global water challenges, the effectivity of environmental engineering techniques, and the vast, versatile, and wide global arena of global water solution procedures.[31]

19.13   VISIONARY SCIENTIFIC RESEARCH PURSUIT IN THE FIELD OF MEMBRANE SCIENCE AND TECHNOLOGY

Scientific research pursuit in our present day human civilization are veritably linked with the global cause of provision of basic human needs such as water. Membrane science research and development are gaining high grounds in scientific scenario. The challenge and the vision in the field of membrane science and technology are growing toward a newer eon in environmental protection. Environmental protection and the wide world of desalination are invariably linked with membrane science and technology. Scientific research endeavor in this treatise is widely observed, and the causes and concerns of environmental catastrophes are veritably addressed.

Nanofiltration (NF), microfiltration, and ultrafiltration (UF) comprise the most advanced and visionary technologies in the wide domain of membrane science and technology. Vision of science is witnessing drastic challenges at each step of human scientific research pursuit. The challenge and future vision of science and technology are inspiring the future scientific innovation. In this section, the author, with strong scientific rigor, delineates some significant work in the field of NF and other membrane technologies.

Gomes et al.[15] delineate, in a well-researched article, integrated NF and upflow anaerobic sludge blanket (UASB) treatment of textile wastewater for in-plant reuse. The filtration characteristics of simulated dyeing effluents containing Acid Orange 7, sodium sulfate, and a pH buffer made of acetic acid and sodium acetate is described using a commercially available NF membrane. The prevailing sources of color in most textile wastewaters are azo dyes. The present work intends to demonstrate the environmental and economic benefits arising from jointly using membrane technology, namely NF, and the UASB reactor for the elimination of textile pollutants concentrated in the retentate and freshwater saving through recycling of the membrane permeate.

Ogunlaja et al.[16] evaluated the efficiency of a textile wastewater treatment plant located in Oshodi, Lagos, Nigeria. The major and the important part of textile processes is based on chemical reactions carried out in aqueous heterogeneous systems which eventually generates effluents of extremely variable composition and constitute an environmental disaster of major and unavoidable concern. This study pointedly focuses on the investigation of characteristics of effluent from textile industry which generates about 1577 m^3 waste per day and the effectiveness of the wastewater treatment practiced.

Central Pollution Control Board Report[17] delineates in detail advanced methods for treatment of textile industry effluents. The challenges in Indian scenario are immense and versatile. The gamut of industrial wastewater treatment depends on science and society. This report targets different advanced methods for treating textile industry effluents as well as for recovery of water/salt in the process. Their economic analysis, benefits, and constraints are discussed in details.

Hassani et al.[18] discussed in minute details NF process on dye removal from simulated textile wastewater. The challenge and vision of membrane science are awesome. Dyes cannot be removed by biological processes; so, comes the need of membrane separation processes such as NF. In this study, NF process was used for removal of different dyestuffs from solution. The rate of dye removal by spiral wound NF membrane in film thin composite MWCO=90 Daltons was evaluated for four classes of dyes: acidic, disperse, reactive, and direct in red and blue dyes medium.

Avlonitis et al.[19] discussed with cogent insight simulated cotton dye effluents treatment and reuse by NF. The textile industry uses enormous quantities of water, which in many cases is disposed to the environment without proper regulations. The effluent contains high salts and organic concentrations and here arises its need for treatment. The challenge of this research was the excellent performance of TRISEP membrane. Dye degradation is significant and salt concentration is decreased with the application of this membrane.

Pazdzior et al.[20] reviewed the integration of NF and biological processes for textile wastewater treatment. The effective implementation of the biological anaerobic–aerobic system in separated reactors to the NF concentrate treatment was presented. Different chemical, physical, and biological methods can be applied to remove dyes from wastewater but there was little research on dye concentrate treatment. The biological methods are the only alternative. Complete mineralization and low costs are this treatment's advantages.

Ramesh Babu et al.[21] discussed cotton textile processing and its waste generation and effluent treatment. The success and vision remains unparalleled in this type of wastewater treatment. This review discusses cotton textile processing and the visionary future of textile wastewater treatment.

Torabian et al.[22] with deep scientific precision dealt with efficiency evaluation of textile basic dye removal from water by NF. NF is effective for treatment of organic and inorganic pollutants in surface water and groundwater resources. The efficiency of textile dye removal by a nanofilter NF90 (Dow FilmTec) is studied. Dye rejection was studied using basic dye. The membrane process which meets the environmental challenges and the wide world of environmental regulations is NF. The authors deeply delve into dye removal by NF in details.

Fersi et al.[23] dealt with deep comprehension on textile plant effluent by UF and NF for water reuse. Membrane fouling stands as a major impediment as membrane flux declines. This challenge needs reenvisioning. Membrane process efficiency can be affected by membrane pore blocking and cake formation. This scientific research pursuit combines two membrane separation techniques. The UF was used as a pretreatment for NF process.

19.14 CHALLENGE OF HUMAN CIVILIZATION, REALIZATION OF ENVIRONMENTAL SUSTAINABILITY, AND HEAVY METAL GROUNDWATER CONTAMINATION

Today, human civilization stands in the midst of deep introspection. Environmental protection and successful realization of environmental sustainability is the absolute need of the hour. Global water crisis is today in the state of immense comprehension. Chemical process engineering and its applications needs reenvisioning with each step of human scientific endeavor in environmental sciences. Global caution, deep environmental concern, and the visionary roads of science will veritably go a long way in the true emancipation of environmental and energy sustainability.[30–33]

The contamination of groundwater by heavy metal, originating either from natural soil sources or from anthropogenic sources, is a matter of immense and feasible concern to the public health.[29] The challenge and the vision for the future needs restructuring in the field of groundwater remediation. Remediation of contaminated groundwater is of highest priority since billions of people all over the world use it for drinking purpose.[29] Selection of a suitable and innovative technology for contamination remediation at a particular site is one of the challenging jobs due to extremely complex soil chemistry and aquifer characteristics, and there remains no thumb rule in solving it.[29] Groundwater contamination will remain a burning global issue with the passage of human history and time. Scientific candor, scientific fortitude, and immense scientific understanding are needed in pursuing the greatness of scientific research in heavy metal remediation.[29] Keeping the sustainability issues and environmental ethics in human foresight, the technologies encompassing natural chemistry, bioremediation, and biosorption are highly recommended to be adopted in appropriate cases.[29]

19.15 DESALINATION SCIENCE, GLOBAL WATER CRISIS, AND THE STATUS OF ENVIRONMENT

In today's human civilization, desalination science has turned out to be a viable alternative to global water crisis in many parts of the planet. The challenge, the definitive vision, and the scientific cognizance are changing the face of scientific research pursuit in the field of desalination science. In today's scientific realm, desalination is a viable alternative to global water shortage crisis in many parts of the world. The status of environment today is in a state of immense distress. Scientific vision, scientific adjudication, and strong scientific forbearance will be the torchbearers toward a greater visionary tomorrow. The human planet needs to restructure and revamp itself in order to realize the true vision of environmental sustainability. Desalination science and technology are in the forefront of global water technology vision. The author repeatedly points out the success of application of novel separation processes and nontraditional environmental engineering techniques in the avenues of scientific endeavor. Environmental sustainability is no more a mirage but a true vision toward future.[30–33]

19.16 VISIONARY RESEARCH PURSUIT IN DESALINATION SCIENCE

Desalination science is moving from one paradigm to another. Global water initiatives are changing the face of human civilization. Groundwater quality is another facet of global water challenge. Science is a visionary colossus without a definite will of its own. Technological vision, the immense scientific envisioning, and the validation of science are extremely necessary in the future forays in the wide world of desalination science.

The engineering and science of desalination in today's scientific realm are scaling visionary heights. The author brings before the scientific horizon the major work done in the field of desalination science.

McCutcheon et al.[24] widely observed a novel ammonia–carbon dioxide forward (direct) osmosis desalination process. The process uses an ammonium carbonate draw solution to extract water from saline-feed water across a semipermeable polymeric membrane. Freshwater scarcity is a growing issue in many regions of the world. Unchecked population growth and stressed freshwater resources cause many countries around the world to move toward ocean as a source of freshwater. Current desalination technologies are immensely expensive and energy intensive. Reverse osmosis (RO) is highly expensive. The need of the hour should be toward alternative desalination technologies. Forward (or direct) osmosis is a process that may be able to desalinate saline water resources at a notably reduced cost. In forward osmosis, like RO, water transports across a semipermeable membrane that is impermeable to salt. However, instead of using a hydraulic pressure to create the driving force for water transport through the membrane, the forward osmosis process utilizes an osmotic pressure gradient.

Fritzmann et al.[25] reviewed state-of-the-art RO desalination. Water scarcity throughout the world has urged the scientific community to venture forward for newer scientific innovations. Water scarcity and water shortage are a threat to human society and progress of human civilization. The most commonly used desalination technologies are RO and thermal processes such as multistage flash (MSF) and multieffect distillation (MED). This review discusses the current state of the art of RO desalination, dealing not only with the RO stage, but also with the entire process from raw water to posttreatment of product water. The discussion of process fundamentals, membranes and membrane modules, and current and future trends of research and development initiatives are delineated in minute details.

Lattemann et al.[26] discussed environmental impact and impact assessment of seawater desalination in detail. The environmental impact assessment of desalination is the imminent need of the hour. Today, technological and scientific validation stands as a major support in the progress of membrane science and desalination. Key issues are the concentrate and chemical discharges to the marine environment, the emission of air pollutants, and the immediate concern of energy demands. This review pointedly focuses on the important concerns of desalination and the keyways and innovations to mitigate environmental concerns.

McGinnis et al.[27] discuss with deep and cogent insight energy requirements of ammonia–carbon dioxide forward osmosis desalination. The energy requirements of ammonia–carbon dioxide forward osmosis desalination are predicted by the use of chemical process modeling software (HYSYS). The innovative forward osmosis process is modeled and simulated using single or multiple distillation columns to separate draw solution solutes from the product water for solute recycling within the forward osmosis system.

Khawaji et al.[28] discussed advances in seawater desalination technologies. Seawater desalination is energy intensive as well as costly alternative to other membrane-based technologies. The authors discuss in minute details important technologies based on the MSF distillation and RO processes. This treatise gives an overview of critical research and development initiatives in the field of seawater desalination.

19.17 VISIONARY FORAYS INTO THE WORLD OF NONTRADITIONAL ENVIRONMENTAL ENGINEERING TECHNIQUES

The challenge of science and technology, the scientific urge to excel, and the global vision are the hallmarks toward a renewed scientific vision and scientific understanding in decades to come. These earlier decades of 21st century are decades of immense scientific challenges and scientific forbearance. Nontraditional environmental engineering techniques such as AOPs in today's visionary world are changing the face of environmental engineering science. The other huge area of scientific vision is the area of heavy metal groundwater contamination and arsenic groundwater remediation. The true vision and the challenge of global water initiatives are facing newer scientific barriers with the rising concern of arsenic and other heavy metal groundwater contamination.[34]

19.18 VISION OF SCIENCE, TECHNOLOGICAL VALIDATION, AND THE CHALLENGE OF RESEARCH AND DEVELOPMENT

Vision of science, technological validation, and the success of research and development initiatives are changing the face of human civilization and environmental engineering science in particular.[34] The alternatives of science and engineering need to be reenvisioned at each step of scientific vision in the field of global water shortage and arsenic groundwater contamination. Developed as well as developing nations are entangled with this burning and vexing issue of heavy metal groundwater contamination. Science has no answer to the world's greatest environmental crisis—arsenic groundwater contamination and subsequent remediation. This treatise stresses repeatedly on the contribution of chemical process engineering to the alleviation of global water crisis and the successful realization of environmental sustainability.

19.19 FUTURISTIC VISION AND FUTURE TRENDS IN RESEARCH ENDEAVOR IN ENVIRONMENTAL ENGINEERING SCIENCE

The futuristic vision of chemical process engineering and the application of membrane separation phenomenon are changing the scientific scenario of global water initiatives. Human scientific advancements today stand in the midst of immense challenges and global environmental concerns. Future research trends in environmental engineering science needs reenvisioning and restructuring with the passage of human history and time. The future research and development initiatives should be targeted toward solutions to global water crisis and the success of heavy metal groundwater remediation.

19.20 CONCLUSION

The challenge and the true vision of chemical process engineering and environmental engineering science are immense and versatile. Human scientific endeavor is witnessing monstrous challenges. Global water initiatives and the success of sustainable development are the immediate need of the hour. Human planet needs reenvisioning with respect to worldwide concern of provision of clean drinking water. This treatise widely encompasses the application of novel separation processes such as membrane science in environmental protection. The success of human civilization lies in the provision of basic human needs such as water. In such a crucial juxtaposition, chemical process engineering is scaling newer and visionary heights. The futuristic trends of global water initiatives enliven the immediate concerns of global water crisis. A new chapter opens up in the avenues of scientific endeavor in water engineering in years to come. Global water initiatives will frame the future of environmental science on human planet.

ACKNOWLEDGMENT

The author wishes to acknowledge the contributions of Chancellor, Vice-Chancellor, Faculty, and students of University of Petroleum and Energy Studies, Dehradun, India, without whom this writing project would not have been completed. Their support is invaluable.

KEYWORDS

- environment
- advanced
- oxidation
- membranes
- bioremediation
- ultraviolet
- separation
- ozonation

REFERENCES

1. Cheryan, M. *Ultrafiltration and Microfiltration Handbook*; Technomic Publishing Company, Inc: Lancaster, 1998.
2. Gogate, P. R.; Pandit, A. B. A Review of Imperative Technologies for Wastewater Treatment I: Oxidation Technologies at Ambient Conditions. *Adv. Environ. Res.* **2004,** *8,* 501–551.
3. Stasinakis, A. S. Use of Selected Advanced Oxidation Processes (AOPs) for Wastewater Treatment—A Mini-Review. *Global NEST J.* **2008,** *10*(3), 376–385.
4. Oller, I.; Malato, S.; Sanchez-Perez, J. A. Combination of Advanced Oxidation Processes and Biological Treatments for Wastewater Decontamination—A Review. *Sci. Total Environ.* **2011,** *409,* 4141–4166.
5. Chakinala, A. G.; Gogate, P. R.; Burgess, A. E.; Bremner, D. H. Treatment of Industrial Wastewater Effluents Using Hydrodynamic Cavitation and the Advanced Fenton Process. *Ultrason. Sonochem.* **2008,** *15,* 49–54.
6. Comninellis, C.; Kapalka, A.; Malato, S.; Parsons, S. A.; Poulios, I.; Mantzavinos, D. Perspective, Advanced Oxidation Processes for Water Treatment: Advances and Trends for R&D. *J. Chem. Technol. Biotechnol.* **2008,** *83,* 769–776.
7. Gogate, P. R. Cavitational Reactors for Process Intensification of Chemical Processing Applications—A Critical Review. *Chem. Eng. Process.* **2008,** *47,* 515–527.

8. Sires, I.; Brillas, E.; Oturan, M. A.; Rodrigo, M. A.; Panizza, M. Electrochemical Advanced Oxidation Processes: Today and Tomorrow. A Review. *Environ. Sci. Pollut. Res.* **2014,** *21,* 8336–8367.

9. Shemer, H.; Kunukcu, Y. K.; Linden, K. G. Degradation of the Pharmaceutical Metronidazole via UV, Fenton and Photo-Fenton Processes. *Chemosphere* **2006,** *63,* 269–276.

10. Saharan, V. K.; Pandit, A. B.; Kumar, P. S. S.; Anandan, S. Hydrodynamic Cavitation as an Advanced Oxidation Technique for the Degradation of Acid Red 88 Dye. *Ind. Eng. Chem. Res.* **2012,** *51,* 1981–1989.

11. Mendez-Arriaga, F.; Torres-Palma, R. A.; Petrier, C.; Esplugas, S.; Gimenez, J.; Pulgarin, C. Mineralization Enhancement of a Recalcitrant Pharmaceutical Pollutant in Water by Advanced Oxidation Hybrid Processes. *Water Res.* **2009,** *43,* 3984–3991.

12. Gogate, P. R.; Shirgaonkar, I. Z.; Sivakumar, M.; Senthilkumar, P.; Vichare, N. P.; Pandit, A. B. Cavitation Reactors: Efficiency Assessment Using a Model Reaction. *AIChE J.* **2001,** *47,* 2526–2538.

13. Klavarioti, M.; Mantzavinos, D.; Kassinos, D. Removal of Residual Pharmaceuticals from Aqueous Systems by Advanced Oxidation Processes. *Environ. Int.* **2009,** *35,* 402–417.

14. Azbar, N.; Yonar, T.; Kestioglu, K. Comparison of Various Advanced Oxidation Processes and Chemical Treatment Methods for COD and Color Removal from a Polyester and Acetate Fibre Dyeing Effluent. *Chemosphere* **2004,** *55,* 35–43.

15. Gomes, A. C.; Goncalves, I. C.; De Pinho, M. N.; Porter, J. J. Integrated Nanofiltration and Upflow Anaerobic Sludge Blanket Treatment of Textile Wastewater for In-plant Reuse. *Water Environ. Res.* **2007,** *79,* 498–506.

16. Ogunlaja, O. O.; Aemere, O. Evaluating the Efficiency of a Textile Wastewater Treatment Plant Located in Oshodi, Lagos. *Afr. J. Pure Appl. Chem.* **2009,** *3*(9), 189–196.

17. *Central Pollution Control Board Report*; Ministry of Environment and Forests, Advanced Methods for Treatment of Textile Industry Effluents, April 2007.

18. Hassani, A. H.; Mirzayee, R.; Nasseri, S.; Borghei, M.; Gholami, M.; Torabifar, B. Nanofiltration Process on Dye Removal from Simulated Textile Wastewater. *Int. J. Environ. Sci. Technol.* **2008,** *5*(3), 401–408.

19. Avlonitis, S. A.; Poulios, I.; Sotiriou, D.; Pappas, M.; Moutesidis, K. Simulated Cotton Dye Effluents Treatment and Reuse by Nanofiltration. *Desalination* **2008,** *221,* 259–267.

20. Pazdzior, K.; Sojka-Ledakowicz, J.; Klepacz-Smolka, A.; Zylla, R.; Ledakowicz, S.; Mrozinska, Z. Integration of Nanofiltration and Biological Processes for Textile Wastewater Treatment. *Environ. Prot. Eng.* **2009,** *35*(2), 97–104.

21. Ramesh Babu, B.; Parande, A. K.; Raghu, S.; Prem Kumar, T. Cotton Textile Processing: Waste Generation and Effluent Treatment. *J. Cotton Sci.* **2007,** *11,* 141–153.

22. Torabion, A.; Nabi Bihdendi, G. R.; Ranjbar, P. Z.; Razmkhah, N. Efficiency Evaluation of Textile Basic Dye Removal from Water by Nanofiltration. *Iran J. Environ. Health Sci. Eng.* **2007,** *4*(3), 177–180.

23. Fersi, C.; Dhahbi, M. Treatment of Textile Plant Effluent by Ultrafiltration and/or Nanofiltration for Water Reuse. *Desalination* **2008,** *222,* 263–271.

24. Mc Cutcheon, J. R.; McGinnis, R. L.; Elimelech, M. A Novel Ammonia-Carbon Dioxide Forward (Direct) Osmosis Desalination Process. *Desalination* **2005,** *174,* 1–11.

25. Fritzmann, C.; Lowenberg, J.; Wintgens, T.; Melin, T. State of the Art of Reverse Osmosis Desalination. *Desalination,* **2007,** *216,* 1–76.

26. Lattemann, S.; Hopner, T. Environmental Impact and Impact Assessment of Seawater Desalination. *Desalination* **2008,** *220,* 1–15.

27. McGinnis, R. L.; Elimelech, M. Energy Requirements of Ammonia-Carbon Dioxide Forward Osmosis Desalination. *Desalination* **2007,** *207*, 370–382.
28. Khawaji, A. D.; Kutubkhanah, I. K.; Wie, J-M. Advances in Sea-Water Desalination Technologies. *Desalination* **2008,** *221*, 47–69.
29. Hashim, M. A.; Mukhopadhyay, S.; Sahu, J. N.; Sengupta, B. Remediation Technologies for Heavy Metal Contaminated Groundwater. *J. Environ. Manag.* **2011,** *92*, 2355–2388.
30. Palit, S. Filtration: Frontiers of the Engineering and Science of Nanofiltration—A Far-Reaching Review. In *CRC Concise Encyclopedia of Nanotechnology*; Ubaldo O.-M., Kharissova. O. V., Kharisov. B. I., Eds.; Taylor and Francis: USA, 2016; pp 205–214.
31. Palit, S. Advanced Oxidation Processes, Nanofiltration, and Application of Bubble Column Reactor. In *Nanomaterials for Environmental Protection*, Boris, I. K., Oxana, V. K., Rasika, D. H. V., Eds.; Wiley: USA, 2015; pp 207–215.
32. Palit, S. Advanced Oxidation Processes, Bioremediation and Global Water Shortage—A Vision for the Future. *Int. J. Pharma BioSci.* **2016,** *7*(1):B, 349–358.
33. Palit, S. Microfiltration, Groundwater Remediation and Environmental Engineering Science—A Scientific Perspective and a Far-Reaching Review. *Nat. Environ. Pollut. Technol.* **2015,** *14*(4), 817–825.
34. Axworthy, T. S. The Global Water Crisis: Addressing an Urgent Security Issue. *Papers for the Inter Action Council*, 2011–2012.

CHAPTER 20

FRUCTOSE AND ITS IMPACT ON THE DIFFUSION OF ELECTROLYTES IN AQUEOUS SYSTEMS

ANA C. F. RIBEIRO[1,*], LUIS M. P. VERISSIMO[1,2], M. LUISA RAMOS[1], DANIELA F. S. L. RODRIGUES[1], and MIGUEL A. ESTESO[2]

[1]*Coimbra Chemistry Centre, Department of Chemistry, University of Coimbra, 3004-535 Coimbra, Portugal*

[2]*Departamento de Química Física, Facultad de Farmacia, Universidad de Alcalá, 28871 Alcalá de Henares (Madrid), Spain*

Corresponding author. E-mail: anacfrib@ci.uc.pt

CONTENTS

ABSTRACT

In this chapter, the importance of the characterization of the diffusion in electrolyte solutions is reviewed and studied in detail with particular applications in industry.

20.1 INTRODUCTION

Characterization of the diffusion in electrolyte solutions is important for fundamental reasons, helping to understand the nature of the structure of electrolytes, and for practical application in fields such as corrosion.[1-6] As an example of the wide, and sometimes surprising, horizon of applicable interests, we have been particularly involved in obtaining data on the effect of chemical systems on the oral cavity (e.g., carbohydrates plus metal ions),[7-15] in order to better understand the corrosion problems related to dental restorations in systems where such data are not currently available.

Fructose is a monosaccharide found in *natura* in fruits and honey. However, exponential growth in the use of high-fructose corn syrup (HFCS) as a sweetener in processed foods appeared to be the focus of the recent scientific literature as a main cause of both the obesity epidemics and a growing number of fructose-related diseases and conditions.[16,17] D-Fructose is soluble in water and reaches the equilibrium state, irrespective of the initial state, in spite of the fact that the process depends on temperature. In D_2O, at 20°C, the anomeric composition is 75% of β-D-fructopyranose (β_p), 21% of α-D-fructofuranose (α_f), and 4% of β-D-fructofuranose (β_f)[18] (Scheme 20.1).

In this regard, taking into account that oral restorations involve various dental metallic alloys, and as these systems are not completely understood relative to their wear and related corrosion problems, we are interested in the transport properties of aqueous solutions of fructose alone[14] and in the presence of copper (II) chloride,[12,13] ammonium vanadate[7,10] and, particularly, calcium chloride[15] as the calcium ion is the main constituent of the surface of teeth (97%) and its concentration in saliva is about 1.5×10^{-3} mol dm^{-3}. These quantities are influenced by the amount of carbohydrates in the oral cavity. In spite of this, the characterization of transport properties, particularly diffusion coefficients, of complexes between sugar and calcium ions on teeth and in saliva remains largely unexplored. As far as the authors know, no data on mutual diffusion coefficients of calcium salts in the presence of fructose, at 25°C—relevant data for in vivo pharmaceutical applications—have been published by other authors. Also underexplored is the interaction of fructose with other metal ions having potential importance in dental metallic alloys. Consequently, our research group has reported experimental data intending to fill this gap.

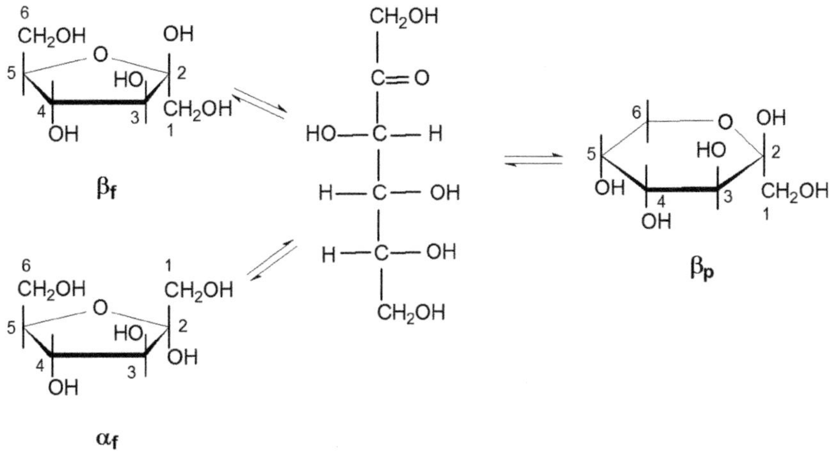

SCHEME 20.1   Representation of different forms of fructose (β_p), (α_f) e (β_f).

More recently, we extended our interest in diffusion to the systems containing fructose and hydrogen phosphate, as phosphates are critical for human life and they fulfill important roles, e.g., ATP/ADP energy management or pH buffering. Mobility of phosphates in and around cells is critical, and degradation of this parameter can seriously compromise health.[19]

Some of these studies on diffusion were complemented by NMR studies. Proton and carbon chemical shifts provide a good indication of the interaction between ligands and metal ions. In particular, broadening or induced shifts of the ^{1}H and ^{13}C signals of the ligand in the presence of the metal ions, when compared with those of the free ligand, can give clear indications of the ligand-to-metal interaction sites. In addition to the information from ^{1}H and ^{13}C chemical shifts, the metal NMR spectra can provide valuable structural details, including the type of metal centers present in these systems, from the characteristic chemical shift. In the present case, [51]V NMR spectra were also used to better characterize the interaction of ammonium vanadate (V) and L-fructose and it was possible to provide further structural insights. In particular, [51]V chemical shifts in vanadium complexes are sensitive to the coordination number and the nature of the ligands. Additional structural information can be gathered from the linewidth of the [51]V signals, which can be related to the symmetry around the metal center.[20–23]

20.2   DIFFUSION

20.2.1   MAIN CONCEPTS

20.2.1.1   MUTUAL AND SELF-DIFFUSION

There are two markedly distinct processes of diffusion, self-diffusion, D^* (also named as intradiffusion, tracer diffusion, single ion diffusion, or ionic diffusion), and mutual diffusion, D, (also known as interdiffusion, concentration diffusion, or salt diffusion).[1-4] Methods such as those based on NMR, polarography, and capillary tube techniques with radioactive isotopes measure self-diffusion coefficients, not mutual diffusion. However, for bulk matter transport, the appropriate parameter is the mutual diffusion coefficient, D. Theoretical relationships derived between self-diffusion and mutual diffusion coefficients, D^* and D, respectively, have had limited success for estimation of D (as well as theoretical models for the calculation of D) and consequently experimental determination of mutual diffusion coefficients is absolutely necessary.

A gradient of concentration in a solution (without convection or migration effects) at constant temperature produces a flow of matter in the opposite direction of the gradient, which arises from random motion of molecules or ions in space. This phenomenon, denominated isothermal diffusion, is an irreversible process. The quantitative measure of this process, the diffusion coefficient, is not a pure kinetic parameter. It depends on kinetic and thermodynamic contributions.[1-4]

The gradient of chemical potential in the solution is the true force producing diffusion. However, for the sake of the experimental determinations, that force can be quantified by the gradient of the concentration at constant temperature. Thus, we may consider the following two approaches to describe the isothermal diffusion: the thermodynamics of irreversible processes and Fick's laws.[1-5]

20.2.1.2   DIFFERENCES BETWEEN FICKIAN AND THERMODYNAMIC DIFFUSION COEFFICIENTS IN BINARY SYSTEMS

The mutual diffusion coefficient, D, in a binary system (i.e., with two independent components) may be defined in terms of the concentration gradient by a phenomenological equation known as Fick's first law (eq 20.1),[1,24]

$$J_i = -D_F \frac{\partial c_i}{\partial x}, \tag{20.1}$$

where J represents the flow of matter of component i across a suitable chosen reference plane per unit area and per unit time, in a one-dimensional system, and c is the concentration of solute in moles per unit volume at the point considered.

However, the expression (20.1) was obtained only because an ideal solution was considered and activity coefficients were ignored. Thus, having in mind a solution that does not behave ideally,

$$J = -D_F \frac{\partial \mu_i}{\partial x}, \tag{20.2}$$

$$\mu_i = \mu_i^0 + RT \ln f_i c_i, \tag{20.3}$$

where μ_i, μ_i^0, and f_i represent the potential and activity coefficient of component i, respectively, with the definition of potential given by eq 20.3.

By manipulating eqs 20.2 and 20.3, we obtain eq 20.4 which relates the thermodynamic coefficient diffusion, D_T, with the Fickian coefficient diffusion, D_F.[1–3]

$$D_T = D_F \left(1 + \frac{d\ln f_i}{d\ln c_i} \right) \tag{20.4}$$

From the last equation, we can see that the diffusion coefficient is not a constant value. As an approach, however, we can assume that the variation of the activity coefficient is not significant for the difference of concentrations causing diffusion (eq 20.5), making D a constant for all practical purposes (and equal to the thermodynamic diffusion coefficient and the Fickian diffusion coefficient) (eq 20.6).

$$\frac{d\ln f_i}{d\ln c_i} \leq 1 \tag{20.5}$$

$$D_T = D_F \tag{20.6}$$

20.2.1.3 MEASUREMENTS OF THE MUTUAL DIFFUSION COEFFICIENTS IN BINARY SYSTEMS

Binary mutual diffusion refers to the fluxes of chemical components produced by gradients in chemical composition. The mutual diffusion coefficient in a two-component solute plus solvent system may be described by Fick's first (eq 20.1) and second laws (eq 20.7),

$$\frac{\partial c}{\partial t} = \frac{\partial}{\partial x}\left(D_F \frac{\partial c}{\partial x} \right). \tag{20.7}$$

In general, the available methods are grouped into two groups: steady- and unsteady-state methods, according to eqs 20.1 and 20.2. In most processes, diffusion is a three-dimensional phenomenon. However, many of the experimental methods used to analyze diffusion restrict it to a one-dimensional process. Also, it is much easier to carry out their mathematical treatments unidimensionally (that may be then generalized to a three-dimensional space).

The resolution of eq 20.2 for a unidimensional process is much easier if we consider D as a constant. This approximation is applicable only when there are small differences of concentration, which is the case in the experimental setups that our team uses, the Lobo's cell open-ended conductimetric technique (Fig. 20.1)[25,26] and in the Taylor technique (Fig. 20.2).[6] In these circumstances, we may consider that all these measurements are parameters with a well-defined thermodynamic meaning.

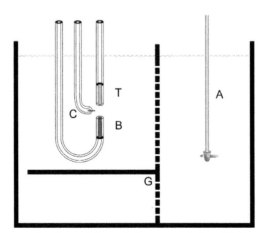

FIGURE 20.1 Schematic view of the Lobo's open-ended capillaries conductimetric cell setup. T and B, top and bottom electrodes (Pt) and capillaries, respectively; C, center electrode (Pt); G, grid and bulkhead; A, stirrer.[25,26]

FIGURE 20.2   Schematic representation of the Taylor dispersion technique.[6]

20.2.1.4   MEASUREMENTS OF THE MUTUAL DIFFUSION COEFFICIENTS IN MULTICOMPONENT SYSTEMS: PSEUDO-BINARY, TERNARY, AND QUATERNARY SYSTEMS

Mutual diffusion in solutions containing two or more solutes is often called multicomponent diffusion. When the interaction between the solute flows is strong, multicomponent diffusion differs significantly from binary diffusion.[6] By using different techniques, it is possible to study the diffusion behavior of multicomponent chemical systems under different conditions.

In the last few years, our group particularly focused on the study of mutual diffusion behavior of binary (e.g., refs. [7–9]) and ternary systems, involving electrolytes in different media (e.g., $CuCl_2$/sucrose/water[13]) or drugs, alone or in combination with cyclodextrins (e.g., HPβ-CD/caffeine/water[27]), and of quaternary systems,[28] helping to understand their structure and their practical applications in fields as diverse as corrosion control or therapeutic optimization, in detail.

Concerning binary systems, we have only two independent components and we need to measure only one diffusion coefficient (D).

Also, the study of the diffusion behavior of pseudo-binary systems such as SDS/cyclodextrin/water,[9] $CuCl_2$/sucrose/water,[13] $CuCl_2$/fructose/water,[13] $CuCl_2$/glucose/water,[13] and NH_4VO_3/carbohydrates/water[10] was conducted. These systems are actually ternary systems, and we really have been measuring only the main diffusion coefficients (D_{11}). However, from experimental conditions, we may consider those systems as pseudo-binary ones, and, consequently, take the measured parameters as binary diffusion coefficients, D.

For ternary systems, mutual diffusion in component (1) + component (2) solutions is described by the coupled Fick eqs 20.8 and 20.9 as follows:

$$J_1 = -D_{11}\nabla C_1 - D_{12}\nabla C_2,$$

(20.8)

$$J_2 = -D_{12}\nabla C_1 - D_{22}\nabla C_2. \tag{20.9}$$

J_1 and J_2 are the molar fluxes of component (1) and component (2) driven by the solute concentration gradients ∇C_1 and ∇C_2. Main diffusion coefficients D_{11} and D_{22} represent the fluxes of the solutes produced by their own concentration gradients. Cross-diffusion coefficients D_{12} and D_{21} describe the coupled flux of each solute driven by a concentration gradient in the other solute. A positive D_{ik} cross-coefficient ($i \neq k$) indicates co-current coupled transport of solute i from regions of higher to lower concentrations of solute k; whereas a negative D_{ik} value indicates a counter-current coupled transport of solute i from regions of lower to higher concentration of solute k.

Diffusion in an aqueous quaternary system is described by the following diffusion equations (eqs 20.10–20.12), which, for brevity, we will denote as ijk (not including the solvent, component 0).

$$-J_1 = {}^{123}(D_{11})_v \frac{\partial c_1}{\partial x} + {}^{123}(D_{12})_v \frac{\partial c_2}{\partial x} + {}^{123}(D_{13})_v \frac{\partial c_3}{\partial x}, \tag{20.10}$$

$$-J_2 = {}^{123}(D_{21})_v \frac{\partial c_1}{\partial x} + {}^{123}(D_{22})_v \frac{\partial c_2}{\partial x} + {}^{123}(D_{23})_v \frac{\partial c_3}{\partial x}, \tag{20.11}$$

$$-J_3 = {}^{123}(D_{31})_v \frac{\partial c_1}{\partial x} + {}^{123}(D_{32})_v \frac{\partial c_2}{\partial x} + {}^{123}(D_{33})_v \frac{\partial c_3}{\partial x}, \tag{20.12}$$

where J_i ($i = 1, 2, 3$) represents the molar flux of solute i in the volume-fixed frame and the ${}^{ijk}D_{ij}$ ($i, j = 1, 2, 3$) are the quaternary diffusion coefficients. The main diffusion coefficients ${}^{123}D_{ii}$ (i.e., ${}^{123}D_{11}$, ${}^{123}D_{22}$, and ${}^{123}D_{33}$) give the flux of solute i produced by its own concentration gradient. The cross-diffusion coefficients ${}^{123}D_{ij}$ (i.e., ${}^{123}D_{12}$, ${}^{123}D_{13}$, ${}^{123}D_{21}$, ${}^{123}D_{23}$, ${}^{123}D_{31}$, and ${}^{123}D_{32}$) give the coupled flux of solute i driven by a concentration gradient in another solute j.

20.3 INFLUENCE OF KINETIC AND THERMODYNAMIC PARAMETERS ON THE DIFFUSION OF ELECTROLYTES IN AQUEOUS SOLUTIONS WITH AND WITHOUT FRUCTOSE IN PSEUDO-BINARY SYSTEMS

On the basis of the Onsager–Fuoss model (eq 20.13),[9] the mutual diffusion coefficient, D, can be considered a product of both kinetic, F_M (or molar mobility coefficient of a diffusing substance) and thermodynamic, F_T (or gradient of the free energy) factors. Thus, two different effects can control the diffusion process: the ionic mobility and the gradient of the free energy,

$$D = F_M \times F_T \qquad (20.13)$$

where

$$F_T = c\frac{\partial \mu}{\partial c} = \left(1 + c\frac{\partial \ln \gamma}{\partial c}\right) \qquad (20.14)$$

and

$$F_M = (D^0 + \Delta_1 + \Delta_2) \qquad (20.15)$$

where μ and γ represent the chemical potential and the thermodynamic activity coefficient of the solute, respectively. Δ_1 and Δ_2 represent the first- and second-order electrophoretic terms and are defined by eq 20.16

$$\Delta n = k_B T A_n \frac{\left(z_1^n t_2^0 + z_2^n t_1^0\right)^2}{|z_1 z_2| a^n} \qquad (20.16)$$

where kB is the Boltzmann's constant, T is the absolute temperature, A_n is a function of the dielectric constant, of the solvent viscosity, of the temperature, and of the dimensionless concentration-dependent quantity (ka), k being the reciprocal of average radius of the ionic atmosphere, and t_1^0 and t_2^0 are the limiting transport numbers of the cation and anion, respectively. Thus, two different effects can control the diffusion process: the ionic mobility and the gradient of the chemical potential. In particular cases, such as systems containing copper chloride at different concentrations and fructose at 0.001 mol dm^{-3},[13] these effects can compensate each other by contributing in opposite directions to the diffusion coefficients, such that there is no net change in diffusion coefficients (Table 20.1). However, for this system, we observed an increase in the conductivity with increasing concentration (Fig. 20.3). Thus, we can conclude that there is an increase in its mobility factor, F_M, which appears to be affected by the presence of the ion–ion interactions, offering less resistance to motion through the liquid. On the other hand, considering that D is not a pure kinetic parameter (eq 20.13), and that this carbohydrate does not affect D, we can conclude that the activity of the solute as compared with a fully dissociated electrolyte is reduced, and consequently leading to lower values of F_T with the concentration (Table 20.2). Also, the diffusion of ammonium vanadate in aqueous solutions of fructose does not appear to be affected. However, in this case, there is a decrease in the conductivity of this electrolyte for the same range of concentrations.[10] In these circumstances, the mobility of this electrolyte, F_M (eq 20.15), also decreases and consequently we can conclude that the activity of this solute increases.

TABLE 20.1 Experimental Mutual Diffusion Coefficient, D, of NH_4VO_3 at 25°C in aqueous solutions of 0.001 mol dm^{-3} fructose at different concentrations, c.[19]

c (mol dm^{-3})	$D \pm S_D{}^a$ (10^{-9} m^2 s^{-1})	$\Delta D/D^b$ (%)
0.001	1.260 ± 0.001	-0.3
0.003	1.200 ± 0.001	-0.2
0.005	1.173 ± 0.002	-0.1
0.008	1.145 ± 0.001	-0.2
0.010	1.130 ± 0.001	-0.4
0.020	1.115 ± 0.001	$+0.4$
0.030	1.100 ± 0.001	$+0.9$

$^a D$ and S_D represent the mean diffusion coefficient of three experiments and the standard deviation of that mean, respectively.

$^b \Delta D/D$ is the deviation between the diffusion coefficients of NH_4VO_3 at the specified concentration in 0.001 mol dm^{-3} fructose and that of the NH_4VO_3 in water.[7]

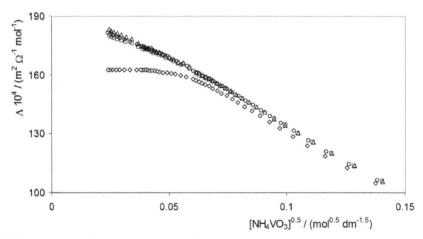

FIGURE 20.3 Molar conductivity, Λ, of ammonium monovanadate in aqueous solutions (O), and in aqueous solutions of fructose (◊) 1 × 10^{-3} mol dm^{-3}, saccharose ( ) 1 × 10^{-3} mol dm^{-3}, and glucose (Δ) 1 × 10^{-3} mol dm^{-3}.[10]

[51]V NMR spectroscopy suggests the complexation in small extension between vanadate and sugar molecules, such as glucose, sucrose, and fructose. There are also indications, in this last case, of oxidation of the carbohydrate and reduction of vanadium.[10]

TABLE 20.2   Experimental Mutual Diffusion Coefficient, D, of $CuCl_2$ at 25°C in aqueous solutions of 0.001 mol dm^{-3} sucrose, fructose, and glucose at different concentrations, c.[13]

Carbohydrates	c (mol dm^{-3})	$D \pm S_D{}^a$ (10^{-9} m^2 s^{-1})	$\Delta D/D^b$ (%)
Sucrose	0.005	1.245 ± 0.002	−0.8
	0.010	1.216 ± 0.002	+1.4
Fructose	0.005	1.236 ± 0.001	−1.4
	0.010	1.218 ± 0.001	+1.6
Glucose	0.005	1.263 ± 0.001	+2.3
	0.010	1.203 ± 0.001	+0.3

^{a}D and S_D represent the mean diffusion coefficient of three experiments and the standard deviation of that mean, respectively.

^{b}$\Delta D/D$ is the deviation between the diffusion coefficients of $CuCl_2$ at a determined concentration in 0.001 mol dm^{-3} sucrose, fructose, and glucose, respectively, and the diffusion coefficients of $CuCl_2$ in water.[12]

In other cases, the fructose molecules exert a significant influence on the diffusion of some aqueous electrolytes. For example, for potassium hydrogen phosphate solutions, a marked decrease in the diffusion coefficients of K_2HPO_4 in aqueous solutions of fructose was observed, compared with those obtained in pure water (Tables 20.3 and 20.4).[19] This is supported by ^{1}H and ^{13}C NMR spectra for D-fructose and for the mixture of D-fructose with sodium hydrogen phosphate salt, in aqueous solution (Figs. 20.4 and 20.5). The ^{1}H NMR and ^{13}C spectra of the anomeric species of D-fructose in the presence of K_2HPO_4 salt display significant shifts for the respective carbon nuclei, suggested in *Journal of Chemical and Engineering Data* (volume 50, pp. 1986–1990) as the proton signals. These effects suggest some noncovalent interaction of hydrogen phosphate anion with D-fructose, possibly involving hydrogen bonding, similar to what has been observed with this sugar and DNA.[29]

Considering now the ternary diffusion measurements for system $CaCl_2$/fructose/water,[15] and comparing these results with those obtained for binary systems at the same temperatures and with the same technique, it is evident that the diffusion behavior of calcium chloride in aqueous solutions at 25°C and 37°C is affected by the presence of these carbohydrate molecules. Having in mind that main diffusion coefficients D_{11} and D_{22} give the molar fluxes of CaCl$_2$ (1) and fructose (2) components, driven by their own concentration gradient, it is observed that, at 25°C, in general, these values are lower than those obtained from binary systems (deviations between 1% and 9%), may be because the diffusion of CaCl$_2$ in aqueous solutions is possibly affected by the eventual presence of new, different species resulting from various equilibria (e.g., several water–sugar hydrogen bonds).

TABLE 20.3 Experimental Mutual Diffusion Coefficient, D, of Aqueous Systems Containing Fructose at Different Concentrations (m) and 0.1 mol kg^{-1} K$_2$HPO$_4$ at 25°C.[19]

m^a (mol kg^{-1})	$D \pm S_D^b$ (10^{-9} m^2 s^{-1})	$\Delta D/D^c$ (%)
0.001	1.025 ± 0.003	−1.4
0.005	1.017 ± 0.010	−2.2
0.010	0.995 ± 0.004	−4.3
0.020	0.971 ± 0.005	−6.6
0.050	0.875 ± 0.004	−15.9
0.101	0.847 ± 0.005	−18.6

[a]Standard uncertainties u are $u_r(m) = 0.03$, $u(T) = 0.01$°C, and $u(P) = 2.03$ kPa. The work solutions were prepared at 25°C in our laboratory.

[b]D and S_D represent the mean diffusion coefficient of three experiments and the standard deviation of that mean, respectively.

[c]$\Delta D/D$ represents the deviations between the diffusion coefficients of the system (fructose + K$_2$HPO$_4$ + H$_2$O) indicated in this table, and the diffusion coefficients of the aqueous solutions of K$_2$HPO$_4$ at 0.1 mol kg^{-1} at the same temperature.[19]

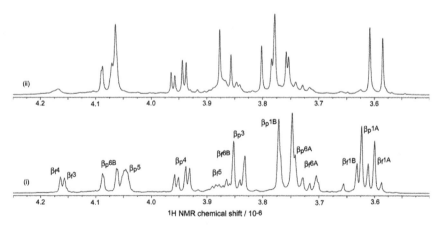

FIGURE 20.4 ^{1}H NMR spectra of D$_2$O solutions of (i) D-fructose 0.10 mol dm^{-3}, pH* 7.2 and (ii) D-fructose/HPO$_4^{2-}$ 0.10:0.10 mol dm^{-3}, pH* 9.0 at 25°C.[19]

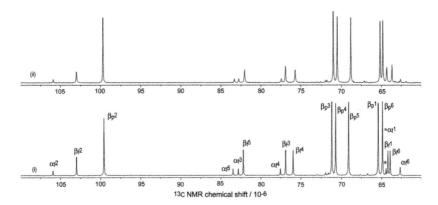

FIGURE 20.5 ^{13}C NMR spectra of D_2O solutions of (i) D-fructose 0.10 mol dm^{-3}, pH* 7.2 and (ii) D-fructose/HPO$_4^{2-}$ 0.10:0.10 mol dm^{-3}, pH* 9.0 at 25°C.[19]

TABLE 20.4   Experimental Ternary Diffusion Coefficients, D_{11}, D_{12}, D_{21}, and D_{22}, for Aqueous Calcium Chloride (1) + Fructose (2) Aqueous Solutions and the Respective Standard Deviations, S_D, at 25°C.[15]

c_1 (mol dm^{-3})	c_2 (mol dm^{-3})	$D_{11} \pm S_D$ (10^{-9} m^2 s^{-1})	$D_{12} \pm S_D$ (10^{-9} m^2 s^{-1})	$D_{21} \pm S_D$ (10^{-9} m^2 s^{-1})	$D_{22} \pm S_D$ (10^{-9} m^2 s^{-1})
0.0000	0.0010				0.686^c
0.0000	0.0250				0.680^c
0.0000	0.0500				0.674^c
0.0000	0.1000				0.661^c
0.0010	0.0000	1.268^a			
0.0250	0.0000	1.173^a			
0.0500	0.0000	1.115^a			
0.1000	0.0000	1.110^a			
0.0010	0.0010	1.159 ± 0.005 (−8.6%)^b	0.013 ± 0.008	0.023 ± 0.008	0.678 ± 0.004 (−1.1 %)^d
0.0250	0.0250		−0.045 ± 0.040	0.090 ± 0.030	
0.0500	0.0500	1.087 ± 0.005 (−7.3%)^b	−0.046 ± 0.011	−0.041± 0.030	0.657 ± 0.002 (−3.3 %)^d
0.1000	0.1000		0.086 ± 0.009	−0.032 ± 0.071	
		1.083 ± 0.028 (−2.9 %)^b			0.660 ± 0.023 (−2.2 %)^d
		1.117 ± 0.045 (0.6 %)^b			0.619 ± 0.023 (−6.2 %)^d

^{a}Our experimental binary D values for aqueous CaCl$_2$.[15]
^{b}The values in parentheses are the relative deviations between the experimental values of D_{11} and the binary values for the same concentration, D.
^{c}Our experimental binary D values for aqueous fructose.[14]
^{d}The values in parentheses are the relative deviations between the experimental values of D_{22} and the binary values for the same concentration, D.

20.4   CONCLUSIONS

For all systems, the variation in D is due mainly to the variation of F_T (attributable to the nonideality in thermodynamic behavior) and, to a lesser extent, the electrophoretic effect in the mobility factor, F_M.

The diffusion of some electrolytes (e.g., $CuCl_2$ and NH_4VO_3) in aqueous solutions at 25°C does not appear to be affected by any association or by aggregate formation between cation or anion and fructose molecules. One possible explanation for this behavior can be due to the fact that if such interactions exist within this region, the lack of effect on diffusion arises from two opposing effects. That is, for $CuCl_2$ and NH_4VO_3, we verified the following statements:

i)   Increase in the mobility of $CuCl_2$
ii)   Decrease in the gradient of the chemical potential with concentration of $CuCl_2$
iii) Decrease in the mobility of NH_4VO_3
iv)   Increase in the gradient of the chemical potential with concentration of NH_4VO_3

However, in other cases, we conclude that fructose interferes significantly with the diffusion of some electrolytes in aqueous solutions (e.g., K_2HPO_4 and $CaCl_2$), leading to a decrease in the diffusion coefficients for the same interval of concentration.

ACKNOWLEDGMENT

The authors are grateful to the Coimbra Chemistry Centre for funding. The Coimbra Chemistry Centre is supported by the Fundação para a Ciência e a Tecnologia (FCT), Portuguese Agency for Scientific Research, through the projects UID/QUI/UI0313/2013 and COMPETE.

KEYWORDS

- electrolytes
- fructose
- diffusion
- mutual diffusion coefficient
- aqueous solutions

REFERENCES

1. Robinson, R. A.; Stokes, R. H. *Electrolyte Solutions*, 2nd ed.; Butterworths: London, 1959.
2. Harned, H. S.; Owen, B. B. *The Physical Chemistry of Electrolytic Solutions*, 3rd ed.; Reinhold Pub. Corp.: New York, 1964.
3. Tyrrell, H. J. V.; Harris, K. R. *Diffusion in Liquids: A Theoretical and Experimental Study*; Butterworths: London, 1984.
4. Lobo, V. M. M. *Handbook of Electrolyte Solutions*; Elsevier: Amsterdam, 1990.
5. Onsager, L.; Fuoss, R. M. *J. Phys. Chem.* **1932**, *36*, 2689–2778.
6. Callendar, R.; Leaist, D. G. Diffusion Coefficients for Binary, Ternary, and Polydisperse Solutions from Peak-Width Analysis of Taylor Dispersion Profiles. *J. Solut. Chem.* 2006, *35*, 353–379.
7. Ribeiro, A. C. F.; Lobo, V. M. M.; Azevedo, E. F. G. Diffusion Coefficients of Ammonium Monovanadate in Aqueous Solutions at 25°C. *J. Solut. Chem.* **2001**, *30*, 1111–1115.
8. Ribeiro, A. C. F.; Lobo, V. M. M.; Natividade, J. J. S. Diffusion Coefficients in Aqueous Solutions of Cobalt Chloride at 298.15 K. *J. Chem. Eng. Data* **2002**, *47*, 539–541.
9. Ribeiro, A. C. F.; Lobo, V. M. M.; Azevedo, E. F. G.; Miguel, M. da G.; Burrows, H. D. Diffusion Coefficients of Sodium Dodecylsulfate in Aqueous Solutions and in Aqueous Solutions of β-Cyclodextrin. *J. Mol. Liq.* **2003**, *102*, 285–292.
10. Ribeiro, A. C. F.; Valente, A. J. M.; Lobo, V. M. M.; Azevedo, E. F. G.; Amado, A. M.; da Costa, A. M. A.; Ramos, M. L.; Burrows, H. D. Interactions of Vanadates with Carbohydrates in Aqueous Solutions. *J. Mol. Struct.* 2004, *703*, 93–101.
11. Ribeiro, A. C. F.; Lobo, V. M. M.; Oliveira, L. R. C.; Burrows, H. D.; Azevedo, E. F. G.; Fangaia, S. I. G.; Nicolau, P. M. G.; Guerra, F. A. D. R. A. Diffusion Coefficients of Chromium Chloride in Aqueous Solutions at 298.15 K and 303.15 K. *J. Chem. Eng. Data* **2005**, *50*, 1014–1017.
12. Ribeiro, A. C. F.; Esteso, M. A.; Lobo, V. M. M.; Valente, A. J. M.; Simões, S. M. N.; Sobral, A. J. F. N.; Burrows, H. D. Diffusion Coefficients of Copper Chloride in Aqueous Solutions at 298.15 K and 310.15 K. *J. Chem. Eng. Data* **2005**, *50*, 1986–1990.
13. Ribeiro, A. C. F.; Esteso, M. A.; Lobo, V. M. M.; Valente, A. J. M.; Simões, S. M. N.; Sobral, A. J. F. N.; Burrows, H. D. Interactions of Copper (II) Chloride with Sucrose, Glucose and Fructose in Aqueous Solutions. *J. Mol. Struct.* **2007**, *826*, 113–119.
14. Ribeiro, A. C. F.; Ortona, O.; Simões, S. M. N.; Santos, C. I. A. V.; Prazeres, P. M. R. A.; Valente, A. J. M.; Lobo, V. M. M.; Burrows, H. D. Binary Mutual Diffusion Coefficients of Aqueous Solutions of Sucrose, Lactose, Glucose and Fructose in the Temperature Range 298.15 K to 328.15. *J. Chem. Eng. Data* **2006**, *51*, 1836–1840.
15. Ribeiro, A. C. F.; Barros, M. C. F.; Lobo, V. M. M.; Sobral, A. J. F. N.; Fangaia, S. I. G.; Nicolau, P. M. G.; Guerra, F. A. D. R. A.; Esteso, M. A. Interaction between Calcium Chloride and Some Carbohydrates as Seen by Mutual Diffusion at 25°C and 37°C. *Food Chem.* **2011**, *124*, 842–849.
16. DiNicolantonio, J. J.; Lucan, S. C. Is Fructose Malabsorption a Cause of Irritable Bowel Syndrome? *Med. Hypotheses* 2015, *85*, 295–297.
17. Dong, B.; Singh, A. B.; Azhar, S.; Seidah, N. G.; Liu, J. High-Fructose Feeding Promotes Accelerated Degradation of Hepatic LDL Receptor and Hypercholesterolemia in Hamsters via Elevated Circulating PCSK9 Levels. *Atherosclerosis* 2015, *239*, 364–374.

18. Jaseja, M.; Perlin, A. S.; Dais, P. Two-Dimensional NMR Spectral Study of the Tauto-
    meric Equilibria of D-Fructose and Related Compounds. *Magn. Reson. Chem.* 1990,
    28, 283–289.
19. Verissimo, L. M. P.; Teigão, J. M. M.; Ramos, M. L.; Burrows, H. D.; Esteso, M. A.;
    Ribeiro, A. C. F. Influence of Fructose on the Diffusion of Potassium Hydrogen Phos-
    phate in Aqueous Solutions at 25°C. *J. Chem. Thermodyn.* **2016**, *101*, 245–250.
20. Caldeira, M. M.; Ramos, M. L.; Cavaleiro, A. M.; Gil, V. M. S. Multinuclear NMR
    Study of Vanadium (V) Complexation with Tartaric and Citric Acids. *J. Mol. Struct.*
    1988, *174*, 461–466.
21. Justino, L. L. G.; Ramos, M. L.; Caldeira, M. M.; Gil, V. M. S. Peroxovanadium(V)
    Complexes of L-Lactic Acid as Studied by NMR Spectroscopy. *Eur. J. Inorg. Chem.*
    2000, *2000*, 1617–1621.
22. Justino, L. G.; Ramos, M. L.; Caldeira, M. M.; Gil, V. M. S. Complexes of Glycolic
    Acid as Studied by NMR Spectroscopy. *Inorganica Chimica Acta* **2000**, *311*, 119–125.
23. Ramos, M. L.; Justino, L. L. G.; Fonseca, S. M.; Burrows, H. D. MR, DFT and Lumi-
    nescence Studies of the Complexation of V(V) Oxoions in Solution with 8-Hydroxy-
    quinoline-5-Sulfonate. *N. J. Chem.* **2015**, *39*, 1488–1497.
24. Bockris, J. O.; Reddy, A. K. N. *Modern Electrochemistry. An Introduction to an Inter-
    disciplinary Area*; Plenum Press: New York, 1977, Vol. 1, 6th Printing.
25. Lobo, V. M. M. *Diffusion and Thermal Diffusion in Solutions of Electrolytes*; Disser-
    tação de Doutoramento: Cambridge, UK: University of Cambridge, 1971.
26. Agar, J. N.; Lobo, V. M. M. Measurement of Diffusion Coefficients of Electrolytes by
    a Modified Open-Ended Capillary Method. *J. Chem. Soc. Faraday Trans.* **1975**, *171*,
    1659–1670.
27. Ribeiro, A. C. F.; Santos, C. I. A. V.; Lobo, V. M. M.; Cabral, A. M. T. D. P. V.; Veiga, F. J.
    B.; Esteso, M. A. Diffusion Coefficients of the Ternary System (2-Hydroxypropyl-Beta-
    Cyclodextrin Plus Caffeine Plus Water) at T=298.15 K. *J. Chem. Thermodyn.* **2009**, *41*,
    1324–1328.
28. Santos, C. I. A. V.; Esteso, M. A.; Lobo, V. M. M.; Ribeiro, A. C. F. Multicomponent
    Diffusion in Cyclodextrin-Drug-Salt-Water Systems (2-Hydroxypropyl-β-Cyclodextrin
    (HP-βCD) + KCl + Theophylline + Water, and β-Cyclodextrin (βCD) + KCl + Theoph-
    ylline + Water). *J. Chem. Thermodyn.* **2012**, *59*, 139–143.
29. Pelmore, H.; Eaton, G.; Symons, M. C. R. Binding of Sugars to DNA. An NMR Study
    of D-Fructose. *J. Chem. Soc. Perkin Trans.* **1992**, *2*, 149–150.

CHAPTER 21

CHEMINFORMATICS—THE PROMISING FUTURE: MANAGING CHANGE OF APPROACH THROUGH ICT EMERGING TECHNOLOGY

HERU SUSANTO[1,2,*]

[1]Department of Information Management, College of Management, Tunghai University, Taichung, Taiwan

[2]Department of Computational Science, The Indonesian Institute of Sciences, Jakarta, Indonesia

[*]Corresponding author. E-mail: heru.susanto@lipi.go.id

CONTENTS

ABSTRACT

In today's society, technology plays an important role, not only in unlimited access for communication purposes but also in education in the area of cheminformatics. Hence, the birth of cheminformatics, which is the combination of chemistry and informatics, opens up a new field of education, although there are countless educational areas that exist today. Cheminformatics has numerous and distinct applications and databases. Cheminformatics contributes by predicting or designing drugs. Immunoassays are commonly used to detect or perform screening test of drug abuse on individuals through their urine or other body fluids. Immunoassays screening commonly points out classes of drugs such as amphetamines, cannabinoids, cocaine, methadone, and opiates. Drug identifications can be carried out in specific detail by applying mass spectrometry methods such as liquid and gas chromatography. This study emphasizes the new approach of applied chemistry through ICT emerging technology that gives information about drug discovery, active compound, and related areas.

21.1 INTRODUCTION

Nowadays information technology is improved very fast, and plays an important role, in various area such as communication, businesses, and education. Here, cheminformatics is one part of information technology improvement. Cheminformatics, is the combination of chemistry and informatics. Initially, this field of chemistry did not have a name until 1998, when the term cheminformatics was first coined by Frank K. Brown. His definition is the "… mixing of those information resources to transform data into information and information into knowledge for the intended purpose of making better decisions faster in the area of drug lead identification and optimization."

Cheminformatics refers to the application of computer technology that is associated before the development of computerized data storage for chemical compounds; we have the traditional documentation system using books and documents. One of the earliest encyclopedias on chemical compounds is Beilstein's *Handbuch der Organischen Chemie* (Beilstein's *Handbook of Organic Chemistry*), of which first edition was published in the 18th century, consisting of two volumes, with more than 2000 pages and registering 1500 compounds. This comprehensive encyclopedia of organic structures covers chemical literature from 1771 to date.[8] ICT concept is one of the integral parts of cheminformatics.[1-5]

This study emphasizes the new approach of applied chemistry through ICT emerging technology revealing drug discovery, active compound, and related areas. This chapter is organized as follow: in the next section, we describe cheminformatics and its application. In Section 21.3, the ICT emerging technology to support cheminformatics is discussed, followed by managing change of cheminformatics to find out the promising future of drug discovery in Section 21.4, and finally conclusion is provided in Section 21.5.

21.2   CHEMINFORMATICS AND ITS APPLICATION

An effective data mining system helps to create and study new chemical objects in which it allows authentication and checking of physical and chemical characteristics among a large collection of described compounds. Getting all this information had always been a problem in chemistry. Chemists should be able to get access to accurate data to their desks whenever it is needed. Therefore, from the beginning, chemists have been developing documentation systems on chemical compounds. The earliest and oldest journal on chemistry is *Chemisches Zentralblatt*, which appeared as early as 1830; another example is, as mentioned above, the Beilstein's *Handbuch der Organischen Chemie* (*Handbook of Organic Chemistry and Chemical Abstracts*) which has been published since 1907. One more notable example is March's *Advanced Chemistry*. The problem is these data storage systems are studied in basic chemical handbooks.

Chemical information branches wanted to benefit from computers as it is much more effective and efficient to save information or data on the computer than just on the desk. Therefore, computations and chemical information are two important components that made up cheminformatics. Polanski[8] stated that Johann Gasteiger, a famous German chemist, supported this statement with the fact that in 1975, *Journal of Chemical Documentation* (a journal specializing in chemical information compiled by American Chemical Society, one of the most authoritative providers of chemistry-related information) changed its name to *Journal of Chemistry and Computer Science*. Similarly, we can use the same title to show recent developments in this field of chemistry, since the journal had just changed the name to *Journal of Chemical Information and Modeling* in 2004. Thus, the journal's history briefly illustrates the scope of the discipline of cheminformatics.

21.2.1 USABILITY OF CHEMINFORMATICS

The main purpose of cheminformatics is to preserve and allow access to tons of data and information that are related to chemistry; moreover, integrating information needed on specific tasks or studies. Another purpose is to aid in the discovery of new drugs. Possible use of information technology is to plan intelligently and to automate the processes associated with chemical synthesis of components of the treatment is a very exciting prospect for chemists and biochemists. One example of the most successful drug discoveries is penicillin. Penicillin is a group of β-lactam antibiotic used in the treatment of infectious diseases caused by bacteria, usually Gram-positive manifold. The way to discover and develop drugs is the result of chance, observation, and many intensive and slow chemical processes. Until some time ago, drug design is considered labor intensive, and the test process always failed.[6–11]

In today's society, technology plays an important role, not only on unlimited access for communicating purposes but also educational. Cheminformatics opens up a new field of education, although there are countless educational areas that exist today. Cheminformatics has numerous and distinct applications and database. Cheminformatics can be learned and taught online. This can give the society easy access about learning the disciplines and methods of cheminformatics. Universities have been partnering up with companies that offer higher education learnings such as Coursera, Udacity, and edX where world experts can connect up with society with an internet connection and a computer for the purpose of tutoring or other educational purposes.[12–18]

The use of information system is important to science educators. It can help them to do their task by exploring the technology that already exists. It can also make science educators to use information system in their higher education learning. It is important to note that the insertion of information system and technology into education or other courses will make a difference in higher education as it has in the simple insertion into public school. Information system and technology is not a vehicle for change, but technology is simply a tool used by its user to do certain objectives.[18–20]

Nowadays, drug abuse is becoming a major issue around the world. Cheminformatics contributes to this issue by predicting or designing drugs. Immunoassays are commonly used to detect or perform screening test of drug abuse on individuals through their urine or other body fluids. Immunoassays screening commonly points out classes of drugs such as amphetamines, cannabinoids, cocaine, methadone, and opiates. Moreover, drugs identifications can be carried out in specific details by applying mass spectrometry methods such as liquid and gas chromatography. In cheminformatics, there is a good relationship between chemistry and technology. The development of information technology has evolved considerably over time. The information technology has developed in conjunction with a variety of disciplines and is applied in various fields. Advances in information technology make the information that can be accessed quickly and precisely. Information technology has changed the way we do science. Surrounded by a sea of data and phenomenal computing capacity, methodologies and approaches to scientific issues developed into a better relationship between theory, experiment, and data analysis.

Consequently, cheminformatics is related to the application of computational methods to solve chemical problems, with special emphasis on manipulation chemical structural information. As mentioned before, the term was introduced by Frank K. Brown in 1998 but there has been no universal agreement about the correct term for this field. Cheminformatics is also known as cheminformatics, chemoinformatics and chemical informatics. Many of the techniques used in cheminformatics are actually rather well established, by the result of years if not decades of research in academic, government, and industrial laboratories. The main reason for its inception can be traced to the need for dealing with large amounts of data generated by the new approach to drug discovery, such as high-throughput screening and combinatorial chemistry. Increase in computer power, especially for desktop engine, has provided the resources to handle this flood. Many other aspects of drug discovery make use of the techniques of cheminformatics, from design of new synthetic route by searching a database of reactions known through development of computational models such as quantitative structure–activity relationship (QSAR), which associates something that is observed through the biological activity of chemical structures through the use of molecular docking program to predict the three-dimensional (3D) structure protein–ligand complexes and then chosen from a set of compounds for screening. One common characteristic is that this method of cheminformatics must apply to a large number of molecules. Cheminformatics, which seams together chemistry and informatics, is obviously linked to computer applications. But, not all chemical branches that relied on computers should automatically be included in the field. Although this term was introduced in the 1980s, it has a long history with its roots going back more than 40 years. The principles of cheminformatics are used in chemical representation and search structure, quantitative structure–activity relationships, chemometrics, molecular modeling, and structural elucidation of computer-aid design and synthesis. Each area of chemistry from analytical chemistry to drug design can benefit from the methods of cheminformatics.

Moreover, ICT applications in the field of molecular science have spawned the field of Biotechnology. Biotechnology is a branch of science that studies the use of living organisms (bacteria, fungi, viruses, etc.) as well as products from living organisms (enzyme, alcohol) in the production process to produce goods and services. This study is increasingly important, because the development has been encouraging and impactful on the field of medicine, pharmacy, environment, and others. These fields include the application of methods of mathematics, statistics, and informatics to answer biological problems, especially with the use of DNA and amino acid sequences as well as the information related to it. Insistence of the need to collect, store, and analyze biological data from a database of DNA, RNA, or protein acts as a spur to the development of bioinformatics. There are nine branches of biotechnology, and one of them is cheminformatics. Cheminformatics is one of biotechnology's disciplines that is a combination of chemical synthesis, biological filtering, and data mining, and is used for drug discovery and development.

21.2.2 DRUG DISCOVERY

Cheminformatics has developed ever since the time of its establishment although throughout the decades when computers and technologies had also developed. Researchers and scientists have been developing a way of assembling information and data within computers. These innovations on computers and technologies have the ability of storing and obtaining chemical information.

The meaning of these disciplines mentioned above is the identification of one of the most popular activities compared with various fields of study that may exist below this field. The way to discover and develop drugs is the result of the agreement, observation, and many intensive and slow chemical processes. Until some time ago, considered drug design should always use a labor-intensive process and the test fails (the process of trial and error). The possibility of the use of IT to plan intelligently and to automate processes associated with chemical synthesis components of the treatment is a very exciting prospect for chemists and biochemists. Award to produce a drug that can be marketed faster is huge, so the target is at the core of cheminformatics.

The academic scope for chemical is very broad. Examples of areas of interest include: planning synthesis, reactions and structure retrieval, modelling, 3D structure retrieval, computational chemistry, visualization tools, and utilities.

Cheminformatics can help chemists and other scientists to produce and manage information. *In silico* analysis using cheminformatics techniques can actually reduce the risks in drug development. Techniques such as virtual screening, library design, and docking figures go into the analysis. Physical properties that may have an impact on whether a substance has the potential to be developed as a drug are often examined by one of the cheminformatics features which makes comparison among a large number of substances. Example is clogP, a measure of the amount of molecular obesity in the system. Sometimes, conclusions can be drawn about a set of associated properties, such as when Chris Lipinski, an experienced medicinal chemist, formulated the famous Rule of Five which says that compounds such as drugs tend to have five hydrogen bond donors or fewer and 10 maximum hydrogen bond acceptors, the calculated logP should be less than or equal to 5, and have a molecular mass of up to 500 daltons. The compound, which showed greater values than these criteria, tends to have poor absorption or permeation, meaning the drug is not an orally active drug in humans.

Research done by Department of Medicine, University of Columbia, had correctly identified additional activity of peptides with 94% accuracy among the top ranked 50 peptides chosen from an in silico library of approximately 100,000 sequences by adopting genetic algorithm. These methods let a radically increased capability to recognize antimicrobial peptide candidates. A genetic algorithm is well suited for difficulties involving string-like data method for search-and-approximation problems. Implementing an iterative method where computational and experimental methodologies are used to find new improved starting point for beginning of genetic algorithms is a more effective tactic. Training the machine to learn algorithms using the new data improves the ability to predict peptide activity. Based on the research, it has been reported that there is several peptides that are active against pathogens of clinical importance, despite these limitations on overview of prediction.

Many compounds kept in databases have already been explored for multiple aims as portion of drug discovery programs. Excavating this information can provide experimental evidence useful for structuring pharmacophores to determine the main pharmacological groups of the compound. Predictor model and DrugBank predictor model dataset were built using data collected from the ChEMBL database by means of a probabilistic method. The model can be used to forecast both the primary target and off-targets of a compound based on the circular fingerprint methodology, one of the technologies developed by cheminformatic method. The study of off-target connections is now known to be as important as to recognize both drug action and toxicology. These molecular structures are drug targets in the treatment neurological diseases such as Alzheimer's disease, obsessive disorders, and Parkinson's disease and depression. In future, developing these multitargeted compounds with selection and chosen ranges of cross-reactivity can report disease in a more subtle and effective ways and will be a key pharmacological concept in future.

21.2.3 CHEMINFORMATICS DATABASE IN FOOD SCIENCE

Computer databases nowadays have become important tools in biological sciences. Bioinformatics tools support the identification of chemical compounds example peptides and proteins by mass spectrometry which is the most reliable tool for applications of the type. Cheminformatics databases are generally used in biological and medical sciences and play an increasingly significant role in modern science. Emphasizing the developing character of data mining and management techniques in animal breeding and food technology made computer databases the most extensive resource for finding and processing such information. One of the main goals of cheminformatics is to clarify life from the chemical outlook. The biological activity of chemical compounds thus falls into both bioinformatics and cheminformatics; for example, food scientists and researcher will need to access databases to find information about the biological activity and behavior of several food components.

21.3 EMERGING TECHNOLOGY

In this section, applications of cheminformatics will be briefly highlighted. The applications included are the most commonly used and contribute on the field of cheminformatics.

21.3.1 SCREEN ASSISTANT 2

Screen Assistant 2 (SA 2) is an open-source JAVA software dedicated to the storage and analysis of small- to grand-size chemical libraries. SA 2 contains information and data on molecules in a MySQL database. SQL, which stands for Structured Query Language, is the common language that is used for the purpose of adding, collecting, regulating, and operating the content in a database. It is much preferred by a whole lot because of its quick response and processing that satisfy a lot of users.

21.3.2 BIOCLIPSE

In cheminformatics, there is a need on applications which can help users with an extensible tool in the obtaining and calculating process of what cheminformatics has to offer. Bioclipse contains 2D editing, 3D visualization, converting files into various formats, and calculation of chemical properties. All these combined into a user-friendly application, where to prepare and edit are easy such as copy and paste, dragging and drop, and to redo and undo process. Bioclipse is in the form of Java and based on Eclipse Rich Client Platform. Bioclipse has advantages over other systems as it can be used in anyway based on the field of cheminformatics.

21.3.3 CINFONY

Toolkits such as RDKit, CDK, and OpenBabel function are very similar with each other but the difference is that each supports various sets of data formats. Although they have complementary features, operating these toolkits on similar programs is challenging because they run on different languages, different chemical models, and have different application programming interfaces (APIs).

Cinfony is a Python module that introduces all three toolkits in an interface which make it easier for users to integrate methods and outcomes from any of those three toolkits. Cinfony makes it easier to perform common tasks in cheminformatics, tasks that include calculating and reading.

21.3.4 KNIME-CDK: WORKFLOW-DRIVEN CHEMINFORMATICS

KNIME (Konstanz Information Miner) is one of the modern, open-source workflow-driven cheminformatics platforms for data analytics. It is a fully open library shared with and accessible to the community. One of the features available is a plug-in feature which is better suited for efficient and easy use, and it enables researchers to automate the routine task and data analysis and also enables building additional nodes; data analysis pipelines from defined components that work well, combined with the existing molecule presentation. KNIME allows you to execute complex statistics and data mining by using tools, such as clustering and machine learning and even plotting and chart tools on the data to examine trends and forecast possible results.

One of the standard roles includes data manipulation tools to manage data in tables, for example, joining, filtering, and partitioning as well as executing these molecule transformations according to common formats. Other tools to manage data use are substructure searching, signatures generating, and fingerprints for a molecular properties. KNIME also uses target prediction tool which can predict the effects of existing drug in terms of their toxicity by giving suggestion on what molecular mechanism is observed behind the undesirable side effects and repurposing by exposing the new uses of the current existing drugs.

21.3.5   CHEMOZART: VISUALIZER PLATFORM AND A WEB-BASED 3D MOLECULAR STRUCTURE EDITOR

Chemozart provides the ability to create 3D molecular structure. This application tool, which is web based, is also used for viewing and editing of these molecular structures. As modern technology evolved, Chemozart which has flexible core technologies can be accessed easily via a UR. The platform is independent and compatible and has been deliberately created in a way that it is compatible with the latest devices, that is, mobile. This application also enables the process of teaching given that it works on mobile devices. Since it is a user-friendly interface, it gives benefit to students. As they easily understand the concept of stereochemical molecules when constructing, drawing, and viewing 3D structures, it is therefore used for educational purpose. With the help of this web-based platform, user can simply create as well as modify or edit or just view the structures of the molecular compounds with just rearranging the position of atoms as simple as dragging them around or use keyboard or (now we can use) any touch screens devices.

21.3.6   OPEN3DALIGN: SOFTWARE FOCUSES ON UNSUPERVISED LIGAND ALIGNMENT

One of the classical tasks of cheminformatics is unsupervised alignment of a structurally varied series of biologically active ligands which leads to various ligand-based drug design methodologies. The most important ligand-based drug design methods are pharmacophore elucidation and 3D quantitative structure–activity relationship studies. Open3DALIGN together with its scriptable interface had the capability of carrying out both conformational searches and multiunsupervised conformational alignment of 3D rigid-body molecular structures which makes automated cheminformatics workflows an ideal component of high throughput. Now different algorithms have been applied to perform single and multiconformation superimpositions on one or more templates. Alignments which contain two operations, feature matching and conformational search, can be achieved by corresponding pharmaco-phores and heavy atoms or any mixed of the two. Feature matching can be accomplished through field-based, pharmacophore-based, and atom-based methods approaches whether to find a same matching molecular interaction fields or searching a collection of pharmacophoric points or heavy atom pairs. Finding the best suited conformer for each ligand which may be mined from prebuilt libraries is the strategy used in conformational search and can be achieved by following rigid alignment on the template and candidate ligands which may also be easily adaptable and aligned on the template. Regardless of the methods and approaches, great computational performance has been achieved through well-organized parallelization of the code features.

21.3.7 CDK-TAVERNA: OPEN WORKFLOW ENVIRONMENT SOLUTION FOR THE BIOSCIENCES

Computational process and analysis of small molecule are among the essentials for both cheminformatics and structural bioinformatics, for example, in drug discovery application. The goals of CDK-Taverna 2.0, through combination of unsimilar open source projects, are structuring a freely available open-source cheminformatics pipelining solution and becoming a progressively influential tool for the biosciences. CDK-Taverna was effectively applied and tested and verified in academic and industrial environments with sea of data of small molecules. Combined with workflows from bioinformatics, statistics, and images analysis, CDK-Taverna supports the process of varied sets of biological data by constructing complex systems biology-oriented workflows. In old days, insufficiencies like workarounds for iterative data reading are removed by sharing the previously accessible workflows developed by a lively community and available online which enables molecular scientists to quickly compute, process, and analyze molecular data as typically found in, for example, today's systems biology scenarios. Graphical workflow editor is currently maintained and is being supported by design and manipulation of workflows. The features are considerably enhanced by the combinatorial chemistry-related reaction list. Implementing the identification of likely drug candidates is one of the additional functionalities for calculating score for a natural product similarity for small molecules. The CDK-Taverna project is recent, constantly updated, and allows multiusage by paralleled threads. It now enables carrying out analysis of large sets of molecules and it is faster in memory processing.

21.3.8 OPEN DRUG DISCOVERY TOOLKIT

Drug discovery has become a significant element supplementing classical medicinal chemistry and high-throughput screening results for many computational chemistry methods were developed to aid them in learning capable drug candidates. Open Drug Discovery Toolkit (ODDT) provides open-source player in the drug discovery field which aims to fulfill the need for comprehensive and open-source drug discovery software because there has been enormous progress in the open cheminformatics field in both methods and software development. Sadly, there has only been little effort to combine them in only one package.

Structure-based methods are the most general and successful methods in drug discovery which are commonly actively used to screen large small-molecule datasets, that is, online databanks or smaller sets, that is, tailored combinatorial chemistry libraries. These methods are crucial for decision-making. Today, much effort is focused toward machine learning which is most valuable in clarifying both nonlinear and trivial correlations in data, respectively.

21.3.9 INDIGO: UNIVERSAL CHEMINFORMATICS API

Indigo is an open-source library which allows developers and chemists to solve many cheminformatics tasks. During the past years, we have enormous development through a collection of more specific tools provided by the universal portable library. Popular programming languages as well as some GUI and command-line tools, which are useful for scientists, are provided. Performance and important chemical features are the core of this C^{2+} cheminformatics library. Among the chemical features of Indigo are that it supports popular chemistry formats and *cis–trans* stereochemistry.

21.4   MANAGING CHANGE OF CHEMINFORMATICS: THE PROMISING FUTURE

Information system has changed the way we do things nowadays. The processes of storing, finding the exact molecule by indexing, searching, retrieving information, and applying information about chemical compounds are made easy with modern technology. Cheminformatics uses computer and informational techniques and applies it to a range of problems in the field of chemistry including chemical problems from making chemical analysis and biochemistry to pharmacology and drug discovery. Cheminformatics contributions to decision-making are known to help certain scientists and chemist, for example, to extract right combination of density and structures from a database containing thousands of molecular structures data that are most likely to provide a specific function or healing effect. These are made available in the database to support research for better chemical decision-making by storing and integrating data in maintainable ways.

Improvement in technology assists cheminformatics in a way that now it has the potential to simulate protein complexes in a solution, for example, pharmacophore analysis/visualization/pattern recognition, and also the most complex biological networks that was not possible with the use of pen and paper. Faster working capability of computer and high-speed networks connection has developed the quality of algorithms and eventually data sources have significantly increased as well. A recent publication by Shaw et al. shows that now millisecond simulations of drug docking using molecular dynamics are possible. This information can enhance knowledge worldwide. According to Glen Robert, there are altogether three areas that had to advance and develop in order to realize the possible growth of computational chemistry.

The first essential area is finding the most accurate and relevant data available. Collaboration among researcher, organization involved in chemical research, and relevant parties who are interested in discovery of new development in chemistry would make it possible that the relevant as well as credible and supported data and information are more open and available for anyone. These could be made possible with crowdsourcing and group of community sharing forum network as well as developing social system for sharing specific area of interest. Too much information could eventually become one of the main challenges faced. Today, in order to face the main challenge, language processing has developed even the most complex complicated documents made by researchers which can be processed by robots to extract and find the most useful information. These robots will assist in finding the best suited data and then filter and organize them in a way they can cater to our area of interest. Another possible solution will be developing network that is self-organized and able to search and navigate the pool of data faster and relevant to our area of interest.

The second area of concern is how cheminformatics will be presented in computers whether the highly professional researcher in chemistry willing to share information or they were bind by the confidentiality and secrecy of their work. These researchers tend to manipulate data by using complex language or symbols to present their research and chemistry concepts although the changes of name and description detail and easier term had been come up by three most influential scientists in chemistry field, namely, Berzelius, Archibald Scott-Couper, and Frederick Beilstein, almost 200 years ago. Now the trend has slowly changed the name form from complex to much simpler and the complex description has become more easy to understand, the complicated details attached to its function are made more simpler, yet there is present the accurate and relevant information with much more details. The description of the information is not easily accessible where we can easily target the most relevant data that we are mostly curious in. More so, the symbols used in presenting molecules and structures are not easily made and understood. Proper indexing had also been made although until now it had not been fully incorporated and much improvement is still needed.

Third area is to replicate reality in the simulation. It is one of the main aims in computational chemistry. With modern technology, nowadays, petabyte computing is offered. Their capacity can simulate real biological systems within millisecond with hardware precisely designed for simulation. With vastly increased development of hardware, the complexity of simulation would be made easier and user-friendly. There will be much improvement in accuracy also. It is also dependent on the capability of the hardware design in future to cater to the need for the development in simulation. When computing capacity is no longer limited in terms of its capacity, extracting, filtering, and evaluating as well as recalculating could be done in more straightforward form and the complexity no longer exists. IBM is now developing chips known as neural chips and with the emergence of cloud computing can improve further the area of cheminformatics. Superhigh-speed connection, vast memory capability, and environmental virtualization had made it possible to change the current cheminformatics in future.

With the modern technology where everything is done by technology, it is possible that according to Robert Glen, these machines will ask and think on their own to assist human in future.

Compounds will not be classified at molecular level anymore but rather in genetic and clinical effects term. Modern technology has been one of the major driving forces of the development of cheminformatics. It is hard to predict the future of cheminformatics but rather these reports will cater for the possible future of cheminformatics and how the information system can support them.

21.4.1 AN INCREASE IN ENVIRONMENTAL CHEMINFORMATICS

Currently, the awareness for the society to study the relationship between environmental exposure and human health had increased from time to time. Nowadays, human beings are exposed to a wide range of environmental chemicals, including pollution present in our air and water, integrally existing in our food, medicines, cosmetics, and many others.

Regardless these chemicals still exist at a low concentration of risk below the toxicological concern, these matters should be taken into serious attention. These combinations of chemicals can potentially affect human health as fast as lightning. Integration of assessable data monitored studies should be cautiously taking into account not just the concentration of chemicals but also exposure of time and life period through a collaborative framework. The expectation of chemical–chemical collaboration tools (here referring to their effects in living organisms, not to chemistry and physics) would be tremendously valued in risk assessment.

On the other hand, to support decision makers minimize risk of exposure, develop legislative and societal demands, a forecast support system could be used in evidence-based medicine to assess environmental threat and chemical warfare agents. The future system may assist with optimizing the selection of active medical ingredients during the process of formulating new drug mixtures. The system will have the capability to handle combinations of drug mixtures of cheminformatics at the "fixed" and "unspecified" mixture level. These technologies of computer-assisted chemical production might become universal in our determination to lessen the synthesis cost of making drugs that are approved and to reduce the impact on environmental chemical reagents.

21.4.2   *EXPANSION OF KNOWLEDGE AND OPEN-ENDED SCIENCE*

Cheminformatics can assist the expansion of knowledge. The problem faced currently is that the vast amounts of molecular property and sea of data often lead to difficulty in making choices and assumptions about the data essentially related to the quality and type of data input that are no longer relevant. Scientists with lack of experience are less likely to investigate multiple diverging options thoroughly while conducting their research. These situations are called epistemological and they are likely to occur when mining difficult data. This epistemological situation is proved and supported by anecdotal behavior; for example, most internet users prefer to take top 10 hits offered by search engines and do not bother to scroll down below 10 hits.

Developing integrative tools and approach based on knowledge within the Cytoscape framework will enable users to visualize molecular interaction networks which provide a basic feature for data integration and analysis and share them via a public setting. The example of public setting is crowdsourcing. Crowdsourcing can help knowledge sharing as it is easier for information to be gathered. This tool will make possible to share and explore alternative hypotheses and multiple situations given the available limitations as well as building high confidence in prediction data and models in future, unlikely for the occurrence of epistemological situation.

21.4.3   *CHEMINFORMATICS INNOVATION AND THE IMPACT ON SOCIETY*

Combining cheminformatics with bioinformatics and other computational systems is expected to give benefit to the developments in proteomics, metabolomics, and metagenomics as well as other sciences. Now in future, we will not need to estimate the bioactivity profile of a chemical at the molecular level but rather we will study biomedical information with the addition of inherent polymorphisms and clinical effects.

These developments in proteomics, metabolomics, and metagenomics as well as other sciences can be supported by developing tools for integrated chemical–biological data acquirement, filtered and processed by taking into account relevant information related to collaborations between proteins and small molecules as well as possible metabolic alterations. These tools will be integrated into the virtual physiological human.

Cheminformatics contributes to the community in many different ways. One of the ways is that it helps on discovering new drugs and also to predict drugs. Cheminformatics helps researchers to experiment on chemical compounds and molecules through computers rather than to practically conduct the experiments. This can reduce costs on the process of the experiment itself.

Although cheminformatics contributes to strong impacts on modern society and science, there are also many obstacles faced by eager and anxious users on learning cheminformatics. This could easily be a disadvantage for those keen people in learning cheminformatics. The most common difficulty encountered is time limitation. Cheminformatics has various types of databases and software that are installed in the computer. Amateurs will need to spend time on the way the cheminformatics works, its database, and software. Thus, most of the people that learn cheminformatics have a shallow knowledge about it due to shortage of time.

Financial limitations, although experimenting on new drugs and discovering new drugs, do not require high expenses but taking the course in learning cheminformatics requires high price of admission among learners in the field of cheminformatics. There are also challenges faced by the implementation of cheminformatics. The world nowadays promotes the practice of going green and environmental friendly. It is crucial to search for the right chemicals which are the ones that have low toxicity and low environmental threat properties. Cheminformatics is highly relied on computers for prediction of drugs on chemical combinations. Other challenges faced by cheminformatics are that it should not be focused on chemistry alone, this is because chemicals have also influences and impacts on cellular functions which trickles a way to biological field.

21.5 CONCLUSION

Cheminformatics has a history almost as long as the computer itself. It is the application of computer technology and methods for the chemical-related field that deals with molecular modeling and computational chemistry. It is without a doubt that cheminformatics helps humanity in many ways, especially in the drug-making activity and in discovery of new drugs. Furthermore, it also opens up new area for research development and also increases opportunities for education.

Information system and technology plays a major and crucial part in cheminformatics especially to store the tremendous amount of data related to chemical compounds and other chemical information. Information system and technology also enables easy accessibility of these data because they can be easily obtained, whereas in the old days, all these data and information were only kept in handbooks which made it hard to get the data when needed. Without information system and technology, chemists would not be able to create new drugs easily resulting in a huge setback on the drug discovery process. In the beginning, cheminformatics engineering has found particular application especially in the pharmaceutical industry, but it is now beginning to penetrate into other areas of chemistry.

KEYWORDS

- **computer-aided design**
- **computer-aided manufacturing**
- **information and communication technology**
- **decision support system**
- **managing change**

REFERENCES

1. Biesken, S.; Mein, T.; Wiswedel, B.; Figueiredo, L.; Berthold, M.; Steinback, C. KNIME-CDK: Workflow Driven Cheminformatics. *BMC Bioinform.* **2013,** *14*, 257.
2. Brown, F. Editorial Opinion: Chemoinformatics—A Ten Year Update. *Curr. Opin. Drug Discov. Devel.* **2005,** *8*(3), 296–30.
3. Churinov, A.; Savelyev, A.; Karulin, B.; Rybalkin, B; Pavlov, D. Indigo: Universal Cheminformatics API. *J. Cheminform.* **2011,** *3*(Suppl 1), P4.
4. Ekins, S; Gupta, R.; Gilford, E.; Bunin, B.; Waller, C. Chemical Space: Missing Pieces in Cheminformatics. *J. Cheminform.* **2010***, 27*, 2035–2039.
5. Fjell, C.; Jenssen, H.; Cheung, W.; Hancock, R.; Cherkasov, A. Optimization of Antibacterial Peptides by Genetic Algorithms and Cheminformatics. *J. Cheminform.* **2010,** *77,* 48–56.
6. Fourches, D. Cheminformatics: At the Crossroad of Eras. *J. Cheminform.* **2014,** *8*, 16.
7. Glen, R. Computational Chemistry and Cheminformatics: An Essay on the Future. *J. Cheminform.* **2011,** *26*, 47–49.
8. Polanski, J. *Cheminform.* **2009***, 4*, 14.
9. Le Guilloux, V. Mining Collections of Compounds with Screening Assistant 2. *J. Cheminform.* **2012,** *4*, 20.
10. Micinski, J.; Minkieicz, P. Biological and Chemical Databases for Research into the Composition of Animal Source Foods. *Food Rev. Int.,* **2013,** *29*, 321–351.

11. Nikolic, K.; Mavridis, L.; Oscar, M.; Ramsay, R.; Agbaba, D.; Massarelli, P.; Rossi, I.; Stark, H.;Contelles, J.; John, B.; Mitchell, O. Predicting Targets of Compounds Against Neurological Diseases Using Cheminformatics Methodology. *J. Cheminform.* **2014**, *29*, 183–198.

12. O'Boyle, N. M.; Hutchison, G. R. Cinfony—Combining Open Source Cheminformatics Toolkits behind a Common Interface. *Chem. Cent. J.* **2008**, *2*, 24.

13. Oprea, T; Taboureau, O.; Bologa, C. Of Possible Cheminformatics Futures. *J. Cheminform.* **2011**, *26*, 107–112.

14. Sajadi, F.; Mohebifar, M. Chemozart: A Web-Based 3D Molecular Structure Editor and Visualizer Platform. *J. Cheminform.* **2015**, *7*, 56.

15. Siedlecki, P.; Zielenkiewicz, P.; Wojcikowski, M. Open Drug Discovery Toolkit (ODDT): A New Open Source Player in the Drug Discovery Field. *J. Cheminform.* **2015**, *7*, 26.

16. Shiri, F.; Balle, T.; Tosco, P. Open3DAlign: An Open Source Software Aimed at Unsupervised Ligand Alignment. *J. Cheminform.* **2011**, *25*, 777–783.

17. Spjuth, O. Bioclipse: An Open Source Workbench for Chemo- and Bioinformatics. *BMC Bioinform.* **2007**, *8*, 59.

18. Truszkowski, A.; Jayaseelan, K.; Neumann, S.; Willighagen, E. L.; Zielesny, A.; Steinback, C. New Developments on the Cheminformatics Open Workflow Environment CDK-Taverna. *J. Cheminform.* 2011, 3, 54.

19. Wild, J. Grand Challenges for Cheminformatics. *J. Cheminform.* **2009**, *1*, 1.

20. Wild, J. Cheminformatics for the Masses: A Chance to Increase Educational Opportunities for the Next Generation of Cheminformaticians. *J. Cheminform.* **2013**, *5*, 32.

CHAPTER 22

APPLICATION OF EVOLUTIONARY MULTIOBJECTIVE OPTIMIZATION IN DESIGNING CHEMICAL ENGINEERING AND PETROLEUM ENGINEERING SYSTEMS AND ITS WIDE TECHNOLOGICAL VISION: A CRITICAL OVERVIEW AND A BROAD SCIENTIFIC PERSPECTIVE

SUKANCHAN PALIT*

Department of Chemical Engineering, University of Petroleum and Energy Studies, Energy Acres, Bidholi via Prem Nagar, Dehradun 248007, Uttarakhand, India

Corresponding author. E-mail: sukanchan68@gmail.com, sukanchan92@gmail.com

CONTENTS

ABSTRACT

Science and engineering are moving drastically and steadfastly toward newer realm and newer visionary frontiers. Chemical engineering and applied mathematics are facing drastic challenges. Chemical engineering process modeling and application of multiobjective optimization are the forlorn needs toward a greater emancipation of science, engineering, and technology. Vision of science in chemical process engineering at such a crucial juncture stands in the midst of deep comprehension and immense optimism. Process engineering and optimization in today's world have an unsevered umbilical cord. In such a situation, petroleum engineering science needs to be revamped with respect to petroleum refining paradigm. Techno-logical vision needs to be revamped with the passage of history and time. Depletion of fossil fuel sources is of immediate and utmost concern. Science of petroleum refining is ushering in a new era and a newer realm. The author with cogent insight and immense vision delineates the greatness of applica-tion of evolutionary optimization in petroleum refining process modeling and chemical engineering systems. Multiobjective optimization is robust and far-reaching. Vision of science, scientific cognizance, and scientific understanding will go a long way in the true emancipation of application of evolutionary optimization in designing and modeling chemical engineering systems. Today, the world of optimization has an umbilical cord with chem-ical and petroleum engineering systems. A scientist's prowess as well as a man's vision needs to be reassessed with every step of scientific research pursuit. The author critically reviews mainly chemical engineering and petroleum engineering systems with utmost importance and grave concern as it is the need of the hour with the depletion of fossil fuel resources. The vision of science and engineering will witness a new dawn with the passage of human history and time.

22.1 INTRODUCTION

Engineering science is moving slowly and steadily toward a new visionary era. Technological vision is ushering in a new revolutionary scientific generation. Mankind's prowess and ultimate vision has urged scientists to devise and discover newer and visionary technologies. Depletion of fossil fuel resources has resulted science and engineering, especially petroleum engineering science, to move toward newer vistas. Scientific frontiers need to be surpassed and the question of sustainable development needs to be reassessed. This treatise elucidates on the wide and varied world of evolutionary multiobjective optimization (EMO) and its application in modeling and designing chemical engineering and petroleum engineering systems.

Science and technology are moving fast toward a newer scientific horizon. Scientific sagacity and scientific validation are reframing the future of application of multiobjective optimization in chemical and petroleum engineering systems. The need of the hour is the application of mathematics in designing petroleum engineering systems. Modeling of fluidized catalytic cracking unit is one such visionary vista of science. Engineering science in today's world is surpassing wide, versatile, and visionary frontiers. The author with deep, instinctive, and cogent insight elucidates on the versatile applications of multiobjective optimization in designing chemical and petroleum engineering systems.

This concise treatise elucidates upon the recent scientific work in the field of multiobjective optimization particularly in genetic algorithm (GA). The author with deep comprehension and lucidity delineates the doctrine of EMO and GA and their application in chemical process engineering, petroleum engineering science, and polymer science. EMO has vast and versatile application domain. Chemical process engineering and petroleum refining technology have an unsevered umbilical cord with GA and EMO.[18,19]

22.2 VISION OF THE PRESENT TREATISE AND THE FUTURISTIC AIM AND MISSION

Multiobjective optimization is an area of multiple-criteria decision making that is concerned with mathematical optimization problems involving more than one objective function to be optimized simultaneously. Technological vision in the field of optimization needs to be reassessed and readjudicated at each step of scientific endeavor. Multiobjective optimization has been applied in many fields of science including engineering, economics, and logistics where optimal decisions need to be taken in the presence of trade-offs between two or more conflicting objectives. Minimizing cost while maximizing comfort while buying a car and maximizing performance while minimizing fuel consumption and emission of pollutants of a vehicle are examples of multiobjective optimization problems involving two or three objectives, respectively. In practical problems, there can be more than three objectives.[18,19]

For a nontrivial multiobjective problem, there does not exist a single solution that simultaneously optimizes each objective. In that case, the objective functions are said to be conflicting, and there exists a (possibly infinite) number of Pareto infinite solutions. A solution is called nondominated, Pareto optimal, Pareto efficient, or noninferior, if none of the objective functions can be improved in value without degrading some of the other objective values. Vision of science and engineering are moving toward a newer realm and newer frontier. The arena of multiobjective optimization is drastically moving toward a visionary era as regards its application. Without additional subjective preference information, all Pareto optimal solutions are considered equally good (as vectors cannot be ordered completely). Researchers delve into multiobjective problems from different viewpoints and thus, there exist different solution philosophies and goals when setting and solving them.[18,19]

Science and technology in our present day human civilization are moving toward a newer visionary frontier and a glorious eon. The purpose of this treatise is vast and versatile. The author with cogent insight and pragmatic approach deals with application of multiobjective optimization in chemical engineering systems and petroleum engineering science. These two glorious examples of optimization science and EMO are unfolding a visionary domain of technology.

22.3 VISION OF MULTIOBJECTIVE OPTIMIZATION AND SCIENTIFIC COGNIZANCE BEHIND IT

Very often real-world applications have several multiple conflicting objectives. Recently, there has been a growing interest in EMO optimization algorithms which combines two major disciplines: evolutionary computation and theoretical frameworks of multiple-criteria decision making. The author elucidates on the fundamental concepts of multiobjective optimization emphasizing the motivation and advantages of using evolutionary algorithms (EAs). The vision of science and engineering is wide and far-reaching in today's world. Multiobjective optimization is opening up new frontiers of applied mathematics and engineering science. Scientific cognizance and scientific sagacity are surpassing veritable barriers and immense difficulties. GA is another avenue of scientific research pursuit.[18,19]

Science and engineering are moving fast in today's human civilization. Mathematical tools, chemical process engineering, and chemical process modeling are the backbones of the future of petroleum engineering science and engineering science as a whole. The world of vicious challenges, the scientific urge to excel and the wide avenues of scientific rigor are all the torchbearers of tomorrow's science. In today's human civilization, science and technology are immense colossus without a definite will of its own. In such a crucial juncture of history of science, validation of science and engineering are of utmost importance. The science and engineering of multi-objective optimization are surpassing vast and visionary frontiers. Applied mathematics and the world of chemical process engineering need to be re-envisioned at every step of scientific research pursuit. Chemical process modeling and optimization techniques are gearing toward a newer dimension of scientific understanding and immense scientific forbearance. The challenge needs to be unfolded and the visionary frontiers surpassed in this crucial juncture of history and time.

22.4 WHAT IS MULTIOBJECTIVE OPTIMIZATION?

Even though some real world problems can be reduced to a matter of single objective very often it is hard to define all the aspects in terms of a single objective. Defining multiple objectives often gives a better idea of the task. Multiobjective optimization has been available for about two decades, and recently its application in real world problems is continuously increasing. In contrast to the plethora of techniques available for single-objective optimization, relatively few techniques have been developed for multiobjective optimization.[18,19]

Multiobjective optimization and its applications are surpassing vast and versatile scientific frontiers. The vision of science, the immense challenges behind it, and the difficulties and barriers will go a long long way in the true emancipation of mathematical tools in designing chemical and petroleum engineering systems. Multiobjective optimization is passing through different and drastic phases in its application procedure in the present decade. Applications are massive yet barriers are many.

22.5 WHAT IS EVOLUTIONARY MULTIOBJECTIVE OPTIMIZATION?

Multiple, often conflicting objectives arise naturally in most real world optimization scenarios. As EA s possess several characteristics that are desirable for this type of problem, this domain of search strategies has been used for multiobjective optimization for more than a decade. At the same time, EMO has become established as a separate subdiscipline combining the fields of evolutionary computation and the classical multiple-criteria decision making.[18,19]

The term EA stands for a class of stochastic optimization methods that simulate the process of natural evolution. The origin of EAs can be traced back to the late 1950s, and since the 1970s several evolutionary methodologies have been proposed, mainly GAs, evolutionary programming (EP), and evolutionary strategies.[18,19]

Although the underlying mechanisms are simple, these algorithms have proven themselves as a general, robust, and powerful search mechanism. In particular, they possess several characteristics that are desirable for problems involving: (1) multiple conflicting objectives and (2) intractably large and highly complex search spaces. As a result, numerous algorithmic variants have been proposed and applied to various problem domains since the mid-1980s.[18,19]

22.5.1 EXAMPLES OF MULTIOBJECTIVE OPTIMIZATION

In economics, many problems involve multiple objectives along with constraints on what combinations of these objectives are attainable. For example, consumer's demand for various goods is determined by the process of maximization of the utilities derived from these goods, subject to a constraint based on how much income is available to spend on those goods and on the prices of those goods. This constraint allows more of one good to be purchased only at the sacrifice of consuming less of another good; therefore, the various objectives (more consumption of each good is preferred) are in conflict with each other. A common method for analyzing such a problem is to use a graph of indifference curves, representing preferences, and a budget constraint, representing the trade-offs that the consumer is faced with. Technological vision and the application of optimization are in the forefront of our present day scientific civilization. Examples of multiobjective optimization are wide, versatile, and visionary. Human scientific progress and scientific validation stands in the midst of immense optimism. The author deeply relates in this treatise the vision behind application of multiobjective optimization and its widening technological scenario.

22.6 GA: AN INTRODUCTION

GA, bio-inspired optimization, and multiobjective optimization are opening up new doors of scientific imagination. Vision of science and engineering in today's world needs to be restructured with the growing concern of energy sustainability and depletion of fossil fuel resources. Human scientific ingenuity and greater scientific forbearance are challenged with the passage of scientific history and time.

GA mimics the principle of natural genetics and natural selection to constitute search and optimization procedures. Simulated annealing mimics the cooling phenomenon of molten metals to constitute a search procedure. Today's world of multiobjective optimization totally encompasses GA and bio-inspired optimization. The science and engineering of GA still remains unexplored.

Professor John Holland of the University of Michigan, Ann Arbor, USA envisaged the concept of these algorithms in the mid-60s and published his seminal work. Thereafter, a number of his students and other researchers have contributed to envision and develop this field.

22.6.1 HISTORY AND SCIENTIFIC DOCTRINE OF GA

GA and bio-inspired optimization in today's scientific world are in the path of newer discoveries. The challenge and the vision need to be revisited and re-envisioned. The scientific doctrine and the immense scientific cognizance in the field of GA still need to be unfolded. Computational time stands as a major backbone in the vast avenue of scientific endeavor in the field of multiobjective optimization. Scientific candor, scientific astuteness, and the world of scientific validation are the torchbearers toward a greater visionary future. EMO in today's world is the path of newer regeneration and a newer revamping.

22.7 EVOLUTIONARY ALGORITHMS

EA is a wide and visionary domain. In artificial intelligence, an EA is a subset of evolutionary computation, a generic population-based metaheuristic optimization algorithm. An EA uses mechanisms inspired by biological evolution, such as reproduction, mutation, recombination, and selection. Candidate solutions to the optimization problem play the role of individuals in a population, and the fitness function determines the quality of the solutions. Evolution of the population then takes place after the repeated application of the above operators. EAs often perform well approximating solutions to all types of problems because they ideally do not make any assumption about the underlying fitness landscape; this generality is shown by successes in fields as diverse as engineering, art, biology, economics, marketing, genetics, operations research, robotics, social science, and a gamut of basic and fundamental sciences.

22.7.1 SWARM INTELLIGENCE

Swarm intelligence (SI) is the collective behavior of decentralized and self-organized natural or artificial system. The concept is employed in work on artificial intelligence. The expression was introduced by Gerardo Beni and Jing Wang in 1989, in the context of cellular robotic systems.

SI systems consist typically of a population of simple agents or Boids interacting locally with one another and with their environment. The inspiration generates from nature, especially biological systems.

22.7.2 OTHER EVOLUTIONARY COMPUTING ALGORITHMS

Other evolutionary computing algorithm which needs special mention are particle swarm optimization, ant colony optimization (ACO), artificial bee colony algorithm, artificial SI, differential evolution, the bees algorithm, artificial immune systems, Bat algorithm, glowworm swarm optimization, gravitational search algorithm, and so on.

ACO, introduced by Dorigo in his doctoral dissertation, is a class of optimization algorithms modeled on the actions of an ant colony. ACO is a probabilistic technique useful in problems that deals with finding better paths through graphs.

22.8 TECHNOLOGICAL VISION AND VALIDATION OF SCIENCE

The world of engineering science along with process engineering and petroleum engineering today stands in the midst of deep comprehension and introspection. The challenge and vision are immense. The author veritably takes the example of design of petroleum refinery particularly Fluidized Catalytic Cracking unit and various other chemical engineering systems. In such a scenario, scientific sagacity and scientific validation stands high above others. Validation of science and advancement of science and technology are in the path of new scientific rejuvenation and newer scientific sagacity. Validation of science with respect to application of multiobjective optimization is the need of the hour. The author repeatedly stresses on this fact of engineering science. GA and multiobjective optimization is ushering in a new era in the field of chemical process engineering.

22.9   EVOLUTIONARY MULTIOBJECTIVE OPTIMIZATION AND PROGRESS OF ENGINEERING

EMO is reaching new heights. Scientific research pursuit is in the avenue of new glory. Vision of science, technological prowess, and the future progress of engineering are the burning and vexing issues of tomorrow's science. Chemical engineering systems and petroleum engineering systems are moving toward a newer scientific civilization and scientific vision. Chemical process engineering in today's progress of human civilization needs to be re-envisioned. EMO and its application to chemical process engineering and chemical process modeling are moving toward a newer future dimension.

22.10   CHALLENGE, SCIENTIFIC VISION, AND THE ROAD AHEAD

Scientific vision and scientific challenges in today's world are witnessing immense challenges and deep and effective introspection. EMO and visionary and ground-breaking environmental engineering tools are reshaping the deep scientific horizon. In today's world of petroleum engineering science and chemical process engineering, the grave concern and scientific challenges of depletion of fossil fuel resources are changing the global engineering scenario. Global petroleum crisis, the scientific devastation and the visionary road ahead are changing the face of scientific vision of tomorrow.[18,19]

22.11   APPLICATION AND VISION OF MULTIOBJECTIVE OPTIMIZATION IN PROCESS OPTIMIZATION

Multiobjective optimization has also been increasingly employed in chemical engineering. The multiobjective generic algorithm (MOGA) was used to optimize the design of the pressure swing adsorption process (cyclic separation process). The design problem considered in that work involves the bi-objective maximization of nitrogen recovery and nitrogen purity. The results obtained in that work provided a good approximation of the Pareto frontier with acceptable trade-offs between the objectives. A multiobjective chemical process problem for the thermal processing of food was solved by some researchers. In that work two case studies (bi-objective and triple objective problems) with nonlinear dynamic models were tackled. A hybrid approach consisting of the weighted Tchebycheff and the normal boundary intersection approaches were utilized. This novel hybrid approach was successful in constructing a Pareto optimal set for the thermal processing of foods. Similar multiobjective problems were encountered and solved for applications involving chemical extraction and bioethanol production processes. Technological visions in today's scientific and engineering world are surpassing wide and versatile frontiers. Technological prowess and vision have an umbilical cord with scientific validation and deep scientific understanding. Sustainable development is the backbone of human scientific progress today. Application of multiobjective optimization in process optimization is of utmost importance and is a visionary imperative to the progress of human civilization.[18,19]

22.12   SOLVING A MULTIOBJECTIVE OPTIMIZATION PROBLEM

As there usually exists multiple Pareto optimal solutions for multiobjective optimization problems, what it means to solve such a problem is not as straightforward as it is for a conventional single-objective optimization problem. Therefore, different researchers have defined the term "solving a multiobjective problem" in various ways. This versatile treatise summarizes some of them and the contexts in which they are used. Many methods convert the original problem with multiple objectives into a single-objective optimization problem. This is called a scalarization problem. If scalarization is done carefully, Pareto optimality of the solutions obtained can be guaranteed.[18,19]

Solving a multiobjective optimization is sometimes understood as approximating or computing all or a representative set of Pareto optimal solutions.

When decision making is emphasized, the objective of solving a multiobjective optimization problem is referred to supporting a decision maker (DM) in finding the most preferred Pareto optimal solution according to his/her subjective preferences. The underlying assumption is that, one solution to the problem must be identified to be implemented in practice. Here, a human DM plays an important role. The DM is expected to be an expert in the problem domain.

The most preferred solution can be found using different philosophies. Multiobjective optimization methods can be divided into four classes. In so-called no preference methods, no DM is expected to be available, but a neutral compromise solution is identified without preference information. Science and engineering of multiobjective optimization is complex and far-reaching.[18,19]

22.13 WHY MULTIOBJECTIVE OPTIMIZATION?

While multidisciplinary design can be associated with the traditional disciplines such as aerodynamics, propulsion, structures, and controls there are also the lifecycle areas of manufacturability, supportability, and cost which require consideration.[18,19]

After all, it is the balanced design with equal or weighted treatment of performance, cost, manufacturability, and supportability which has to be the ultimate goal of multidisciplinary system design optimization. Vision of science and scientific sagacity are in the path of a new realm and vitalized glory. Human scientific understanding and scientific vision needs to be revamped in today's competitive arena. Scientific validation and eon of technology is ushering in a new frontier of science.[18,19]

22.14 APPLICATION OF EVOLUTIONARY MULTIOBJECTIVE OPTIMIZATION IN CHEMICAL PROCESS ENGINEERING AND PETROLEUM ENGINEERING SCIENCE

Chemical process engineering and petroleum engineering science are moving toward a new phase of human scientific endeavor. The challenge, the immense vision, and the scientific urge to excel are the hallmarks toward a newer and visionary scientific regeneration. The author in this section deeply comprehends the success of research work in the field of application of multiobjective optimization in chemical process engineering and petroleum engineering science. Science is veritably moving toward a new generation. The author delineates the immense scientific work in the field of design of fluidized catalytic cracking reactors with the help of multiobjective optimization and GA.

Hugget et al.[1] lucidly delineates the global optimization of a dryer by using neural networks and GAs. For any optimum design problems, the objective function is the result of a complex numerical code and may not be differentiable and explicit. The science of optimization is complex and rigorous. The first aim was to propose a way of solving such complexity on an example problem. A novel and global strategy involving artificial neural networks (ANN) and a GA is presented and validated for an industrial convective dryer.

Nandi et al.[2] discussed with deep comprehension artificial neural-network assisted stochastic process optimization strategies. This article presents two hybrid robust process optimization approaches integrating ANN and stochastic optimization formalisms—GA and simultaneous perturbation stochastic approximation (SPSA). An ANN-based process model was developed solely from process input-output data and then its input space comprising design and operating variables was optimized by employing either the GA or the SPSA methodology. ANN and GA stands as veritable backbones to the field of process optimization in today's chemical engineering world.

Sundaram et al.[3] delineated in details, the design of fuel additives using neural networks and EAs. A fuel additive is a substance added to gasoline in small quantities to provide improved performance or to correct deficiencies.

Giovanoglou et al.[4] dealt with lucid details and deep insight optimal solvent design for batch separation based on economic performance. A mixed-integer dynamic optimization (MIDO) framework for solvent design in batch processes is presented. Performance measures reflecting process economics and computed on the basis of process dynamics are used to validate candidate solvent structures built from the UNIQUAC Functional-Group Activity Coefficients (UNIFAC) molecular groups. Science and engineering of optimization is at its helm in this research work and scientific rigor needs to be restructured at each step of endeavor and pursuit.

Capon-Garcia et al.[5] dealt with cogent insight multiobjective evolutionary optimization of batch process scheduling under environmental and economic concerns. The simultaneous considerations of economic and environmental objectives in batch process scheduling today is a subject of major concern. This work presents a hybrid optimization strategy based on rigorous local search and GA to deal efficiently with industrial batch scheduling problem.

Technological vision of optimization and GA is entering a new phase of scientific understanding. The challenge of science and technology is unimaginable in today's scientific world. The author gleans through the research output in the field of application of evolutionary multiobjective in designing chemical and petroleum engineering systems. The main purpose of this treatise is wide and versatile. The upshot of this treatise surpasses visionary frontiers.

Routray et al.[6] discussed with cogent insight kinetic parameter estimation for a multiresponse nonlinear reaction model. The kinetic parameters are successfully estimated for the propane oxidative dehydrogenation (ODH) reaction over vanadia-alumina catalyst under steady state conditions. The 5% V_2O_5/Al_2O_3 catalyst is synthesized using an incipient wetness impregnation technique.

Sarkar et al.[7] delineated with deep comprehension multiobjective optimization of semi-batch reactive crystallization processes. The determination of the optimal feed profiles for a reactive crystallizer is an important dynamic optimization problem, as the feed profiles offer a significant control over quality of the product crystals. Crystallization processes typically have multiple performance objectives and optimization using different objective functions leads to significantly different optimal operating conditions.

Corriou et al.[8] did a well-informed and well-observed study on optimization of a pulsed operation of gas separation by membrane. Science of optimization and GA are breaking visionary scientific horizon. A pulsed cyclic membrane process, originally proposed by Paul for gas separations, has been studied through a simulation and optimization study. According to the study, for carbon dioxide hydrogen separation, it is shown that cyclic operation based on an already reported material, could potentially compete with the most selective, still virtual, polymers, both in terms of selectivity and productivity.

Majdalani et al.[9] dealt lucidly on reactive transport parameter estimation with a comparison between GAs versus Monte Carlo approach. This treatise observes with deep vision of the transport of Cadmium (Cd) and tributyltin (TBT) in column experiments as test cases. The reactive transport model is described by a set of chemical reactions with equilibrium constants as the major adjustable parameters.

Elnashaie et al.[10] delved deeply on the modeling, simulation, and optimization of industrial fixed bed catalytic reactors. This widely informative treatise discusses mathematical modeling of industrial fixed bed catalytic reactors which are of utmost importance to the petrochemical and petroleum refining industries. The challenge and the vision of science and engineering of modeling and simulation of fixed bed catalytic reactors are eye openers toward a new eon of petroleum engineering science.

22.15   REVIEW WORK DONE ON EVOLUTIONARY MULTIOBJECTIVE OPTIMIZATION

Fonseca et al.[11] discussed with cogent insight EAs in multiobjective optimization. The application of EAs in multiobjective optimization is currently receiving growing interest from researchers with various backgrounds. In this review, current multiobjective evolutionary approaches are discussed, ranging from the conventional analytical aggregation of the different objectives into a single function to a number of population-based approaches and the more recent ranking schemes based on the definition of Pareto-optimality. The challenge of optimization particularly bio-inspired optimization is witnessing a new eon of scientific vision and scientific understanding. From the treatise, directions for future research in multiobjective fitness assignment and search strategies are veritably identified.

Back et al.[12] described with immense lucidity an overview of EAs for parameter optimization. Three main streams of EAs, probabilistic optimization algorithms based on the model of natural evolution are compared in this article: evolutionary strategies (EAs), EP, and GAs. The comparison is deeply performed and well comprehended with respect to certain characteristic components of EAs: the representation scheme of object variables, mutation, recombination, and the selection operator.

A report[13] by ETH, Zurich gave to the scientific domain a comparison of multiobjective evolutionary algorithms (MOEAs). A systematic comparison of various evolutionary approaches to multiobjective optimization using six carefully chosen test functions is done with due introspection. Each test function involves a particular feature that is known to cause difficulty in the evolutionary optimization process, mainly in converging to the Pareto optimal front (e.g., multimodality and deception). The suitability of the technique is deeply understood through this treatise.

Veldhuizen et al.[14] discussed deeply MOEAs and analyzing the state of the art. Solving optimization problems with multiple objectives is generally a difficult goal. Science of optimization is unfolding new chapters and a new eon as human scientific endeavor moves from one decade over another. This discussion's aim and objective deals with defining multiobjective optimization problems and certain related concepts, present an MOEA classification scheme, and evaluate the variety of contemporary MOEAs.

Zhou et al.[15] deeply comprehend in a review MOEAs and a survey of the state of art. A multiobjective optimization problem involves several conflicting objectives and has a set of Pareto optimal solutions. By evolving a population of solutions, MOEAs are able to approximate the Pareto optimal set in a single run. This study surveys the development of MOEAs primarily during the last decade.

22.16 GA AND FUTURE OF SCIENCE AND ENGINEERING

GA is a popular stochastic optimization technique, often used to solve complex large scale optimization problems in many fields. GAs have a wide range of applications and have a multitude of visionary avenues. Human civilization in today's scientific world is moving toward a newer visionary trend. Future directions and future dimensions in the application domain of GA are wide, vast, and versatile. Scientific cognizance and scientific challenges are the hallmark of a newer scientific generation for the future. Scientific progress of application of mathematical tools in petroleum engineering science is of utmost importance with the burning issue of depletion of fossil fuel resources. GA and application of multiobjective optimization are changing the face of chemical process modeling and petroleum engineering science. There is also a burning need of scientific validation in the veritable pursuit of science and engineering.[16–19]

22.17 DOCTRINE OF GA AND SCIENTIFIC/TECHNOLOGICAL COGNIZANCE

Application of GA in designing chemical process plants and petroleum refineries are surpassing visionary frontiers. Human intelligence, scientific cognizance, and academic rigor are at its helm. GA paradigm is moving through one visionary phase over another. Scientific and technological cognizance today stands in the crossroads of deep insight and intense comprehension. Energy sustainability and the grave concerns of depletion of fossil fuel resources has plunged human civilization to an unimaginable crisis. A scientist's vision as well as an engineer's forays into the world of unknown has propelled science and engineering to move toward newer innovation. The challenge of human vision and the success of scientific rigor are changing the face of human scientific research pursuit in petroleum engineering science and chemical process modeling.[16–19]

22.18 SUCCESS OF CHEMICAL PROCESS MODELING AND APPLICATION OF GA

Chemical process modeling, simulation, and optimization of chemical processes today stand in the midst of deep and cogent insight. GA paradigm is changing the chemical process modeling scenario. Bio-inspired optimization is a novel technique in the field of multiobjective optimization. Engineering science and the world of vicious scientific challenges is changing the process modeling scenario. Chemical process modeling, simulation, optimization, and control are moving from one phase to another in the wide and vast avenue of scientific vision and scientific forbearance. The challenge is enormous but the wide vision is far-reaching. Man's visions as well as a scientist's definitive prowess are in today's civilization in the path of glory and scientific fortitude.[16–19]

22.19 EVOLUTIONARY MULTIOBJECTIVE OPTIMIZATION, GA, AND THE WIDE FUTURE OF CHEMICAL PROCESS MODELING

EMO and the related field of GA are surpassing visionary boundaries. History of optimization science and the vision and urge to excel are the forerunners to the newer areas of technologies of mathematical modeling. Science is moving fast toward a newer realm and newer regeneration. Human scientific research pursuit is at its deepest peril with the concerns of ecological imbalance and depletion of fossil fuel resources. Future of chemical process modeling and simulation are in the path of definitive vision and immense scientific introspection.[16,17]

22.20 CHEMICAL ENGINEERING SYSTEMS AND PETROLEUM ENGINEERING SCIENCE

Chemical engineering systems and petroleum engineering science are in the threshold of a new era of scientific regeneration. The challenge and vision needs to be reassessed and readjudicated. Chemical engineering and petroleum engineering systems have far-reaching capabilities. The world of technological vision needs to be revamped with the progress of human civilization. Application of mathematics and application of multiobjective optimization is ushering in a new scientific era of science and engineering. In today's world, scientific validation is the ultimate pillar toward advancement of technology and engineering. A scientist's world of knowledge and mankind's prowess are emboldened with every stride of scientific research pursuit. Chemical engineering and petroleum engineering science are in the threshold of a vibrant and visionary era. In such a crucial juncture, application of mathematics and optimization is of utmost importance to the future strides in design and process modeling in chemical engineering. Chemical engineering systems and its application are in the threshold of a new revolutionary era. Depletion of fossil fuel resources is reframing and rebuilding the present and future scientific generation. Man's visions as well as civilisation's prowess are emboldened with every step of human history and time. The world of challenges in the domain of chemical engineering science is instinctive and equally intuitive. Mankind's prowess and civilization's greatness are witnessing veritable challenges. Depletion of fossil fuel resources has re-envisioned mankind and scientific research pursuit. Environmental engineering science and ecological balance are the pillars of success of futuristic human vision.[18,19]

22.21 EVOLUTIONARY MULTIOBJECTIVE OPTIMIZATION AND ITS SCIENTIFIC PERSPECTIVES

Evolutionary techniques have been used for the purpose of single-objective optimization for more than three decades. But gradually people discovered that many real world problems are naturally posed as multiobjective. Many instinctive questions arise in this arena. Scientific perspectives in today's world are challenged and surpassing arduous frontiers. EMO is in the throes of a new scientific generation and innovative scientific order. Scientific and engineering perspectives are ushering in a new dawn of human civilization and a new scientific vision and order. EMO is in the path of new glory and a visionary era. Application and visionary perspectives are the pillars and pallbearers of today's science.[18,19]

22.22 APPLICATION OF EVOLUTIONARY MULTIOBJECTIVE OPTIMIZATION IN CHEMICAL AND PETROLEUM ENGINEERING SYSTEMS

Application of EMO to the field of science and engineering is vast, versatile, and far-reaching. The world of scientific validation will go a long way in the true emancipation of application of optimization to chemical engineering and petroleum engineering systems. Chemical engineering and petroleum engineering systems are robust and ground-breaking. Human scientific research pursuit and progress of science today stands in the midst of deep comprehension and unimaginable disaster. Petroleum engineering science needs to be re-envisioned with each step of human scientific endeavor. [18,19]

22.23   SCIENTIFIC ENDEAVOR IN THE FIELD OF MULTIOBJECTIVE OPTIMIZATION

Scientific endeavor in the field of multiobjective optimization needs to be reassessed and rebuilt with the passage of scientific history and time. Scientific sagacity and scientific cognizance in today's world are in the forefront of scientific understanding and advancement of technology. The world of technology is witnessing a drastic and rational challenge in the field of multiobjective optimization. Human scientific endeavor is in the midst of immense changes and veritable challenges. Scientific endeavor is visionary in today's world of applied mathematics, chemical and petroleum engineering.

22.24   SCIENTIFIC CHALLENGES AND FUTURE DIRECTIONS IN THE FIELD OF MULTIOBJECTIVE OPTIMIZATION

Challenges of science in the field of multiobjective optimization are immense and visionary. Struggles in science and engineering are the epitomes of the future directions of multiobjective optimization. Technological advancements, the vision of science, and the progress of civilization ahead will all go a long way in the realization of sustainability in chemical engineering and petroleum engineering science.

Multiobjective optimization and GA are surpassing wide and visionary frontiers in its application in chemical process modeling and process engineering. Vision of science, progress of chemical process engineering, and the wide world of GA will go a long and visionary way in the true emancipation of application of mathematics of tomorrow. Scientific challenges, scientific forbearance, and deep scientific understanding in the field of EMO are the pallbearers of greater understanding of GA of tomorrow.[18,19]

22.25   FUTURE VISION OF EVOLUTIONARY MULTIOBJECTIVE OPTIMIZATION AND FUTURE TRENDS IN SCIENTIFIC RESEARCH PURSUIT

Science and technology in today's world are moving fast surpassing visionary frontiers. Future vision and future objectives in application of multiobjective optimization needs to be re-addressed and re-envisioned. Present and future trends in research in this domain are ground-breaking and veritably far-reaching. Global sustainability research is at its zenith today and our human civilization is witnessing drastic and dramatic changes. Energy and environmental sustainability and holistic sustainable development are the pallbearers toward a new generation of scientific hope and optimism. Petroleum engineering science in a similar vein is gearing toward newer vision and challenges with the passage of human history and time. At such a crucial and eventful juncture, mathematical tools and chemical process modeling are the backbones of scientific research pursuit in petroleum engineering science. The future vision of EMO and the scientific rigor in research pursuit are targeted toward a newer visionary scientific horizon. Science, in this present century, is gearing toward a newer futuristic horizon and opening up wide windows of deep introspection and veritable comprehension.[16-19]

22.26 CONCLUSION

Science and technology are moving toward a visionary direction. The present concern of today's human civilization is scientific validation and the grave future of sustainability. Sustainable development is the veritable pillar of human scientific progress. Chemical engineering science and petroleum engineering science are ensconced by the grave concern of depletion of fossil fuel resources. The present treatise gives a glimpse of multiobjective optimization. Future recommendations of the study are ground-breaking and visionary. Today, scientific challenges stand in the midst of deep insight, optimism, and irrevocable comprehension. This treatise widely views and elucidates the frontiers of mathematics of optimization and its vicious challenges. Civilization's visions are today opening new doors of intuitive science in years to come. The pillars of multiobjective optimization and GA are surpassing wide, vast, and versatile frontiers. Human civilization, in this present century, is yearning toward a new visionary era and a new scientific order. As human civilization moves through one decade over other, energy and environmental sustainability garners immense importance. In such a situation, mathematical tools and chemical process modeling are opening new vistas of research.

The scientific challenge for the future is inspiring and ground-breaking. Human scientific endeavor in chemical process modeling and chemical process engineering are surpassing vast and versatile frontiers. Mathematical tools and optimization techniques are changing the face of human scientific research pursuit. The global concern of petroleum engineering science, the scientific urge to innovate and the future avenues of progress will go a long way in the true emancipation of global technological initiative in the field of application of EMO in design and optimization.

KEYWORDS

- **evolution**
- **multiobjective**
- **optimization**
- **genetic**
- **algorithm**

REFERENCES

1. Hugget, A.; Sebastian, P; Nadeau, J. P. Global Optimization of a Dryer by Using Neural Networks and Genetic Algorithms. *AIChE J.* **1999**, *45*(6), 1227–1238.
2. Nandi, S.; Ghosh, S.; Tambe, S. S.; Kulkarni, B. D. Artificial Neural-Network-Assisted Stochastic Process Optimization Strategies. *AIChE J.* **2001**, *47*(1), 126–141.
3. Sundaram, A.; Ghosh, P.; Caruthers, J. M.; Venkatasubranium, V. Design of Fuel Additives Using Neural Networks and Evolutionary Algorithms. *AIChE J.* **2001**, *47*(6), 1387–1406.
4. Giovanoglou, A.; Barlatier, J.; Adjiman, C. S.; Pistikopoulos, E. N. Optimal Solvent Design for Batch Separation Based on Economic Performance. *AIChE J.* **2003**, *49*(12), 3095–3109.
5. Capon-Garcia, E.; Bojarski, A. D.; Espuna, A.; Puigjaner, L. Multiobjective Evolutionary Optimization of Batch Process Scheduling Under Environmental and Economic Conditions. *AIChE J.* **2013**, *59*(2), 429–444.
6. Routray, K.; Deo, G. Kinetic Parameter Estimation for a Multiresponse Nonlinear Reaction Model. *AIChE J.* 2005, *51*(6), 1733–1746.
7. Sarkar, D.; Rohani, S.; Jutan, A. Multiobjective Optimization of Semibatch Reactive Crystallization Processes. *AIChE J.* **2007**, *53*(5), 1164–1177.
8. Corriou, J-P.; Fonteix, C.; Favre, E. Optimization of a Pulsed Operation of Gas Separation by a Membrane. *AIChE J.* **2008**, *54*(5), 1224–1234.
9. Majdalani, S.; Fahs, M.; Carrayrou, J.; Ackerer, P. Reactive Transport Parameter Estimation: Genetic Algorithm vs. Monte Carlo Approach. *AIChE J.* **2009**, *55*(5), 1959–1968.
10. Elnashaie, S. S. E. H.; Elshishini, S. S. *Modeling, Simulation and Optimization of Industrial Fixed Bed Catalytic Reactors;* Gordon and Breach Science Publishers: Switzerland, 1993.
11. Fonseca, C. M.; Fleming, P. J. An Overview of Evolutionary Algorithms in Multiobjective Optimization. *Evol. Comput.* 1995, *3*(1), 1–16.
12. Back, T.; Schwefel, H-P. An Overview of Evolutionary Algorithms for Parameter Optimization. *Evol. Comput.* 1993, *1*(1), 1–23.
13. Zitzler, E.; Deb, K.; Thiele, L. *Comparison of Multiobjective Evolutionary Algorithms,* TIK-Report No.70, ETH Zurich, Zurich, Switzerland, 1999.
14. Van Veldhuizen, D. A.; Lamont, G. B. Multiobjective Evolutionary Algorithms: Analyzing the State of the Art. *Evol. Comput.* **2000**, *8*(2), 125–147.

15. Zhou, A.; Qu, B-Y; Li, H.; Zhao, S-Z.; Suganthan, P. N.; Zhang, Q. Multiobjective Evolutionary Algorithms: a Survey of the State of the Art. *Swarm Evol. Comput.* **2011**, *1*, 32–49.
16. Palit, S. Application of Evolutionary Multi-Objective Optimization in Designing Fluidised Catalytic Cracking Unit and Chemical Engineering Systems-A Scientific Perspective and A Critical Overview. *Int. J. Comput. Intell. Res.* **2016**, *12*(1), 17–34.
17. Palit, S. The Future Vision of the Application of Genetic Algorithm in Designing a Fluidized Catalytic Cracking Unit and Chemical Engineering Systems. *Int. J. Chemtech. Res.* **2014-2015**, *7*(4), 1665–1674.
18. Palit, S. Modelling, Simulation And Optimization Of A Riser Reactor Of Fluidised Catalytic Cracking Unit With The Help Of Genetic Algorithm And Multi-Objective Optimization- A Scientific Perspective And A Far-Reaching Review. *Int. J. Chem. Eng.* **2016**, 9(1), 89–102.
19. Deb, K. *Multi-objective Optimization Using Evolutionary Algorithm, Wiley Interscience Series in Systems and Optimization;* John Wiley and Sons: USA, 2010.

CHAPTER 23

INTENSIFICATION OF EFFICIENCY OF PROCESS OF GAS CLEANING IN APPARATUSES "ROTOKLON"

R. R. USMANOVA[1] and G. E. ZAIKOV[2,*]

[1]*Department of Chemical Science, Ufa State Technical University of Aviation, Ufa 450000, Bashkortostan, Russia*

[2]*Department of Polymer Science, N. M. Emanuel Institute of Biochemical Physics, Russian Academy of Sciences, Moscow 119991, Russia*

Corresponding author. E-mail: gezaikov@yahoo.com

CONTENTS

ABSTRACT

In this chapter, the intensification of efficiency of process of gas cleaning in apparatuses "Rotoklon" is reviewed and discussed in detail.

23.1   INTRODUCTION

The big group of wet-type collectors in which contact of gases to a liquid is carried out at the expense of blow of a gas stream on a liquid surface refers to apparatuses of impact-sluggish act. The gas stream is passed through holes of various configurations. As a result of such interaction, drops of diameter 300 microns are formed. Feature of apparatuses of percussion is the total absence of mechanism of moving a liquid. Therefore, all energy necessary for creation of a surface of contact is brought through a gas stream. In this connection scrubbers, "rotoklon" sometimes name apparatuses with internal circulation of a liquid. Despite a large variety of builds of apparatuses of this type found application in the industry, especially abroad, the reliable theoretical method of their calculation is still not known.[2]

Apparatuses of impact-sluggish act refer to wet-type collectors with internal circulation of a liquid; therefore, removal of sludge is carried out more often periodically (in process of its accumulation in a bunker part). Continuous removal of sludge is possible, but the basic advantage of impact-sluggish apparatuses—decrease in the specific charge of an irrigation water at the expense of its repeated circulation—in this case is lost. Absence of small holes for dispensation of a liquid and mechanical twirled parts allows to maintain apparatuses of this type by maintaining the firm suspended matter which is in contact with a gas fluid stream.

However, along with advantages they have a row of features to which it is necessary to pay attention at their sampling and maintenance.

For normal maintenance of wet impact-sluggish dedusters, the great value is maintained at a fixed level of liquid in the apparatus. Even insignificant decrease in fluid level (at the expense of its removal with sludge, transpiration, or liquid ablation) can lead to sharp decrease of efficiency and on the contrary—the level increase (at excessive feed) calls growth of a water resistance of the apparatus. Also one of defects of apparatuses of impact-sluggish act is efficiency of trapping of corpuscles of size not less than 4 microns.[1,2,8]

23.2   EXPERIMENTAL INSTALLATION AND THE TECHNIQUE OF REALIZATION OF EXPERIMENT

The rotoklon represents the basin with water on which surface on a connecting pipe of feeding of dusty gas, the dust-laden gas mix arrives. Thus, gas changes a traffic route. The dust containing in gas, penetrates into liquid under the influence of an inertial force. Turning the blades of an impeller is done manually, rather each other on a threaded connection by means of handwheels. The slope of blades was installed at the interval of 25–45° to an axis.

In a rotoklon, three pairs of the blades having profile of a sinusoid are installed. Blades can be controlled for installation of their position. Depending on the cleanliness level of an airborne dust flow, the lower lobes by means of handwheels are installed on an angle defined by operational mode of the device. The rotoklon is characterized by presence of three channels, a formation, the overhead, and bottom blades. And in each following on a run of gas, the bottom blade channels are installed above the previous. Such arrangement promotes a gradual entry of a water gas flow in slotted channels and thereby reduces the device hydraulic resistance. The arrangement of an input part of lobes on an axis with a capability of their turn allows creating a diffusion reacting region. Sequentially slotted channels mounted in a diffusion zone are equipped with a rotation angle lobes and a hydrodynamic zone of intensive wetting of corpuscles of a dust. In process of flow moving through the fluid-flow curtain, the capability of multiple stay of corpuscles of a dust in hydrodynamically reacting region is supplied that considerably raises a dust clearing efficiency and ensures functioning of the device in broad bands of cleanliness level of gas flow.

The construction of a rotoklon with adjustable sinusoidal lobes is developed and protected by the patent of the Russian Federation, capable to solve a problem of effective separation of dust from gas flow.[3] Thus, water admission to contact zones implements as a result of its circulation in the apparatus.

The rotoklon with the adjustable sinusoidal lobes is presented in Figure 23.1 and consists of a body (3) with connecting influent (7) and effluent (5) pipes. Moving of the overhead lobes (2) can be done by screw jacks (6); the lower lobes (1) are fixed on an axis 8 with rotation capability. The rotation angle of the lower lobes is chosen from a condition of a persistence speed of an airborne dust flow. For angle of rotation regulating a handwheel at the output parts of the lower lobes (1) are embedded. Quantity of lobe pair is determined by productivity of the device and cleanliness level of an airborne dust flow that is a regime of a stable running of the device. In the lower part of a body, there is a connecting pipe for draining the slime water 9. Before connecting pipe for gas make (5) the labyrinth drip pan (4) is installed. The rotoklon works as follows. Depending on cleanliness level of an airborne dust flow the overhead lobes (5) by means of screw jacks (6), and the lower lobes (1) by means of handwheels are installed on an angle defined by operational mode of the device. Dusty gas arrives in the upstream end (7) in the upper part of the body (3) apparatuses. Having reached the liquid surface, gas changes the direction and moves to the slot-hole channel formed upper (2) and inferior (1) blades. Thanks to a traffic high speed, gas captures the upper layer of the liquid and atomizes it in small-sized drip and foam. After passage of all slot-hole channels, gas moves to the labyrinth drip pan (4) and is inferred in an atmosphere through the discharge connection (5). The collected dust settles in the loading pocket of a rotoklon and through a connecting pipe for removal of slurry (9), together with a liquid, is periodically inferred from the apparatus.[4]

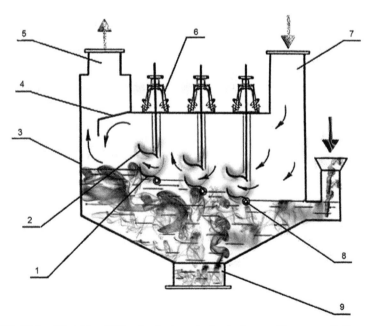

FIGURE 23.1 "Rotoklon." The front view: 1, the bottom guide vanes; 2, the overhead guide vanes; 3, a body; 4, the labyrinth drip pan; 5, connecting pipe for gas; 6, screw jacks; 7, connecting pipe for gas entry; 8, an axis; 9, a connecting pipe for a drain of slurry.

The mentioned structural features do not allow using correctly available solutions on hydrodynamics of dust-laden gas flows for a designed construction. In this connection, for the well-founded exposition of the processes occurring in the apparatus, there was a necessity of realization of experimental researches.

Experiments were conducted on the laboratory-scale plant "rotoklon" and presented in Figure 23.2.

The examined rotoklon had three slotted channels speed of gas with gas speed up to 15 mps. At this speed the rotoklon had a hydraulic resistance 800 Pa. Working in such regime, it supplied efficiency of trapping of a dust with input density 0.5 g/nm^3 and density 1200 kg/m^3 at level of 96.3%.[5]

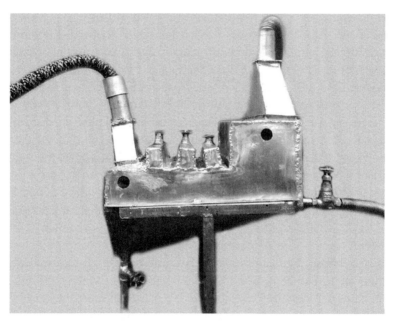

FIGURE 23.2 Experimental installation of "rotoklon."

In the capacity of modeling system air and a dust of talc with a size of corpuscles d = 2–30 micron, white black and a chalk have been used. The apparatus body was filled with water on level h_g = 0.175 m.

Cleanliness level of an airborne dust mix was determined by a direct method.[5] On direct sections of the pipeline before and after the apparatus the mechanical sampling of an airborne dust mix was made. After determination of matching operational mode of the apparatus, gas test were taken by means of tubes for a mechanical. On tubes for researches with various diameter tips have been installed.

23.3 MATHEMATICAL MODELING OF TRAFFIC OF DISPERSION PARTICLES IN BLADE IMPELLERS

Artificial creation of turbulence at the expense of installation of blades of an impeller is necessary for an intensification of processes of concretion.

For the concretion called by turbulence of a stream, it is necessary to observe two cases.

First, if a pulse of corpuscles of an aerosol is approximately same as that at Wednesday, they will move approximately with the same speed, even the sections of air surrounding them. In this case, motion of corpuscles can be presented by means of factor of turbulent diffusion D_t. This factor can matter, in 10^4–10^6 times more, than factors of thermal diffusion. The processes of concretion called by turbulence of a stream, can be observed as usual concretion, but with use of the big factors of diffusion.

Another case of concretion of an aerosol in a turbulent stream is characterized by origination of the inertia and differences between corpuscles of different sizes. Owing to turbulence of a corpuscle are sped up till the various speeds depending on a size, and can then face with each other. For a monodisperse aerosol this mechanism has no value. For unequigranular, an aerosol with known size distribution,[2] the speed of concretion is proportional to a basic speed of a turbulent stream in extents 9, speed of concretion increases very rapidly with increase in speed of a turbulent stream. As very small corpuscles are rapid and sped up, value of this mechanism decreases with decrease of a size of corpuscles, and it is the most momentous for corpuscles, whose diameters make 0.1–1 microns. In all cases, Brownian diffusion when diameters of corpuscles there are less than 0.01 microns[3] predominates. As previously explained, we achieved increase in efficiency of trapping of corpuscles less than 4 microns at the expense of concretion, and also at the expense of capture by larger drops formed as a result of an ejection.

Computer modeling has been applied to studying of hydrodynamic characteristics in the apparatus in program Ansys CFX intended for a problem solving of a mechanics of fluids.

Numerical calculation in the given program complex does not allow to model a torch of the sprayed liquid and to size up its dispersity as in the program the account of a surface tension force of liquid and its interacting with a circumambient is not mentioned. Therefore, result of numerical experiment are the gained values of speed of gas and liquid streams and pressure in any point of the crank chamber which allow to investigate in details process of interacting of streams, to choose an optimum relationship of sizes of a rotoklon and to define a range of its work.

For calculation, the method of final volumes which is more rapid-transfer alternative of a method of final elements simplified, hence, was used. The movement of gas without a discrete phase, in view of the power and thermal affecting of corpuscles on negligible gas phase has been simulated only. This assumption is true in that case when the mass fraction of corpuscles in a two-phase stream does not exceed 30%.[6] The numerical analysis of a current of gas in the great dispatch-vortical apparatus is reduced to the solution of system of mean field Nave–Stoks equations on Reynolds. For short circuit of gas-dynamic Nave–Stoks equations, the standard k - ε model of turbulence was used. Besides turbulence and heat exchange model, it was used volume of fluid for modeling with a free surface of water and a stream of air.[7]

Boundary conditions have been chosen the following (in the capacity of gas has been used):

- air, liquids (on which air—water goes),
- entry temperature (300 K),
- gas pressure (1014 Pa),
- intensity of turbulence on an entry and an apparatus exit (5%).

Calculations of the apparatus with a nozzle and without it have been carried out to learn about nozzle effect on hydrodynamic parametres. Over the range changes of speed from 2 to 12 mps as a result of blow of a gas stream about a liquid surface, spherical deepening is formed, and the expelled liquid forms a ledge. From a surface of a fluid-flow ledge, there is an ejection of films, drops, and streams in a zone of blades. All results presented more low are an operating mode of the apparatus at speed of gas on an exit from a connecting pipe on a water surface.

23.4   THE ANALYSIS OF RESULTS OF MODELING

Gas stream lines (Fig. 23.3) from which it is visible are presented that gas in the apparatus with a nozzle in an ejection zone has numerous uniform eddyings. Formation of a turbulent trace behind shovels of an impeller results from a breakaway of a boundary layer from its surface and increase in frequency of turbulent pulsations.

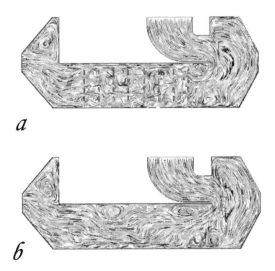

a - the apparatus without a nozzle; *b* - the apparatus with a nozzle

FIGURE 23.3 Paths of gas and liquid in the apparatus: (a) The apparatus without a nozzle and (b) the apparatus with a nozzle.

Also intensity of turbulence (Fig. 23.4) is a momentous parameter for the equipment of clearing of gas emissions.

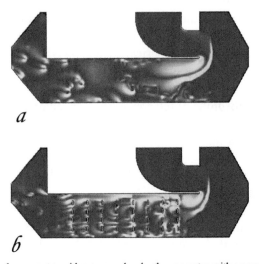

a - the apparatus without a nozzle; *b* - the apparatus with a nozzle

FIGURE 23.4 Frequency of turbulent pulsations: (a) The apparatus without a nozzle and (b) the apparatus with a nozzle.

The field of speed (Fig. 23.5) shows that with nozzle installation the uniform distribution of speed in the apparatus is observed more or less.

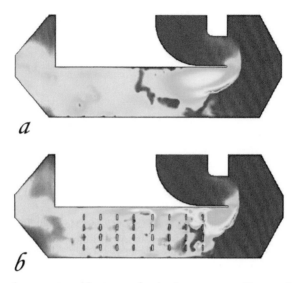

a - the apparatus without a nozzle; *b* - the apparatus with a nozzle

FIGURE 23.5   Field of speeds: (a) The apparatus without a nozzle and (b) the apparatus with a nozzle.

Thus, it is possible to draw deductions that nozzles reinforce turbulence of cleared gas in a zone of an ejection of the apparatus that theoretically should affect increase in efficiency of trapping of small corpuscles.

23.5 CONCLUSIONS

1. By means of modeling, it is defined that at speed of bringing gas of 6 mps and above, active ablation of the liquid is initiated. There is a necessity to make good the loss the liquid.
2. It is established that the basic pressure losses in the apparatus occur in a zone of blow of gas about the liquid. An optimum level of water before equipment start is level to an upstream end edge.
3. It is proved that the rise in water level considerably increases not only the rotoklon water resistance but also there are pulsations of pressure with considerable amplitudes of differences.

KEYWORDS

- rotoklon
- contact channels
- ansys CFX
- nozzle
- efficiency of gas cleaning

REFERENCES

1. Kouzov, P. A.; Malgin A. D.; Skryabin, G. M. *Clearing of Gases and Air of a Dust in the Chemical Industry*; St.-Petersburg, 1993.
2. Shcwidki, V. S.; Ladygichev, M. G. *Clearing of Gases;* Moscow, 2002.
3. Usmanova, R. R.; Zhernakov, V. S.; Panov A. K. Rotoklon a Controlled Sinusoidal Blades R. F. Patent 2317845, February 27, 2008.
4. Usmanova, R. R.; Zaikov, G. E. Clearing of Industrial Gas Emissions: Theory, Calculation, and Practice. In *Experimental Research and Calculation of Efficiency of Sedimentation of Dispersion Particles in a Rotoklon*; AAP: U.S., Canada, 2015.
5. Usmanova, R. R.; Zaikov, G. E.; Stoyanov, O. V.; Klodziuska, E. Research of the Mechanism of Shock-Inertial Deposition of Dispersed Particles from Gas Flow. *Bull. Kazan Technol. Univ.* **2013**, *16*(9), 203–207 (in Russian).
6. Richkov, C. P.; Savelyev, J. A. Application of a Rotoklon "Ural" in the Industry. *Safety of Work in the Industry* **1982**, *8*, 43–45 (in Russian).
7. Hirt, C. W.; Nicholls, B. D. Volume of Fluid (VOF) Method for Dynamical Free Boundaries. *J. Comput. Phys.* **1981**, *39*, 201–225.
8. Patel, V. C.; Rodi, W.; Scheuerer, G. Turbulence Models for Near-Wall and Low Reynolds Number Flows: A Review. *AIAA J.* **1985**, *23*, 1308–1319.

INDEX

Printed and bound by CPI Group (UK) Ltd, Croydon, CR0 4YY

23/10/2024

01777704-0006